Baolong Guo, Juanjuan Zhu
Signals and Systems
Information and Computer Engineering

Information and Computer Engineering

Volume 3

Already published in the series:

Volume 2
Jie Yang, Congfeng Liu, Random Signal Analysis, 2018
ISBN 978-3-11-059536-9, e-ISBN 978-3-11-059380-8,
e-ISBN (EPUB) 978-3-11-059297-9

Volume 1
Beijia Ning, Analog Electronic Circuit, 2018
ISBN 978-3-11-059540-6, e-ISBN 978-3-11-059386-0,
e-ISBN (EPUB) 978-3-11-059319-8

Baolong Guo, Juanjuan Zhu

Signals and Systems

—

DE GRUYTER

Science Press
Beijing

Authors
Prof. Baolong Guo
Xidian University
Xi'an, China

Associate Prof. Juanjuan Zhu
Xidian University
Xi'an, China

ISBN 978-3-11-059541-3
e-ISBN (PDF) 978-3-11-059390-7
e-ISBN (EPUB) 978-3-11-059296-2
ISSN 2570-1614

Library of Congress Control Number: 2018946184

Bibliographic information published by the Deutsche Nationalbibliothek
The Deutsche Nationalbibliothek lists this publication in the Deutsche Nationalbibliografie;
detailed bibliographic data are available on the Internet at http://dnb.dnb.de.

© 2018 Walter de Gruyter GmbH, Berlin/Boston and China Science Publishing & Media Ltd.
Cover image: Prill/iStock/Getty Images Plus
Typesetting: le-tex publishing services GmbH, Leipzig
Printing and binding: CPI books GmbH, Leck

www.degruyter.com

Preface

This book is primarily designed for undergraduate courses in signals and systems. The course is intended for instruction in the field of electrical and computer engineering. It can be a professional basic course for communication, automatic control, information processing and artificial intelligence. Our main motivation in writing the book was to help students master the basic methods of system analysis. The students will not only study essential elements in an engineering program but also expand the scope of actual applications.

1. The content emphasizes three central issues: basic signal, signal decomposition and analysis methods of linear time-invariant systems. The mathematical foundations are strongly emphasized to help students grasp important concepts.
2. The content of the book is arranged in an analogical way. The analysis of continuous-time and discrete-time systems is treated separately. All the knowledge points are evolved from the time domain to the frequency and the transformation domains. The content of the whole book is systematic and logical. It is easy for students to grasp the complete knowledge system of signals and systems analysis.
3. The book adopts a series of strategies such as task-driven and problem-inspired methods, all aiming at stimulating the interest of the students. In each chapter, some questions are proposed to introduce the main problems that the chapter will solve. It can also motivate the readers to ponder deeply and gain great interest in learning the material. Some marginal notes are added to give explanations or enlighten thinking.
4. The applications of signal and system are introduced to focus on basic concepts and characteristics. The course is linked with the following courses: digital signal processing, automatic control and image processing. It provides students with an appreciation for the range of applications of the technique being learned and for directions for further study. From the window of this course, students will have a deep understanding of more related knowledge in the subject of signals and systems.
5. We have extensively used MATLAB to validate our analytical results and to illustrate the design procedures for a variety of problems. To further enhance students' understanding of the main signal processing concepts, MATLAB simulations are illustrated. The introduction and analysis of typical cases in engineering practice are guided by a scientific method to promote students' active inquiry through the use of MATLAB.

The book consists of eight chapters. We begin Chapter 1 by introducing some of the elementary ideas related to the mathematical representation of signals and systems. In particular, we discuss transformations (such as time shifts and scaling) of the inde-

https://doi.org/10.1515/9783110593907-201

pendent variables of a signal. We also introduce basic definitions and classifications of systems. The basic system properties such as causality, linearity and time invariance are discussed. The framework of analytical methods is given to help users have a general and comprehensive understanding of the book. Chapter 2 introduces the time-domain analysis of linear time-invariant continuous-time (LTIC) systems, including the convolution integral used to evaluate the output in response to a given input signal. Chapter 3 provides the time-domain analysis of linear time-invariant discrete-time (LTID) systems, including the convolution sum used to calculate the output of a discrete-time (DT) system.

Chapters 4 through 6 present a thorough description and analysis of transformation methods in both continuous and discrete time. Chapter 4 defines the continuous-time Fourier series (CTFS) as a frequency-domain representation for continuous-time (CT) periodic signals. The continuous-time Fourier transform (CTFT) is then presented to provide an alternative to the convolution integral for the evaluation of the output response. Many important properties of Fourier transforms are detailed. The close relationship between Fourier series and transforms are emphasized and illustrated to compute the response of the system. In the last section in Chapter 4, the sampling theorem is provided.

Chapter 5 then develops the Laplace transform to analyze LTIC systems in the complex frequency domain. The relationship between the Fourier transform and the Laplace transform is given. Chapter 6 gives the method of the Z-transform to deal with LTID systems. We use these transform methods to determine the frequency responses of LTIC or LTID systems described by differential or difference equations. We also provide several examples illustrating how transform domain methods can be used to compute the zero-input and zero-state responses. In the last sections in Chapters 5 and 6, a signal flow graph and system simulation using Mason's rule are analyzed.

Chapter 7 develops the state-space equations to analyze multiple-input multiple-output (MIMO) systems. The method of establishing state equations is illustrated, and the procedures of solving the state and output equations with the Laplace transform and the Z-transform are detailed. The system transfer function and the stability analysis of systems are analyzed in the last section.

Chapter 8 concludes the book by motivating the students with several applications in communication systems, control systems, fast Fourier transform and digital filters in digital signal processing, Kalman filters and image processing. We give a brief introduction to design techniques for IIR and FIR filters. All these applications help engineering students establish a comprehensive framework for current and future developments in the engineering fields.

The book has been designed to be a professional basic course at the sophomore level. A course in electrical circuits, although not essential, would be highly useful as several examples of electrical circuits have been used as systems to motivate the students. For the specialty of electronic information, Chapters 1–6 should be included in the course. In this book, we have used MATLAB to validate our analytical results and also to illustrate the design procedures for a variety of problems. The MATLAB code is provided in each example. Consequently, several MATLAB exercises have been included in the Concept Problem sections.

We express our appreciation to Professor Junli Zheng of Tsinghua University for his ideas and help with the initial Chinese manuscript. We would also show great thanks to Professor Zhigong Wang, Qiao Meng, Houjin Chen and Chen Liu from the Teaching Committee of Electrical and Electronic Courses at Ministry of Education. Special thanks are due to Professor Xianjue Luo of Xi'an Jiaotong University, Professor Zhemin Duan and Lei Guo of Northwestern Polytechnical University for their valuable suggestions. We would also like to express our sincere thanks to Professor Jiandong Li, Xinbo Gao, Guangming Shi, Xiaozi Sun, Changhong Liang, Fenglin Fu and Sanyang Liu of Xidian University for their support. In addition, Professor Yongrui Zhang, Xiaoping Li, Songlin Wang, Wei Sun and Associate Professor Yunyi Yan, Xianxiang Wu and Fanjie Meng from Xidian University contributed significantly during the preparation of the book. Teacher Jinxin Zhang, Wangpeng He and the postgraduate students of ICIE helped a lot in collecting materials. The technical support and help provided by Science Press and the executive editor were crucial in making this edition a reality.

Any suggestions or concerns regarding the book may be communicated to the authors at the email address: blguo@xidian.edu.cn.

Baolong Guo
Juanjuan Zhu
Xidian University
January 2018

Contents

1 Overview of signals and systems

Please focus on the following key questions.
1. What are signals, systems and basic signals?
2. What is a linear time-invariant system?

1.0 Introduction

Information, such as voice, text, images, symbols, etc., is exchanged and transmitted in all fields of human society. Since electricity has been available, transmission technology by electrical signals has developed rapidly. In 1876, A. G. Bell invented the telephone, as shown in Figure 1.1, which converted sound signals into electrical signals to be transmitted along wires. In the late nineteenth century, research was carried out on transmission technology by electromagnetic waves. Now, the transmission of radio signals is spread all over the world, and even to the universe. The development prospect of personal communication technology indicates the fact that anyone is able to communicate with other people in the world at anytime and anywhere. The concepts of signals and systems arise in a wide variety of fields, including communications, aeronautics and astronautics, circuit design, acoustics, speech processing and biomedical engineering.

Input Voice   Output Voice

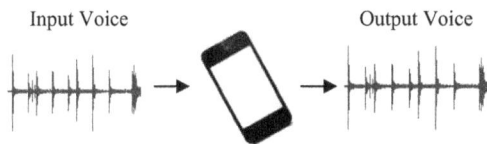

Fig. 1.1: Example of a telephone system

This chapter will introduce basic definitions and classification of signals and systems. The mathematical description and representations of the elementary signals are detailed. The framework of the classical communication system and control system is introduced. The system analysis methods involved in this book will also be briefly given in the last section of this chapter.

https://doi.org/10.1515/9783110593907-001

1.1 Basic definitions and classification of signals

1.1.1 Concepts

Message: All the reports or news are referred to as the message. It will change the receiver's knowledge state. The degree of knowledge state change is decided by the amount of information contained in the message.

Information: The meaningful content of the message is called information. In information theory, the amount of information carried by the message is defined as shown in Equation (1.1):

$$I = -\log P(x) \tag{1.1}$$

where $P(x)$ is the probability of occurrence of an event x and I is the amount of information.

Signal: The carriers of information are referred to as signals [1]. For example, the ringing of a bell is a sound signal, a traffic light is a light signal, and the content received from a TV antenna is an electrical signal. In addition, there are text signals, image signals and bio-electricity signals. Signals need to be transmitted and processed in some particular systems. For instance, the wireless radio modulates voice signals by the carrier signal, which is suitable for long-distance transmission. The carrier signal is then emitted through the antenna. The voice signal is received and demodulated by the radio receiver.

! **Note:** The signal is the carrier of information, which is depicted as a function.

1.1.2 Description of signals

A signal is a function of one or more independent variables, which may describe a wide variety of physical phenomena. In this book, we focus on electrical signals involving a single independent variable. For convenience, we will generally refer to signals as voltage or current functions of time. The descriptions of signals can be functions of time $f(t)$ and their graphical representation.

1.1.3 Classification of signals

1. Deterministic signal and stochastic signals
A *deterministic signal* is a signal with a certain value at any time, and it can be described by a certain function.

A *stochastic signal* is a signal with random values. The probability of a certain value at a certain time may be known. This kind of signal can also be called an uncertain signal.

Thermal fluctuation noise and lightning disturbance signals in electronic systems are two typical stochastic signals.

The study and analysis of deterministic signals can be extended to analyze stochastic signals. Specifically, the input and output of the deterministic signals are replaced by the statistics of the stochastic signals. In this book, only deterministic signals are discussed.

2. Continuous-time signals and discrete-time signals
(1) Continuous-time signals
A continuous-time signal can be defined at all time t in a continuous-time range $(-\infty < t < \infty)$, which is abbreviated as *continuous signal* or *CT signal*. In general, continuous signals with continuous function values are called *analog signals*.

The domain of definition t is continuous and sometimes it may contain break points. However, the range of function values can be continuous or discrete [2]. As shown in Figure 1.2, continuous-time signals have a continuous definitional domain, and their function domains might be discrete [Figure 1.2 (b)].

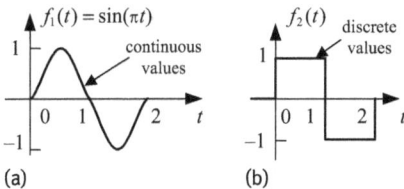

Fig. 1.2: Illustrations of continuous-time signals

(2) Discrete-time signals
The signal defined only at discrete times is referred to as a *discrete-time signal*, which is abbreviated as *discrete signal* or *DT signal*. Discrete signals with discrete function values are called *digital signals*.

The "discrete" in discrete signals means that the definitional domain of functions, i.e., time, is discrete. Specifically, the function values at the defined discrete times exist, while those at the rest of time are undefined.

Note: Do not mistake "undefined value" for "0 value".

(3) Discrete sequence
Figure 1.3 (a) illustrates a signal $f(t)$ defined only at some discrete times t_k ($k = 0$, $\pm 1, \pm 2, \ldots$). The intervals of adjacent discrete points can be equal or not. Given an equal interval T, the discrete-time signal can be represented as $f(kT)$. It is abbreviated as $f(k)$, as shown in Figure 1.3 (b). These equal-interval discrete signals are referred to as *discrete sequences* or *sequences*. The sequence $f(k)$ is defined only at integer numbers of k.

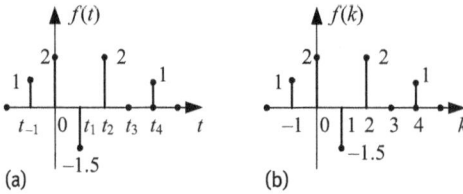

Fig. 1.3: Representation of discrete signals; (a) $f(t)$ defined at some discrete time, (b) discrete sequence $f(k)$

Mathematically, the discrete sequence $f(k)$ is denoted by:

$$f(k) = \begin{cases} 1, & k = -1 \\ 2, & k = 0 \\ -1.5, & k = 1 \\ 2, & k = 2 \\ 0, & k = 3 \\ 1, & k = 4 \\ 0, & \text{else} \end{cases} \tag{1.2}$$

It can also be represented as a sequence,

$$f(k) = \{\ldots, 0, 1, 2, -1.5, 2, 0, 1, 0, \ldots\}$$
$$\uparrow k = 0 \tag{1.3}$$

!   **Note:** Remember to give the label of "$k = 0$".

3. Periodic and nonperiodic signals

Periodic signals change repeatedly at a uniform duration T (or an integer N).
    A periodic continuous-time signal $f(t)$ has the following property:

$$f(t) = f(t + mT), \quad m = 0, \pm 1, \pm 2, \ldots \tag{1.4}$$

Likewise, a periodic discrete-time signal $f(k)$ has the following property:

$$f(k) = f(k + mN), \quad m = 0, \pm 1, \pm 2, \ldots \tag{1.5}$$

In this case, the smallest value of T (or integer N) that satisfies Equation (1.4) [or Equation (1.5)] is called the *fundamental period* of the signal.

!   **Note:** N must be an integer, but T can be a noninteger.

A signal that is not periodic is called an *aperiodic* or *non-periodic signal*.

A familiar example of a periodic CT signal is a sinusoidal function represented mathematically by the following expression:

$$f(t) = A \sin(\omega_0 t + \theta).$$

The sinusoidal signal $f(t)$ has a fundamental period $T = 2\pi/\omega_0$. Although one CT sinusoidal signal is periodic, with the combination of two sinusoidal signals it needs to be verified whether the period exists [3]. We have the following proposition.

Note: The continuous sinusoidal signal is definitely periodic.   !

Proposition: A signal $g(t)$ that is a linear combination of two periodic signals, $f_1(t)$ with fundamental period T_1 and $f_2(t)$ with fundamental period T_2 as follows:

$$g(t) = a f_1(t) + b f_2(t)$$

is periodic if

$$\frac{T_1}{T_2} = \frac{m_2}{m_1} = \text{rational number}.$$

The fundamental period is given by $m_1 T_1 = m_2 T_2$ provided that the values of m_1 and m_2 are chosen such that the greatest common divisor between m_1 and m_2 is 1.

Example 1.1.1. Determine whether or not each of the following signals is periodic. If the signal is periodic, determine its fundamental period:

(1) $f_1(t) = \sin(2t) + \cos(3t)$ (2) $f_2(t) = \cos(2t) + \sin(\pi t)$

Solution:
(1) The sinusoidal signals $\sin(2t)$ and $\cos(3t)$ are both periodic signals with fundamental periods T_1 and T_2, respectively:

$$\omega_1 = 2 \text{ rad/s}, \quad T_1 = 2\pi/\omega_1 = \pi \text{ s}$$
$$\omega_2 = 3 \text{ rad/s}, \quad T_2 = 2\pi/\omega_2 = (2\pi/3) \text{ s}$$

Calculating the ratio of the two fundamental periods yields:

$$\frac{T_1}{T_2} = \frac{3}{2},$$

which is a rational number. Supposing that the fundamental period $T = m_1 T_1 = m_2 T_2$, then the fundamental period is given by $T = m_1 T_1 = 2T_1 = 2\pi$ s. Alternatively, the fundamental period of $f_1(t)$ can also be evaluated from $T = m_2 T_2 = 3T_2 = 2\pi$.

(2) We know that the fundamental periods of $\cos(2t)$ and $\sin(\pi t)$ are $T_1 = \pi$ s and $T_2 = 2$ s, respectively. Calculating the ratio of the two fundamental periods yields:

$$\frac{T_1}{T_2} = \frac{\pi}{2},$$

which is not a rational number. Hence, the signal $f_2(t)$ is not a periodic signal.

Although all CT sinusoidal signals are periodic, the DT sinusoidal sequence $f(k) = A\sin(\beta k + \theta)$ may not always be periodic [4].

❗ Note: The sum signal of two continuous sinusoidal signals is not necessarily periodic.

Example 1.1.2. Determine whether or not the sinusoidal sequence $f(k) = \sin(\beta k)$ is periodic. If the sequence is periodic, determine its fundamental period.

Solution:

$$f(k) = \sin(\beta k) = \sin(\beta k + 2m\pi)$$

$$= \sin\left[\beta\left(k + m\frac{2\pi}{\beta}\right)\right] = \sin[\beta(k + mN)], \quad m = 0, \pm 1, \pm 2, \ldots \quad (1.6)$$

where β is the digital angular frequency, whose unit is in rad.
(1) If $2\pi/\beta$ is an integer, the sinusoidal sequence is periodic. The fundamental period is $N = 2\pi/\beta$.
(2) If $2\pi/\beta$ is a rational number, the sinusoidal sequence is also periodic. The fundamental period is $N = M(2\pi/\beta)$, where M is the smallest integer that results in an integer value for N.
(3) If $2\pi/\beta$ is not a rational number, the sinusoidal sequence is not periodic.

❗ Note: The discrete sinusoidal signal is not necessarily periodic.

Example 1.1.3. Determine whether or not each of the following sequences is periodic. If the sequence is periodic, determine its fundamental period.

$$(1)\ f_1(k) = \sin(3\pi k/4) + \cos(0.5\pi k) \qquad (2)\ f_2(k) = \sin(2k)$$

Solution:
(1) According to the analysis of Example 1.1.2, we have:

$$\beta_1 = 3\pi/4\,\text{rad}, \quad 2\pi/\beta_1 = 8/3$$
$$\beta_2 = 0.5\pi\,\text{rad}, \quad 2\pi/\beta_2 = 4$$

The corresponding fundamental periods are $N_1 = 8$ and $N_2 = 4$, respectively. So, the fundamental period of $f_1(k)$ is $N = 8$, which is the lowest common multiple of periods N_1 and N_2.
(2) We know that the digital angular frequency of $\sin(2k)$ is $\beta = 2$ rad. Since $2\pi/\beta = \pi$ is not an irrational number, $f_2(k) = \sin(2k)$ is not periodic.

❗ Note: The sum signal of two discrete periodic sequences is definitely periodic.

4. Energy signals and power signals
Supposing that a CT signal $f(t)$ represents the voltage or the current across a resistor with resistance $1\,\Omega$, the instantaneous power is $|f(t)|^2$. The total energy over an infinite

time interval $-\infty < t < \infty$ is defined as:

$$E \stackrel{\text{def}}{=} \int_{-\infty}^{\infty} |f(t)|^2 \, dt,$$

(1.7)

and the average power over this time interval is:

$$P \stackrel{\text{def}}{=} \lim_{T \to \infty} \frac{1}{T} \int_{-\frac{T}{2}}^{\frac{T}{2}} |f(t)|^2 \, dt.$$

(1.8)

The signal with $0 < E < \infty$ is defined as the finite energy signal, abbreviated as the *energy signal*.

The signal with $0 < P < \infty$ is defined as the finite power signal, abbreviated as the *power signal*.

Similarly, for DT signals, the signal with $0 < E = \sum_{k=-\infty}^{\infty} |f(k)|^2 < \infty$ is defined as an *energy signal*. Moreover, the signal with $P = \lim_{N \to \infty} 1/(N+1) \sum_{k=-N/2}^{N/2} |f(k)|^2 < \infty$ is defined as a *power signal*.

Note: A signal cannot be both an energy and a power signal simultaneously.

Energy signals have zero average power, whereas power signals have infinite total energy. The signals with limited time interval (i.e., signals with nonzero values over a finite time interval) are definitely energy signals [5]. Some signals, however, for example, the signal $e^t \varepsilon(t)$, is a growing exponential with infinite energy, whose average power cannot be calculated. Such signals are generally of little interest to us. Most periodic signals are typically power signals, while nonperiodic signals may be energy or power signals.

5. One-dimensional and multi-dimensional signals

Mathematically, signals can be represented as a function of one or more variables, which are referred to as one-dimensional or multi-dimensional signals, respectively.

The voice signal is a one-dimensional signal, which can be represented as a sound pressure function $f(t)$ of time. The image signal is a two-dimensional signal, which is a light intensity function $I(x, y)$ of a coordinate's position. In addition, there exist signals of more variables. In this book, mainly one-dimensional signals are studied, whose independent variable is mostly time t or k.

6. Causal and noncausal signals

In practical signal processing applications, input signals start at time $t = 0$. Signals that start at $t = 0$ are referred to as *causal signals*, which satisfy the condition $f(t) = 0, t < 0$. Correspondingly, a signal satisfying the condition $f(t) = 0, t \geq 0$ is referred to as a *noncausal signal*.

The same concept can be extended to DT signals. The DT signals that start at $k = 0$ are referred to as *causal signals*, which satisfy the condition $f(k) = 0, k < 0$. Correspondingly, a signal satisfying the condition $f(k) = 0, k \geq 0$ is referred to as a DT *noncausal signal*.

1.1.4 Representation and plotting of signals with MATLAB

MATLAB provides many tool functions for signal representation and graphic plotting.

Example 1.1.4. Represent and plot the continuous signal $f(t) = 5e^{-0.8t}\sin(\pi t), 0 < t < 5$ with MATLAB.

Solution:

```
b=5; a=0.8;
t=0:.001:5;
x=b*exp(-a*t).*sin(pi*t);
plot (t,x);              % Waveform plotting
```

The resulting waveform is shown in Figure 1.4.

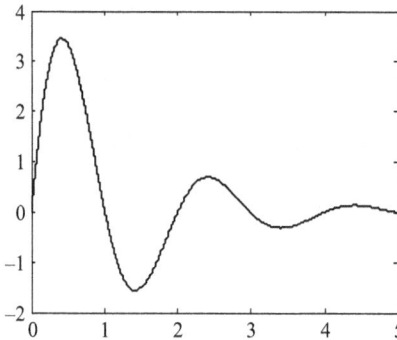

Fig. 1.4: Waveform of the continuous signal in Example 1.1.4

Example 1.1.5. Represent and plot the DT signal $f(k) = 2(0.8)^k, -5 < k < 5$ with MATLAB.

Solution:

```
c=2; d=0.8;
k=-5:5;
y=c*d.^k;               % the symbol ".^" stands for group operation
stem(k,y);
```

The resulting waveform is shown in Figure 1.5.

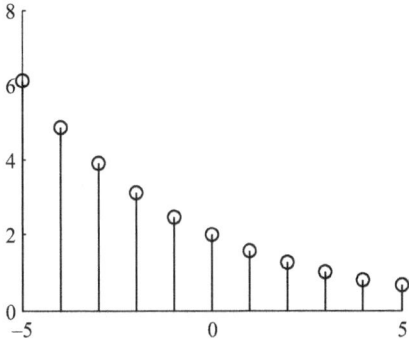

Fig. 1.5: Waveform of the signal in Example 1.1.5

1.2 Basic operations of signals

1.2.1 Operations "+", "−" and "×" of signals

The operations "+", "−" and "×" of two signals $f_1(\cdot)$ and $f_2(\cdot)$ mean addition, subtraction and multiplication, respectively, of the values at each corresponding time.
For example,

$$f_1(k) = \begin{cases} 2, & k = -1 \\ 3, & k = 0 \\ 6, & k = 1 \\ 0, & \text{else} \end{cases} \qquad f_2(k) = \begin{cases} 3, & k = 0 \\ 2, & k = 1 \\ 4, & k = 2 \\ 0, & \text{else} \end{cases}$$

$$f_1(k) + f_2(k) = \begin{cases} 2, & k = -1 \\ 6, & k = 0 \\ 8, & k = 1 \\ 4, & k = 2 \\ 0, & \text{else} \end{cases} \qquad f_1(k) \times f_2(k) = \begin{cases} 9, & k = 0 \\ 12, & k = 1 \\ 0, & \text{else} \end{cases}$$

Note: The symbol $f_1(\cdot)$ means that it can be a function of t or k.

1.2.2 Signal transformations in the time domain

1. Time inversion
When a CT signal $f(t)$ is time reversed, the inverted signal is denoted by $f(-t)$. Likewise, when a DT signal $f(k)$ is time reversed, the inverted signal is denoted by $f(-k)$. As shown in Figure 1.6, we observe that the signal inversion can be performed graphically by simply flipping the signal $f(\cdot)$ about the y-axis.

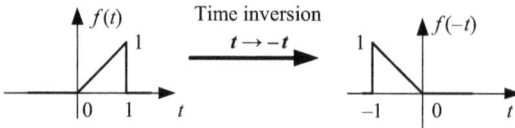

Fig. 1.6: Time inversion

2. Time shifting

The time-shifting operation delays or advances forward the input signal in time. Consider a CT signal $f(t)$, whose time-shifted signal is defined as $f(t - t_0)$. Likewise, the time-shifted signal of $f(k)$ is defined as $f(k - k_0)$. If $t_0 > 0$ (or $k_0 > 0$), the signal is delayed in the time domain. Graphically, this is equivalent to shifting the origin of the signal towards the right-hand side by duration t_0 (k_0) along the time axis. On the other hand, if $t_0 < 0$ (or $k_0 < 0$), the signal is shifted towards the left-hand side.

As shown in Figure 1.7, $f(t - 1)$ is a delayed version of $f(t)$ and $f(t + 1)$ is a time-advanced version. The waveforms are identical to that of $f(t)$ except for a shift of one time unit towards the right-hand and the left-hand side, respectively.

! **Note:** Pay attention to the correct time-shifting direction of $f(t \pm t_0)$.

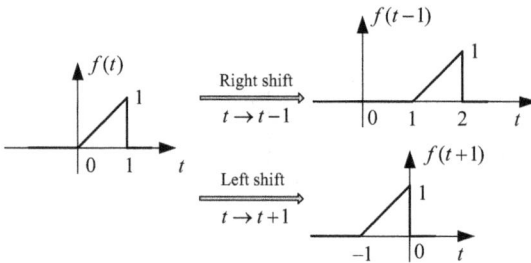

Fig. 1.7: Time-shifting operation of a CT signal

Example 1.2.1. Given a signal $f(t)$, plot the waveform of $f(2 - t)$.

Solution 1. As shown in Figure 1.8, $f(t + 2)$ is first obtained by shifting the given signal $f(t)$ to the left by two time units along the t-axis. Then, $f(-t + 2)$ is obtained by flipping $f(t + 2)$ about the y-axis.

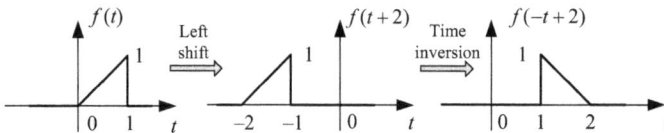

Fig. 1.8: Combined CT operations of time shifting and inversion

Solution 2. As in Figure 1.9, $f(-t)$ is first obtained by flipping $f(t)$ about the y-axis. Then there is a shift $f(-t)$ towards the right-hand side by two time units to obtain $f[-(t-2)] = f(-t+2)$.

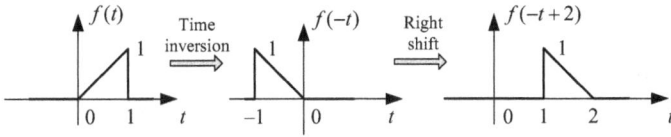

Fig. 1.9: Combined CT operations of inversion and time shifting

Note: Which solution process is better?

3. Time scaling

The time-scaling operation compresses or expands the input signal in the time domain. A CT signal $f(t)$ scaled by a factor a in the time domain is denoted by $f(at)$. If $a > 1$, the signal $f(at)$ is a compressed version of $f(t)$. On the other hand, if $0 < a < 1$, the signal $f(at)$ is an expanded version of $f(t)$. These two operations are illustrated in Figure 1.10, where $f(2t)$ and $f(0.5t)$ are the compressed and expanded versions of $f(t)$, respectively.

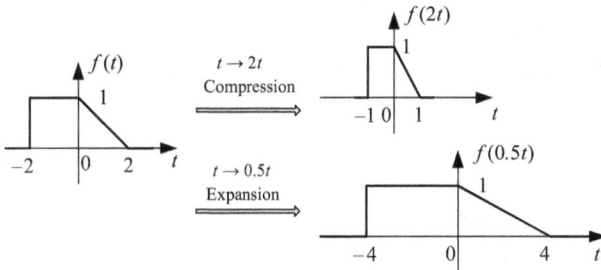

Fig. 1.10: Representations of the time scaling of a CT signal

For a DT signal $f(k)$, some values might be lost when performing the time-scaling operation. For example, if we decimate $f(k)$ by 2, the decimated function [6] $f(2k)$ retains only the alternate samples given by $f(0), f(2), f(4)$ and so on. Compression is, therefore, an irreversible process in the DT domain, as the original sequence $f(k)$ cannot be recovered precisely from the decimated sequence $f(2k)$.

Note: The time scaling of the DT sequence is not discussed here.

Example 1.2.2. Given a signal $f(t)$, sketch the waveform of $f(-4 - 2t)$.

Solution 1. As in Figure 1.11, follow steps (i)–(iii) to obtain $f(-4 - 2t)$.
(i) Shift $f(t)$ towards the right-hand side by four time units to obtain $f(t - 4)$.
(ii) Compress $f(t - 4)$ by a factor of 2 to obtain $f(2t - 4)$.
(iii) Reverse $f(2t - 4)$ to obtain $f(-4 - 2t)$.

! **Note:** This process is the recommended order of operations.

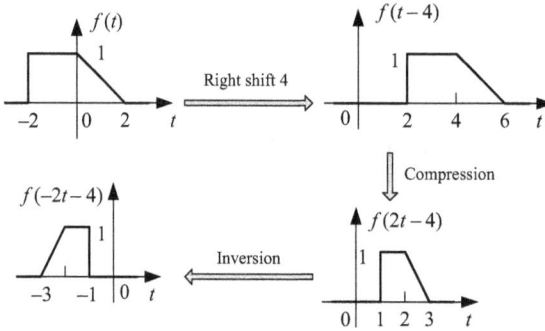

Fig. 1.11: Operations of time shifting, time scaling and inversion

Solution 2. As in Figure 1.12, follow steps (i)–(iii) to obtain $f(-4 - 2t)$.
(i) Compress $f(t)$ by a factor of 2 to obtain $f(2t)$.
(ii) Shift $f(2t)$ towards the right-hand side by two time units to obtain $f[2(t - 2)] = f(2t - 4)$.
(iii) Reverse $f(2t - 4)$ to obtain $f(-4 - 2t)$.

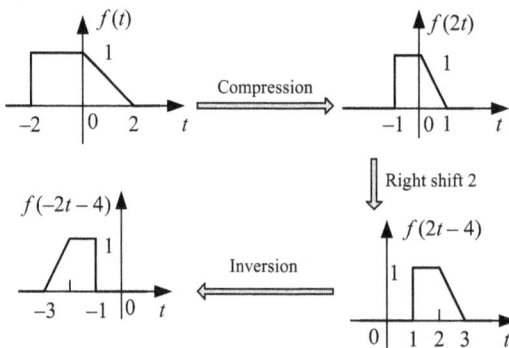

Fig. 1.12: Operations of compression, time shifting and inversion

Example 1.2.3. Given a signal $f(-4 - 2t)$, sketch the waveform of $f(t)$.

Solution: As in Figure 1.13, follow steps (i)–(iii) to obtain $f(t)$.
(i) Reverse $f(-2t - 4)$ to obtain $f(2t - 4)$.
(ii) Expand $f(2t - 4)$ by a factor of 0.5 to obtain $f(t - 4)$.
(iii) Shift $f(t - 4)$ towards the left-hand side by four time units to obtain $f(t)$.

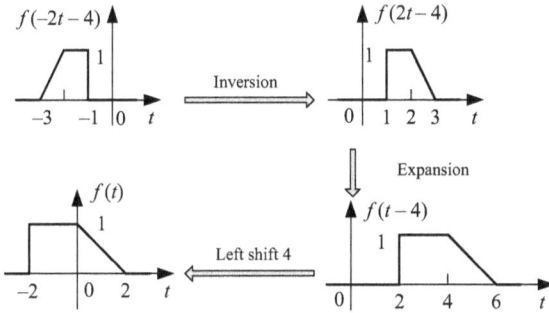

Fig. 1.13: Operations of inversion, expansion and time shifting

1.3 Elementary signals

In this section, we define four elementary functions that will be used frequently to represent more complicated signals [7]. Representing signals in terms of the elementary functions simplifies the analysis and design of linear systems.

1.3.1 The continuous-time unit step function

In Figure 1.14, the unit step function $\varepsilon(t)$ is obtained by calculating the limit of $y_n(t)$, which is illustrated in Figure 1.14. Note that the unit step is discontinuous at $t = 0$:

$$\varepsilon(t) \overset{\text{def}}{=} \lim_{n \to \infty} y_n(t) = \begin{cases} 0 , t < 0 \\ 1 , t > 0 \end{cases} \tag{1.9}$$

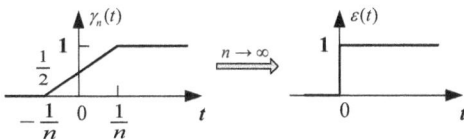

Fig. 1.14: Definition of step function

The properties of the CT unit step function are listed as follows:
(1) It is used to represent certain signals. For example, the signal shown in Figure 1.15 can be represented as:

$$f(t) = 2\varepsilon(t) - 3\varepsilon(t-1) + \varepsilon(t-2) \tag{1.10}$$

Note: How can we write the expression exactly and quickly?

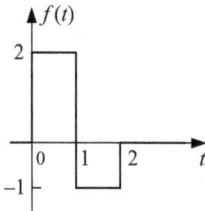

Fig. 1.15: Decomposition representation of signals

(2) It is used to express the time interval of signals. As in Figure 1.16, the interval of the signal in (b) is $t > 0$, and the interval of (c) is $t_1 < t < t_2$.

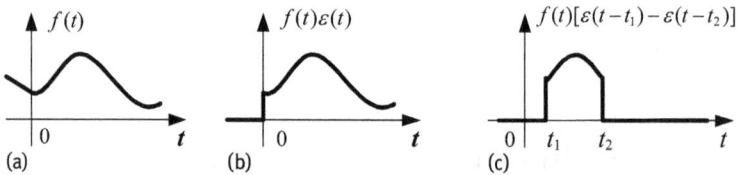

Fig. 1.16: Expression of the time interval of signals by unit step functions

(3) The integral of $\varepsilon(t)$ is calculated as follows:

$$\int_{-\infty}^{t} \varepsilon(\tau)\,d\tau = t\varepsilon(t) \tag{1.11}$$

Note: How do we prove it?

1.3.2 The continuous-time unit impulse function

The CT unit impulse function $\delta(t)$, also known as the Dirac delta function or, simply, the delta function, is defined in terms of two properties as follows:

$$\begin{cases} \delta(t) = 0, & t \neq 0 \\ \int_{-\infty}^{\infty} \delta(t)dt = 1 \end{cases} \tag{1.12}$$

Consider a tall narrow rectangle $p_n(t)$ with width $2/n$ and height $n/2$, as shown in Figure 1.17 (a), such that the area enclosed by the rectangular function equals 1. As $n \to \infty$, the rectangular function converges to the CT impulse function $\delta(t)$ with an infinite amplitude at $t = 0$. However, the area enclosed by CT impulse function is finite and equals 1:

$$\delta(t) \overset{\text{def}}{=} \lim_{n \to \infty} p_n(t) \tag{1.13}$$

Note: The area enclosed by the CT impulse function is $(n/2) \times (2/n) = 1$.

The impulse function is illustrated in Figure 1.17 (b) by an arrow pointing vertically upwards. The height of the arrow corresponds to the area enclosed by the CT impulse function.

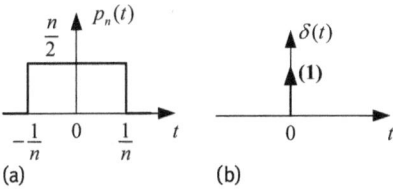

(a)   (b)

Fig. 1.17: Generation of the unit impulse function; (a) Rectangular pulse, (b) Unit impulse

Figure 1.18 depicts the relationships between unit step function and unit impulse function. The unit step function $\varepsilon(t)$ and unit impulse function $\delta(t)$ are the limiting forms of $\gamma_n(t)$ and $p_n(t)$, respectively. Moreover, the rectangular pulse $p_n(t)$ is the derivative of $\gamma_n(t)$. The relationships between $\varepsilon(t)$ and $\delta(t)$ can be described by:

$$\varepsilon(t) \overset{\text{def}}{=} \lim_{n \to \infty} \gamma_n(t), \quad \delta(t) \overset{\text{def}}{=} \lim_{n \to \infty} p_n(t), \quad p_n(t) = \frac{d\gamma_n(t)}{dt} \quad \Rightarrow \quad \delta(t) = \frac{d\varepsilon(t)}{dt} \tag{1.14}$$

$$\varepsilon(t) = \int_{-\infty}^{t} \delta(\tau)\,d\tau \tag{1.15}$$

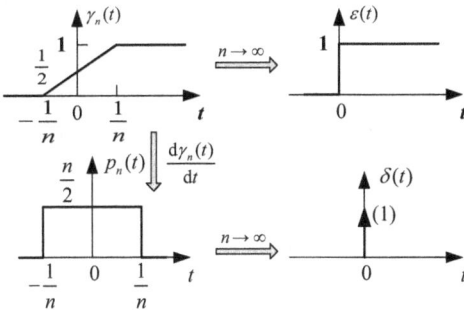

Fig. 1.18: Relationship between unit step function and unit impulse function

Consider the discontinuous signal $f(t)$ in Figure 1.19 (a). Because of the relationship between the CT unit impulse and unit step, we can readily calculate the derivative of this signal. Their expressions are as follows:

$$f(t) = 2\varepsilon(t + 1) - 2\varepsilon(t - 1) \quad f'(t) = 2\delta(t + 1) - 2\delta(t - 1) \tag{1.16}$$

Then we plot the derivative of signal $f(t)$. Specifically, the derivative of $f(t)$ is clearly 0, except at the discontinuities. As in Figure 1.19 (b), the impulse pulse is placed at each discontinuity of $f(t)$. Note, for example, that the discontinuity in $f(t)$ at $t = 1$ has a value of -2, so that an impulse scaled by -2 is located at $t = 1$ in the signal $f'(t)$.

! **Note:** Pay attention to the plot of $-2\delta(t - 1)$.

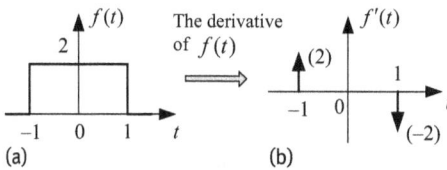

Fig. 1.19: Illustration of derivative of discontinuous signal; (a) The discontinuous signal $f(t)$, (b) The derivative signal

1.3.3 Properties of the CT unit impulse function

1. Sampling property
When an arbitrary function $f(t)$ is multiplied by a unit impulse function, the product is given by:

$$f(t)\delta(t) = f(0)\delta(t) \tag{1.17}$$

When $f(t)$ is multiplied by a shifted impulse function, the product is given by:

$$f(t)\delta(t - a) = f(a)\delta(t - a) \tag{1.18}$$

In other words, multiplication of a CT function and an impulse function produces an impulse function, which has an area equal to the value of the CT function at the location of the impulse. It is referred to as the sampling property [8]. Integrating the multiplication can produce:

$$\int_{-\infty}^{\infty} f(t)\delta(t)\,dt = f(0) \quad \int_{-\infty}^{\infty} f(t)\delta(t - a)\,dt = f(a) \tag{1.19}$$

Example 1.3.1. Simplify the following expressions:

(1) $\sin\left(t + \dfrac{\pi}{4}\right)\delta(t)$

(2) $\displaystyle\int_{-1}^{9} \sin\left(t - \dfrac{\pi}{4}\right)\delta(t)\,dt$

(3) $\displaystyle\int_{-3}^{0} \sin\left(t - \dfrac{\pi}{4}\right)\delta(t-1)\,dt$

(4) $\displaystyle\int_{-1}^{1} 2\tau\delta(\tau - t)\,d\tau$

(5) $\displaystyle\int_{-1}^{t} (\tau - 1)^2\delta(\tau)\,d\tau$

(6) $\dfrac{d}{dt}\left[e^{-2t}\varepsilon(t)\right]$

Solution:

(1) Using Equation (1.17) yields

$$\sin\left(t + \dfrac{\pi}{4}\right)\delta(t) = \sin\left(\dfrac{\pi}{4}\right)\delta(t)|_{t=0} = \dfrac{\sqrt{2}}{2}\delta(t).$$

(2) Using Equation (1.18) yields

$$\int_{-1}^{9} \sin\left(t - \dfrac{\pi}{4}\right)\delta(t)\,dt = \sin\left(t - \dfrac{\pi}{4}\right)\Big|_{t=0} = -\dfrac{\sqrt{2}}{2}.$$

(3) Since the integral interval does not include the impulse location at $t = 1$,

$$\int_{-3}^{0} \sin\left(t - \dfrac{\pi}{4}\right)\delta(t-1)\,dt = 0.$$

(4) Using Equation (1.17) yields

$$\int_{-1}^{1} 2\tau\delta(\tau - t)\,d\tau = \int_{-1}^{1} 2t\delta(\tau - t)\,d\tau = 2t\int_{-1}^{1} \delta(\tau - t)\,d\tau.$$

If the impulse location satisfies $\tau = t \in (-1, 1)$, the integral $\int_{-1}^{1} \delta(\tau - t)\,d\tau$ yields 1; otherwise, it yields 0. So, the result expression is

$$2t \cdot [\varepsilon(t + 1) - \varepsilon(t - 1)].$$

(5) Using Equation (1.17) yields

$$\int_{-1}^{t} (\tau - 1)^2\delta(\tau)\,d\tau = \int_{-1}^{t} (0 - 1)^2\delta(\tau)\,d\tau = \int_{-1}^{t} \delta(\tau)\,d\tau = \varepsilon(t)$$

(6)

$$\dfrac{d}{dt}\left[e^{-2t}\varepsilon(t)\right] = e^{-2t}\delta(t) - 2e^{-2t}\varepsilon(t) = \delta(t) - 2e^{-2t}\varepsilon(t)$$

2. Derivative $\delta'(t)$ of the CT unit impulse function

The derivative of the multiplication of a CT function and an impulse function is given by:

$$[f(t)\delta(t)]' = f(t)\delta'(t) + f'(t)\delta(t) .$$

The expression is converted to:

$$f(t)\delta'(t) = [f(t)\delta(t)]' - f'(t)\delta(t) = f(0)\delta'(t) - f'(0)\delta(t) . \tag{1.20}$$

The definition of derivative $\delta'(t)$ of the CT unit impulse functions is obtained by integrating Equation (1.20):

$$\int_{-\infty}^{\infty} \delta'(t)f(t)\,dt = -f'(0) \tag{1.21}$$

! **Note:** In the deduction process, we used $\int_{-\infty}^{\infty} \delta'(t)\,dt = 0$.

Likewise, the n-order derivative of impulse function $\delta^{(n)}(t)$ is defined as:

$$\int_{-\infty}^{\infty} \delta^{(n)}(t)f(t)\,dt = (-1)^n f^{(n)}(0) . \tag{1.22}$$

Example 1.3.2. Simplify the expression $\int_{-\infty}^{\infty}(t-2)^2\delta'(t)\,dt$.

Solution:

$$\int_{-\infty}^{\infty}(t-2)^2\delta'(t)\,dt = -\frac{d}{dt}[(t-2)^2]\,|_{t=0} = -2(t-2)\,|_{t=0} = 4$$

3. Time-scaling property

The scaled version $\delta(at)$ is given by:

$$\delta(at) = \frac{1}{|a|}\delta(t) . \tag{1.23}$$

When $a = -1$, we have:

$$\delta(-t) = \delta(t) .$$

The impulse function is an even function.

The scaled and time-shifted version $\delta(at-t_0)$ of the unit impulse function is given by:

$$\delta(at - t_0) = \frac{1}{|a|}\delta\left(t - \frac{t_0}{a}\right) \tag{1.24}$$

The n-order derivative $\delta^{(n)}(at)$ is defined as:

$$\delta^{(n)}(at) = \frac{1}{|a|} \cdot \frac{1}{a^n}\delta^{(n)}(t) \tag{1.25}$$

When $a = -1$ and $n = 1$, we have $\delta'(-t) = -\delta'(t)$. The $\delta'(t)$-function is an odd function.

Example 1.3.3. Given the signal $f(t)$ shown in Figure 1.20, sketch the signals of $g(t) = f'(t)$ and $g(2t)$.

Solution: Note that the discontinuity in $f(t)$ at $t = -2$ has a value of 4, so that an impulse is located at $t = -2$ in the signal $f'(t)$. The derivative signal $g(t)$ holds a constant value of -1 in the time interval of $(-2, 2)$. Finally, compress $g(t)$ by a factor of 2 to obtain $g(2t)$ with half the amplitude of the impulse function. The time interval of the constant value -1 compresses to $(-1, 1)$.

Note: Pay attention to the plot of the time scaling of the impulse function. !

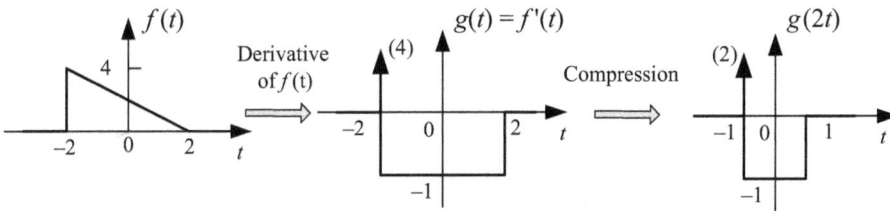

Fig. 1.20: Illustration of derivation and time scaling

1.3.4 The discrete-time unit step and impulse sequences

1. The DT unit impulse sequence
The DT unit impulse sequence is defined by:

$$\delta(k) \stackrel{\text{def}}{=} \begin{cases} 1, & k = 0 \\ 0, & k \neq 0 \end{cases} \tag{1.26}$$

The waveform of $\delta(k)$ is shown in Figure 1.21. Unlike the CT unit impulse function, the DT impulse function has no ambiguity in its definition; it is well defined for all values of k.

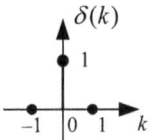

Fig. 1.21: The DT unit impulse sequence

Note: The definition is different to that of the CT impulse function. !

The DT unit impulse sequence has similar sampling properties as follows:

$$f(k)\delta(k) = f(0)\delta(k) \tag{1.27}$$

$$f(k)\delta(k - k_0) = f(k_0)\delta(k - k_0) \tag{1.28}$$

$$\sum_{k=-\infty}^{\infty} f(k)\delta(k) = f(0) \tag{1.29}$$

Example 1.3.4. Simplify the following expressions:

(1) $\displaystyle\sum_{k=-\infty}^{\infty} \delta(k)$ (2) $\displaystyle\sum_{k=-\infty}^{\infty} (k - 5)\delta(k)$ (3) $\displaystyle\sum_{k=-\infty}^{2} (k - 5)\delta(k - 4)$

Solution:

(1)

$$\sum_{k=-\infty}^{\infty} \delta(k) = 1$$

(2)

$$\sum_{k=-\infty}^{\infty} (k - 5)\delta(k) = (k - 5)|_{k=0} = -5$$

(3) Since the summing interval does not include the impulse location at $k = 4$,

$$\sum_{k=-\infty}^{2} (k - 5)\delta(k - 4) = 0 .$$

2. The DT unit step sequence
The DT unit step sequence is defined by:

$$\varepsilon(k) \overset{\text{def}}{=} \begin{cases} 1, & k \geq 0 \\ 0, & k < 0 \end{cases} \tag{1.30}$$

The waveform of $\varepsilon(k)$ is shown in Figure 1.22. We know that the CT unit step function $\varepsilon(t)$ is piecewise continuous with a discontinuity at $t = 0$. However, the DT function $\varepsilon(k)$ has no such discontinuity.

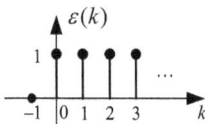

Fig. 1.22: The DT unit step sequence

Note: The value at $k = 0$ is 1, which is different to that of the CT step function.

3. The relationship between $\varepsilon(k)$ and $\delta(k)$

There is a close relationship between the DT unit impulse and unit step sequence. The DT unit impulse is the first difference of the DT step sequence:

$$\delta(k) = \varepsilon(k) - \varepsilon(k-1) \qquad (1.31)$$

Conversely, the DT unit step is the running sum of the unit impulse. That is,

$$\varepsilon(k) = \sum_{i=-\infty}^{k} \delta(i) \qquad (1.32)$$

1.4 Basic definitions and classification of systems

1.4.1 Introduction to systems

The physical structures used to generate, transmit and process signals are referred to as systems. In contexts ranging from signal processing and communications to electromechanical motors, and automotive vehicles, a system can be viewed as a process in which input signals are transformed by the system or cause the system to respond in some way, resulting in other signals as outputs. In other words, a system establishes a relationship between a set of inputs and the corresponding set of outputs. For example, the mobile communication system [9] shown in Figure 1.23 (a) can be viewed as

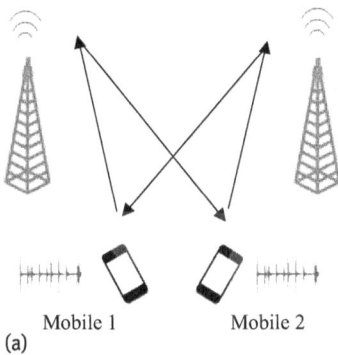

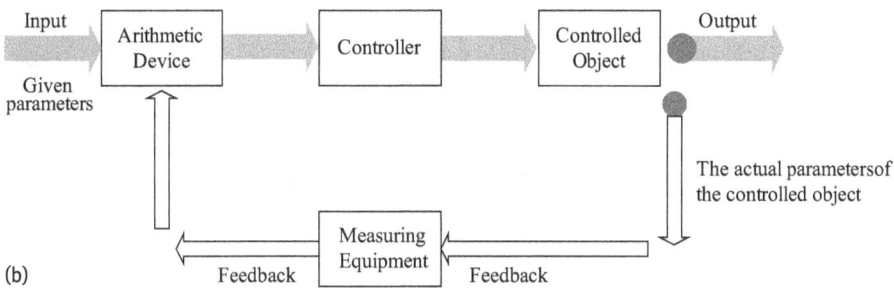

Fig. 1.23: Examples of systems; (a) Block diagram of the mobile communication system, (b) Block diagram of the control system

a system to transmit signals. The control system shown in Figure 1.23 (b) is to adjust the parameters so that the output is the desired value. A system can be viewed as a single machine, such as a cell phone, a television, stabilized voltage supply, etc. In addition, a system can also be viewed as circuit modules with a certain function, such as bleeder circuits, filter circuits, amplifier circuits, etc.

The block diagram representing general schematics of a system is illustrated in Figure 1.24. The system processes or transforms the input signals to result in the output signals. The systems are designed based on certain applications. The most important system is the *linear time-invariant (LTI) system*, which is mainly studied in this book.

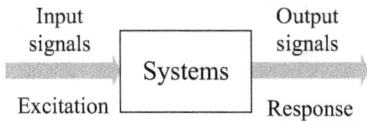

Input signals   Output signals

Systems

Excitation   Response   **Fig. 1.24:** Block diagram of systems

1.4.2 Classifications of systems

In the analysis or design of a system, it is desirable to classify the system according to some generic properties that the system satisfies. In this section, we classify systems into seven basic categories:

(i)   continuous-time and discrete-time systems
(ii) dynamic and instantaneous systems
(iii) single-input single-output and multiple-input multiple-output systems
(iv) linear and nonlinear systems
(v) time-invariant and time-varying systems
(vi) causal and noncausal systems
(vii)stable and unstable systems

1. Continuous-time and discrete-time systems

A *continuous-time system* is a system in which continuous-time input signals are applied to result in continuous-time output signals. Similarly, a *discrete-time system* is a system that transforms discrete-time input signals into discrete-time outputs. In most of this book, we will treat discrete-time systems and continuous-time systems separately but in parallel.

2. Dynamic and instantaneous systems

A *dynamic system* is a system whose output signal at a given time depends both on the present input value and the system's initial conditions. Electrical systems containing dynamic elements, i.e., capacitor, inductor, etc., are dynamic systems. In contrast, a system is called an *instantaneous system* if its output value at a given time depends only on the input signal at that same time. The circuit systems containing only instantaneous elements, i.e., resistors, are instantaneous systems, which do not have dynamic memory.

Note: We focus on dynamic systems. !

3. Single-input single-output and multiple-input multiple-output systems
A system with only one input and one output is called a *single-input single-output system* or *SISO system*. Multiple-input multiple-output (MIMO) systems are often approximated by a combination of several single-input systems [10]. Throughout this book, we will focus our discussion on the analysis and design of SISO systems.

4. Linear and nonlinear systems
Let $y_1(\cdot)$ be the response of a system to an input $f_1(\cdot)$ and let $y_2(\cdot)$ be the output corresponding to the input $f_2(\cdot)$. Then the system is linear if it satisfies the linear property.
(i) The response to $af_1(\cdot)$ is $ay_1(\cdot)$, where a is any complex constant.
(ii) The response to $f_1(\cdot) + f_2(\cdot)$ is $y_1(\cdot) + y_2(\cdot)$.

The first of these two properties is known as the homogeneity property; the second is known as the additive property. They hold in continuous systems and discrete systems. The two properties defining a linear system can be combined into a single statement as follows:

$$\text{Continuous system:} \quad af_1(t) + bf_2(t) \quad \rightarrow \quad ay_1(t) + by_2(t), \quad (1.33)$$
$$\text{Discrete system:} \quad af_1(k) + bf_2(k) \quad \rightarrow \quad ay_1(k) + by_2(k). \quad (1.34)$$

Here, a and b are any complex constants.
    The whole response of a dynamic system is composed of the zero-state response on the input $\{f(\cdot)\}$ and zero-input response on the initial states $\{x(0)\}$ of the system.

Note: The input is the external excitation and the initial state is the internal excitation. !

The whole response:
$$y(\cdot) = T[\{f(\cdot)\}, \{x(0)\}] \quad (1.35)$$

The zero-state response:
$$y_{zs}(\cdot) = T[\{f(\cdot)\}, \{0\}] \quad (1.36)$$

The zero-input response:
$$y_{zi}(\cdot) = T[\{0\}, \{x(0)\}] \quad (1.37)$$

A dynamic system is linear if it satisfies the following three conditions:
(i) Decomposability of the whole response:

$$y(\cdot) = y_{zs}(\cdot) + y_{zi}(\cdot) = T[\{f(\cdot)\}, \{0\}] + T[\{0\}, x\{(0)\}] \qquad (1.38)$$

(ii) Linearity of the zero-state response:

$$T[\{af_1(t) + bf_2(t)\}, \{0\}] = a \cdot T[\{f_1(\cdot)\}, \{0\}] + b \cdot T[f_2\{\cdot\}, \{0\}] \qquad (1.39)$$

(iii) Linearity of the zero-input response:

$$T[\{0\}, \{ax_1(0) + bx_2(0)\}] = a \cdot T[\{0\}, \{x_1(0)\}] + b \cdot T[\{0\}, \{x_2(0)\}] \qquad (1.40)$$

Example 1.4.1. Determine whether the CT systems are linear.
(a) $y(t) = 3x(0) + 2f(t) + x(0)f(t)$
(b) $y(t) = 2x(0) + |f(t)|$
(c) $y(t) = x^2(0) + 2f(t)$

Solution:
(a) The zero-state response is $y_{zs}(t) = 2f(t)$, and the zero-input response is $y_{zi}(t) = 3x(0)$. Obviously, $y(t) \neq y_{zs}(t) + y_{zi}(t)$.
From Equation (1.38), the system does not satisfy decomposability. Therefore, the system is nonlinear.
(b) The whole response is decomposed into $y_{zs}(t) = |f(t)|$ and $y_{zi}(t) = 2x(0)$.
For the zero-state response,

$$T[\{a \cdot f(t)\}, \{0\}] = |a \cdot f(t)| \neq a \cdot |f(t)| = a \cdot y_{zs}(t) .$$

Hence, it is not linear.
(c) The whole response is decomposed into $y_{zs}(t) = 2f(t)$ and the zero-input response $y_{zi}(t) = x^2(0)$.
For the zero-input response:

$$T[\{0\}, \{a \cdot x(0)\}] = [a \cdot x(0)]^2 \neq a \cdot y_{zi}(t) .$$

Hence, it is not linear.

Example 1.4.2. Determine whether or not the following system is linear:

$$y(t) = e^{-t}x(0) + \int_0^t \sin(x)f(x)\,dx$$

Solution: Letting $x(0) = 0$, the zero-state response is $y_{zs}(t) = \int_0^t \sin(x)f(x)\,dx$.
Letting the input $f(t) = 0$, the zero-input response is $y_{zi}(t) = e^{-t}x(0)$.
Obviously, $y(t) = y_{zs}(t) + y_{zi}(t)$, thus the system $y(t)$ is decomposable.

Since:

$$T[\{af_1(t) + bf_2(t)\}, \{0\}] = \int_0^t \sin(x)[af_1(x) + bf_2(x)]\,\mathrm{d}x$$

$$= a \int_0^t \sin(x)f_1(x)\,\mathrm{d}x + b \int_0^t \sin(x)f_2(x)\,\mathrm{d}x$$

$$= aT[\{f_1(t)\}, \{0\}] + bT[\{f_2(t)\}, \{0\}] \,,$$

the zero-state response is linear.

Likewise, the zero-input response satisfies the following linearity:

$$T[\{0\}, \{ax_1(0) + bx_2(0)\}] = e^{-t}[ax_1(0) + bx_2(0)]$$

$$= ae^{-t}x_1(0) + be^{-t}x_2(0)$$

$$= aT[\{0\}, \{x_1(0)\}] + bT[\{0\}, \{x_2(0)\}]$$

Therefore, the system $y(t)$ is a linear system.

Note: If the fact violates any of the three conditions, it is not a linear system. !

5. Time-invariant and time-varying systems

A system is said to be a *time-invariant system* if a time delay or time advance of the input signal leads to an identical time-shift in the output signal. A CT system with $f(t) \rightarrow y_{zs}(t)$ is time-invariant if:

$$f(t - t_d) \rightarrow y_{zs}(t - t_d) \tag{1.41}$$

for any arbitrary time-shift t_d. Likewise, a DT system with $f(k) \rightarrow y_{zs}(k)$ is time invariant if:

$$f(k - k_d) \rightarrow y_{zs}(k - k_d) \tag{1.42}$$

for any arbitrary discrete time-shift k_d.

As shown in Figure 1.25, a time-invariant system responds exactly the same way no matter when the input signal is applied. For a time-delay input $f(t-1)$, its zero-state response $y_{zs}(t - 1)$ is equally time shifted with the identical waveform.

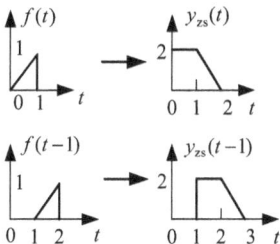

Fig. 1.25: Illustration of the time-invariant system

Example 1.4.3. Determine whether or not each of the following systems is time invariant:

(a) $y_{zs}(k) = f(k) \cdot f(k-1)$
(b) $y_{zs}(t) = t \cdot f(t)$
(c) $y_{zs}(t) = f(-t)$

Solution:

(a) In this system, we have:

$$f(k) \rightarrow f(k)f(k-1) = y_{zs}(k)$$

If the time-shifted signal $f(k - k_d)$ is applied at the system, the new output is given by:

$$f(k - k_d) \rightarrow f(k - k_d)f(k - 1 - k_d) = y_{zs}(k - k_d)$$

Obviously, the system is time invariant.

(b) In this system, we have:

$$f(t) \rightarrow t \cdot f(t) = y_{zs}(t)$$

If the time-shifted signal $f(t - t_d)$ is applied at the system, the new output is given by:

$$f(t - t_d) \rightarrow t \cdot f(t - t_d)$$

The shifted output $y_{zs}(t - t_d)$ is given by:

$$y_{zs}(t - t_d) = (t - t_d)f(t - t_d)$$

Since $t \cdot f(t - t_d) \neq y_{zs}(t - t_d)$, this system is time varying.

(c) In this system, we have:

$$f(t) \rightarrow f(-t) = y_{zs}(t)$$

If the time-shifted signal $f(t - t_d)$ is applied at the system, the new output is given by:

$$f(t - t_d) \rightarrow f(-t - t_d)$$

The shifted output $y_{zs}(t - t_d)$ is given by:

$$y_{zs}(t - t_d) = f[-(t - t_d)]$$

Since $f(-t - t_d) \neq y_{zs}(t - t_d)$, this system is time varying.

! **Note:** If there exists a variant coefficient in front of $f(\cdot)$ or $y(\cdot)$, or time-inversion or time-scaling operations in $f(\cdot)$ or $y(\cdot)$, the system is time varying.

If the system is not only linear but also time invariant, it is an LTI system. The continuous LTI system has the differentiation and integration property as follows:

$$\text{If} \quad f(t) \rightarrow y_{zs}(t) , \quad \text{then} \quad f'(t) \rightarrow y'_{zs}(t) \tag{1.43}$$

$$\text{If} \quad f(t) \rightarrow y_{zs}(t) \quad \text{and} \quad f(-\infty) = 0, y_{zs}(-\infty) = 0 ,$$

$$\text{then} \quad \int_{-\infty}^{t} f(x)\,dx \rightarrow \int_{-\infty}^{t} y_{zs}(x)\,dx \tag{1.44}$$

6. Causal and noncausal systems

A CT system is *causal* if the output at time t_0 depends only on the input $f(t)$ for $t \le t_0$. Likewise, a DT system is causal if the output at time instant k_0 depends only the input $f(k)$ for $k \le k_0$. A system that violates the causality condition is called a noncausal system.

In other words, for a causal CT system, if its input signal $f(t)$ satisfies the condition:

$$f(t) = 0 \quad \text{for} \quad t < t_0 .$$

then, its zero-state response satisfies the condition:

$$y_{zs}(t) = 0 \quad \text{for} \quad t < t_0 .$$

Example 1.4.4.
(a) CT time-delay system $y_{zs}(t) = f(t - 1) \Rightarrow$ causal system
(b) CT time-forward system $y_{zs}(t) = f(t + 1) \Rightarrow$ non-causal system
(c) CT time-scaling system $y_{zs}(t) = f(2t) \Rightarrow$ non-causal system
(d) DT time-advanced system $y_{zs}(k) = f(k - 1) \Rightarrow$ causal system

Example 1.4.5. Let $x(0_-)$ be the initial state of a causal continuous LTI system. When $x(0_-) = 1$, the full response to a causal input signal $f_1(t)$ is:

$$y_1(t) = e^{-t} + \cos(\pi t) , \quad t > 0 .$$

When $x(0_-) = 2$, the full response to an input signal $f_2(t) = 3f_1(t)$ is:

$$y_2(t) = -2e^{-t} + 3\cos(\pi t) , \quad t > 0 .$$

Determine the zero-state response of the system $y_{3zs}(t)$ when the input signal is

$$f_3(t) = \frac{df_1(t)}{dt} + 2f_1(t - 1) .$$

Solution: Assuming that:

$$x(0_-) = 1 \rightarrow y_{zi}(t), \qquad\qquad f_1(t) \rightarrow y_{zs}(t)$$
$$x(0_-) = 2 \rightarrow 2y_{zi}(t), \quad f_2(t) = 3f_1(t) \rightarrow 3y_{zs}(t),$$

the full response is composed of zero-input response and zero-state response. We have:

$$y_1(t) = y_{zi}(t) + y_{zs}(t) = e^{-t} + \cos(\pi t), \quad t > 0 \tag{1.45}$$

and:

$$y_2(t) = 2y_{zi}(t) + 3y_{zs}(t) = -2e^{-t} + 3\cos(\pi t), \quad t > 0 \tag{1.46}$$

Subtracting twice Equation (1.45) from Equation (1.46), we obtain:

$$y_{zs}(t) = -4e^{-t} + \cos(\pi t), \quad t > 0$$

The closed-form analytic solution is rewritten as:

$$f_1(t) \rightarrow y_{zs}(t) = [-4e^{-t} + \cos(\pi t)]\varepsilon(t)$$

! **Note:** Why is the function $\varepsilon(t)$ added?

Using the differentiation property given in Equation (1.43), we have:

$$\frac{df_1(t)}{dt} \rightarrow \frac{dy_{zs}(t)}{dt} = -3\delta(t) + [4e^{-t} - \pi\sin(\pi t)]\varepsilon(t)$$

Using the time-invariance property of LTI systems, we obtain:

$$f_1(t-1) \rightarrow y_{zs}(t-1) = \{-4e^{-(t-1)} + \cos[\pi(t-1)]\}\varepsilon(t-1).$$

Finally, applying the linearity property, we have:

$$f_3(t) = \frac{df_1(t)}{dt} + 2f_1(t-1) \rightarrow y_{3zs}(t) = \frac{dy_{zs}(t)}{dt} + 2y_{zs}(t-1),$$

$$y_{3zs}(t) = -3\delta(t) + [4e^{-t} - \pi\sin(\pi t)]\varepsilon(t) + 2\{-4e^{-(t-1)} + \cos[\pi(t-1)]\}\varepsilon(t-1).$$

7. Stable and unstable systems

Before defining the stability criterion for a system, we define the bounded property for a signal. A CT signal $f(t)$ or a DT signal $f(k)$ is said to be bounded in magnitude if:

CT signal	$\|f(t)\| < \infty,$	$-\infty < t < \infty;$	(1.47)
DT signal	$\|f(k)\| < \infty,$	$-\infty < k < \infty,$	(1.48)

A system is said to be *bounded-input, bounded-output (BIBO) stable* if an arbitrary bounded-input signal $f(\cdot)$ always produces a bounded-output response $y_{zs}(\cdot)$:

CT system	$\|f(t)\| < \infty \rightarrow \|y_{zs}(t)\| < \infty,$	$-\infty < t < \infty;$	(1.49)
DT system	$\|f(k)\| < \infty \rightarrow \|y_{zs}(k)\| < \infty,$	$-\infty < k < \infty.$	(1.50)

The stability property of systems will be discussed and analyzed in the following chapters.

1.5 Framework of analytical methods

1.5.1 Analytical methods for LTI systems

1. Analytical content of signals and systems

Content of study: Given an LTI system (including the initial state) and an input signal, the output signal or response of the system are determined by solving the mathematical equations.

Methods of analysis: The analytical methods of systems include the input–output techniques (also known as the external analysis), and the state-variable methods (also known as the internal analysis). Chapters 2 to 6 belong to external analysis, which includes time-domain analysis in Chapters 2 and 3, Fourier transform analysis in Chapter 4, Laplace transform analysis in Chapter 5 and the Z-transform analysis in Chapter 6. Chapter 7 is about internal techniques, analyzing the internal properties of continuous and discrete systems by state equations.

2. Basic ideas for solving LTI systems

(i) Determine the zero-input response and zero-state response individually.
(ii) Decompose the input signal into the sum of multiple basic signals. According to the additive and homogeneity properties of linear systems, the response is equal to the sum of the individual response of each basic signal.

1.5.2 Key issues to study

In this book, we treat CT and DT systems separately. Chapters 2 focuses on continuous-time domain analysis and Chapter 3 on discrete-time domain analysis. Frequency-domain analysis is discussed in Chapter 4 and S-domain analysis in Chapter 5 aim at analyzing CT systems. Chapter 6 provides the Z-domain methods to analyze DT systems. In brief, three key issues are covered in each chapter: signal decomposition, basic response and systematic methods of LTI systems.

1. Signal decomposition

Figure 1.26 shows the decomposition methods based on each elementary signal in each chapter.

2. Basic response

Figure 1.27 shows the basic response in the time domain and in the transform domain in each chapter.

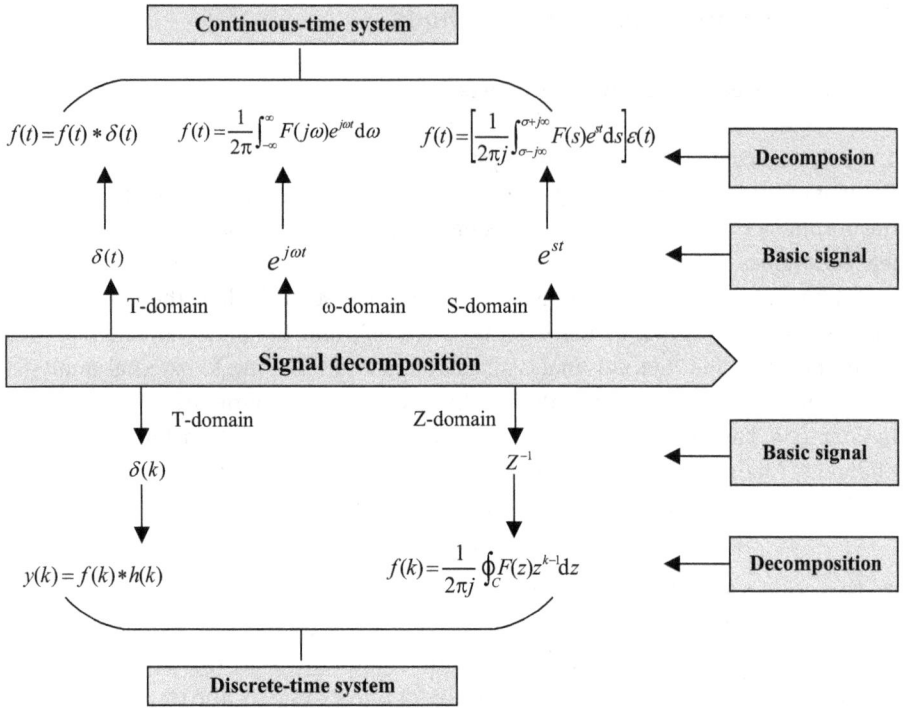

Fig. 1.26: Signal decompositions based on different basic signals

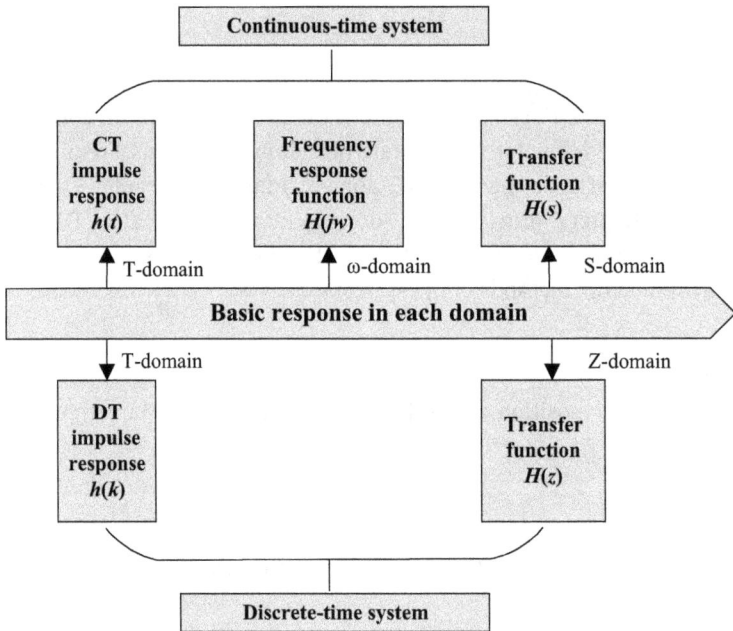

Fig. 1.27: Basic responses in time and transform domains

3. Systematic methods of LTI systems
Figure 1.28 shows the systematic analysis methods of LTI systems in each chapter.

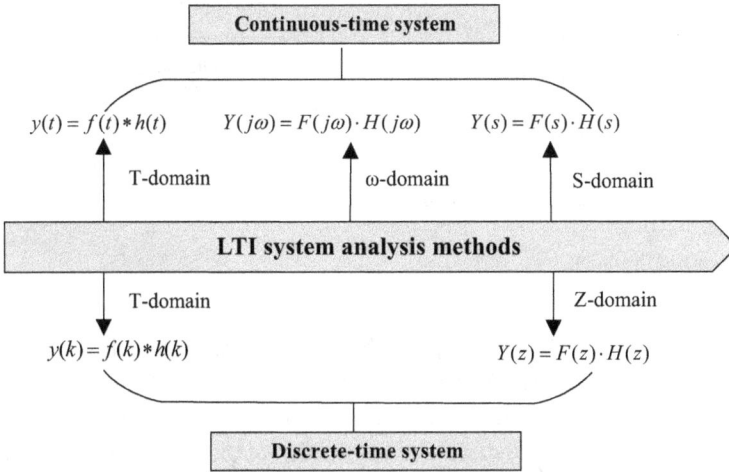

Continuous-time system

$y(t) = f(t) * h(t)$ $Y(j\omega) = F(j\omega) \cdot H(j\omega)$ $Y(s) = F(s) \cdot H(s)$

T-domain   ω-domain   S-domain

LTI system analysis methods

T-domain   Z-domain

$y(k) = f(k) * h(k)$ $Y(z) = F(z) \cdot H(z)$

Discrete-time system

Fig. 1.28: Systematic analysis methods of LTI systems

Note: All methods are unified under a framework.

1.5.3 Framework of all chapters

Figure 1.29 shows the framework of the three parts of this books. The first part is about time-domain analysis in the first three chapters. The second part is about transform-domain analysis in Chapters 4–6, which are the core contents of this book. The third part extends the introductions to the corresponding applications.

Chapter 1 introduces signals and systems, including their mathematical and graphical interpretations. The classifications and properties of systems are detailed.

Chapter 2 focuses on the method of establishing and solving the differential equations of CT systems. The convolution integral operation is discussed to solve the zero-state response in the time domain.

Chapter 3 gives the method of establishing and solving the difference equations of DT systems. The convolution sum operation is discussed to solve the zero-state response in the time domain.

Chapter 4 defines the CT Fourier series and the Fourier transform as a frequency-domain representation. The frequency-response function is applied to analyze systems. The sampling theorem is given to realize the discretization of continuous-time signals.

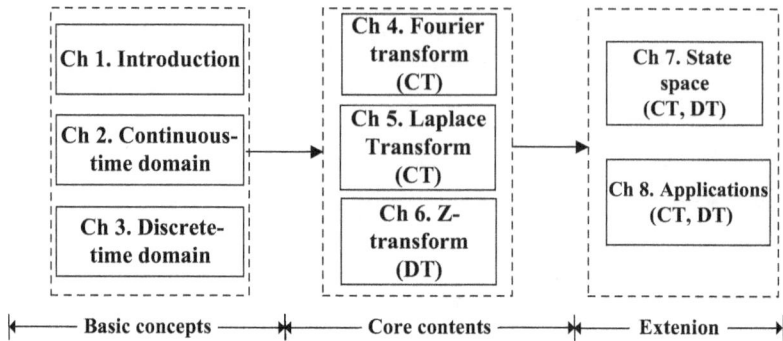

Fig. 1.29: The framework of this book

Chapter 5 details the Laplace transform, transfer function and system flow graph. Chapter 6 is devoted to a discussion of the Z-transform. Chapter 7 presents the methods of establishing the state and output equations. The state-space solution is discussed in the transform domain. Chapter 8 introduces the application of the related concepts in the course of signals and systems, such as control systems, digital signal processing, image processing, communication systems, etc.

1.6 Summary

This chapter introduced the basic concepts of signals and systems. In Section 1.1, we classified signals in six different categories. Section 1.2 presented signal operations of time-shifting, inversion and scaling. The properties of the basic signals $\delta(t)$ and $\delta(k)$ were illustrated in Section 1.3. The requirements of linear time-invariant (LTI) system were detailed in Section 1.4. A system is linear if it satisfies the principle of superposition. A system is time-invariant if a time-shift in the input signal leads to an identical shift in the output signal without affecting the shape of the output.

Chapter 1 problems

1.1 Determine whether the signal $f(t) = \cos 2t + \sin 3t$ is a periodic signal. If yes, give its period.

1.2 For the signal $f(k) = e^{j\omega k}$, determine if it is an energy signal or a power signal.

1.3 Determine whether the sinusoidal DT sequences are periodic:

(1) $f_1(k) = \sin(\pi k/12 + \pi/4)$ (3) $f_2(k) = \sin(3\pi k/10 + \theta)$

(3) $f_3(k) = \cos(0.5k + \varphi)$ (4) $f_4(k) = e^{j(7\pi k/8 + \theta)}$

1.4 Calculate the following integrations:

(1) $\displaystyle\int_{-\infty}^{\infty} \sin\left(t - \frac{\pi}{4}\right)\delta\left(t - \frac{\pi}{2}\right)dt$

(2) $\displaystyle\int_{-\infty}^{\infty} e^{-t}\delta(t - 3)\,dt$

(3) $\displaystyle\int_{-\infty}^{\infty} \varepsilon\left(t - \frac{t_0}{2}\right)\delta(t - t_0)\,dt$

(4) $\displaystyle\int_{-3}^{1} \delta(t - 4)\,dt$

1.5 Simplify the following expressions:

(1) $\dfrac{5 - jt}{7 + t^2}\delta(t)$

(2) $\displaystyle\int_{-\infty}^{\infty} (t + 5)\delta(t - 2)\,dt$

(3) $\displaystyle\int_{-\infty}^{\infty} e^{j0.5\pi t + 2}\delta(t - 5)\,dt$

1.6 Figure P1.1 shows the waveform of $f(t)$. Sketch the waveforms of the following signals:

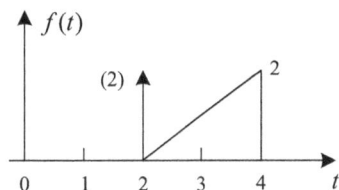

Fig. P1.1: The waveform for Problem 1.6

(1) $f_1(t) = f(2t - 1)$

(2) $f_2(t) = f(2t + 4)$

(3) $f_3(t) = f(-2t - 4)$

(4) $f_4(t) = f\left(\dfrac{1}{2}t - \dfrac{1}{4}\right)$

(5) $f_5(t) = f\left(\dfrac{1}{2}t + \dfrac{1}{4}\right)$

(6) $f_6(t) = f\left(-\dfrac{1}{2}t - \dfrac{1}{4}\right)$

1.7 Figure P1.2 shows the waveform of $f(5 - 2t)$. Sketch the waveform of $f(t)$.

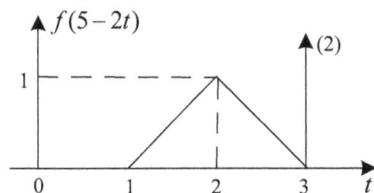

Fig. P1.2: The waveform for Problem 1.7

1.8 Consider three DT systems with the following input–output relationships. Determine whether the systems are (i) linear and (ii) time invariant:

(1) $y(k) = 2^k f(k)$

(2) $y(k + 3) - ky^2(k) = f(k)$

(3) $y(k) = f(k) + 3f(k - 1) + 4f(k - 2)$

1.9 Consider two CT systems with the following input–output relationships. Determine whether the systems are time invariant:

(1) $y(t) = \sin(x(t))$

(2) $y(t) = t\sin(x(t))$

1.10 Determine whether or not each of the following systems is linear and time variant:

(1) $y(t) = g(t)f(t)$　　　　　　　　　　(2) $y(t) = Kf(t) + f^2(t)$

(3) $y(t) = tf(t)\cos t$　　　　　　　　　(4) $y(t)f(t) = 1$

(5) $y(t) = f(t - 1)$　　　　　　　　　　(6) $y(t) = \displaystyle\int_{-\infty}^{t} f(\tau)\cos(t - \tau)\,d\tau$

(7) $y(k) = \displaystyle\sum_{i=0}^{k+2} k^2 f(i),\ (k = 0, 1, 2\ldots)$　(8) $y(k) = \alpha f(k) + \beta f(k - 1) + \alpha f(k - 2)$

1.11 Determine whether the following systems are stable:

(1) $y(t) = 50f(t) + 10$

(2) $y(t) = \displaystyle\int_{-\infty}^{t} f(\tau)\,d\tau$

(3) $y(k) = 50\sin[f(k)] + 10$

(4) $y(k) = e^{f(k)}$

1.12 Determine whether the following systems are causal:

(1) $y(t) = 5f(t - 2)$

(2) $y(t) = f(t + 1)$

(3) $y(k) = f(k - 2)$

(4) $y(k) = f(k - 2) + f(k + 10)$

1.13 Determine whether the following systems are dynamic:

(1) $y(t) = af(t) + b$　　　　　　　　　(2) $y(t) = f(t - 5)$

(3) $y(t) = \sin(f(t)) + 1$　　　　　　　(4) $y(t) = f(2t)$

(5) $y(k) = f(k - 1)$　　　　　　　　　(6) $y(k) = f(k) + 3$

1.14 For an LTIC system, the initial states are $y_1(0)$, $y_2(0)$, and the zero-input response is $y_{zi}(t)$. The system satisfies the following conditions:

$$y_1(0) = 1 , \quad y_2(0) = 0 , \quad y_{zi1}(t) = 2e^{-t} + 3e^{-3t} , \quad t \geq 0$$

$$\text{and} \quad y_1(0) = 0 , \quad y_2(0) = 1 , \quad y_{zi2}(t) = 4e^{-t} - 2e^{-3t} , \quad t \geq 0$$

Determine the zero-input response $y_{zi}(t)$ if the initial states are $y_1(0) = 5$, $y_2(0) = 3$.

1.15 Generate and sketch the following signals using MATLAB:

(1) $f(t) = 5 \sin(2\pi t) \cos(\pi t - 8)$ for $-5 \leq t \leq 5$,

(2) $f(t) = 5e^{-0.2t} \sin(2\pi t)$ for $-10 \leq t \leq 10$

(3) $f(k) = -0.92 \sin\left(0.1\pi k - \dfrac{3}{4}\pi\right)$ for $-10 \leq k \leq 20$

(4) $f(k) = 7(0.6)^k \cos(0.9\pi k)$

2 Time-domain analysis of LTIC systems

Please focus on the following key questions.

1. How can we represent circuit analysis as a problem of continuous system analysis?
2. What is the classical solution to the differential equations of LTIC systems?
3. What is the elementary signal and its response in continuous time-domain analysis?
4. How can we derive a new operation of the convolution integral from continuous-time signal decomposition in the time domain? What is the equation of solving zero-state response in the continuous-time domain?

2.0 Introduction

In this chapter, the linear, time-invariant, continuous-time (LTIC) system [11] is analyzed to solve the response in time domain. In circuit systems, the excitation source is regarded as the input signal, and the branch voltage or current to be solved is regarded as the response. Circuit theorems are used to build the differential equation, and the classic solutions in the time domain are applied to solve the response. This classic solution is applicable to simple inputs, such as direct current and exponential signals. For sinusoidal signals, the calculation is more complicated. If the signal is arbitrary, the classic solution cannot be solved. Therefore, Chapter 2 mainly concerns the response solutions of LTIC systems with any arbitrary input signal. The basic idea is to decompose the signal by the unit impulse function to calculate the output based on convolving the applied input with the impulse response.

This chapter firstly introduces the linear constant-coefficient differential equation to model an LTIC system. In Section 2.2, classical time-domain analysis is used to solve the differential equation to obtain the zero-input response and the zero-state response. In Section 2.3, the unit impulse response is defined as the output of an LITIC system to the unit impulse function applied at the input. Moreover, the unit step response is defined as the output of the system when a unit step function is applied at the input. In Section 2.4, the signal decomposition in time domain and convolution integral are discussed in detail. The chapter is concluded in Section 2.5, with a summary of the important concepts covered in the chapter.

https://doi.org/10.1515/9783110593907-002

2.1 Representation of the LTIC system

2.1.1 Analytical description based on mathematical models

For the electrical circuit system shown in Figure 2.1, the voltage source $u_S(t)$ is the applied input, and the voltage $u_C(t)$ is the response. According to the Kirchhoff voltage law (KVL) and the volt–ampere relationship (VAR), the linear differential equation is described in the following form:

$$\begin{cases} LC\dfrac{d^2u_C}{dt^2} + RC\dfrac{du_C}{dt} + u_C = u_S \\ u_C(0_+), \quad u'_C(0_+) \end{cases} \tag{2.1}$$

where the $u_C(0_+)$, $u'_C(0_+)$ is the initial value of the system.

! **Note:** How can we to establish the differential equation?

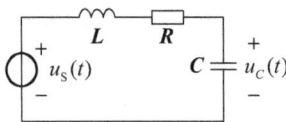

Fig. 2.1: An RLC circuit

The above equation can be abstracted into the following general expression:

$$a_2\frac{d^2y(t)}{dt^2} + a_1\frac{dy(t)}{dt} + a_0y(t) = f(t) \tag{2.2}$$

where the coefficients $a_2 \sim a_0$ are constants. This equation is a linear, second-order differential equation with constant coefficients. In fact, it can be shown that an LTIC system can always be modeled by a linear, constant-coefficient differential equation with appropriate initial conditions.

2.1.2 Description based on the block diagram

For a linear, constant-coefficient differential equation, three basic operations are included [12]: multiplication, differentiation and addition. These basic operations can be expressed by the ideal parts to be connected with each other, and this is drawn as a block diagram. As is shown in Figure 2.2, the basic component units are the integrator, the adder, the and multiplier.

! **Note:** We use the integrator as a basic component instead of differentiator.

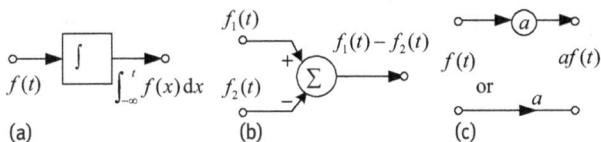

Fig. 2.2: Basic components of the CT system; (a) Integrator, (b) Adder, (c) Multiplier

Example 2.1.1. The linear differential equation of an LTIC system is $y''(t) + ay'(t) + by(t) = f(t)$; draw the block diagram.

Solution:
(i)   Rewrite the equation as $y''(t) = -ay'(t) - by(t) + f(t)$.
(ii) Draw two integrators and let the output of the last integrator be $y(t)$. Then, the input of the first integrator is $y''(t)$, which is the output of the adder.
(iii) As is shown in Figure 2.3, draw three inputs $-ay'(t) - by(t) + f(t)$ of the adder.

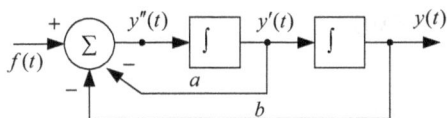

Fig. 2.3: Block diagram of Example 2.1.1

Example 2.1.2. The linear differential equation of an LTIC system is:

$$y''(t) + 3y'(t) + 2y(t) = 4f'(t) + f(t) .$$

Draw the block diagram.

Solution:
(i) Suppose $x(t)$ satisfies $x''(t) + 3x'(t) + 2x(t) = f(t)$; draw the corresponding block diagram in Figure 2.4, which meets the relationship between $x(t)$ and $f(t)$.
(ii) Due to the differentiation property and linearity of the LTIC system, the response satisfies $y(t) = 4x'(t) + x(t)$, which is drawn as the output of the right adder.

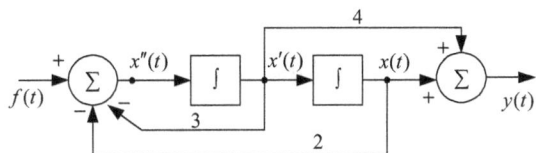

Fig. 2.4: Block diagram of Example 2.1.2

Example 2.1.3. The block diagram of an LTIC system is shown in Figure 2.5. Determine the differential equation of the input $f(t)$ and the output $y(t)$.

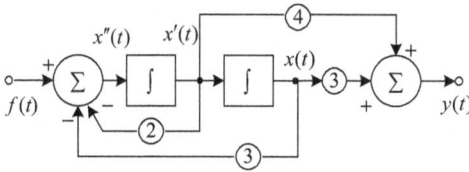

Fig. 2.5: Block diagram of Example 2.1.3

Solution:
(i) Set the auxiliary variable $x(t)$ at the output of the last integrator.
(ii) For the left adder, the input-output representation is:

$$x''(t) = -2x'(t) - 3x(t) + f(t), \tag{2.3}$$

which is rewritten as:

$$x''(t) + 2x'(t) + 3x(t) = f(t). \tag{2.4}$$

(iii) For the right adder, the input–output equation is:

$$y(t) = 4x'(t) + 3x(t). \tag{2.5}$$

(iv) According to the characteristics of the LTIC system, Equation (2.4) and (2.5) are combined to obtain the differential equation [13]:

$$y''(t) + 2y'(t) + 3y(t) = 4f'(t) + 3f(t)$$

! **Note:** Readers can prove it.

2.2 Classical solution of the differential equation

2.2.1 Classical solution of the direct method

Consider the differential equation of the input $f(t)$ and the output $y(t)$ for the LTIC system as follows:

$$y^{(n)}(t) + a_{n-1}y^{(n-1)}(t) + \cdots + a_1 y^{(1)}(t) + a_0 y(t)$$
$$= b_m f^{(m)}(t) + b_{m-1}f^{(m-1)}(t) + \cdots + b_1 f^{(1)}(t) + b_0 f(t)$$

The classical solutions of differential equation meets [14]:

$$y(t) = y_h(t) + y_p(t), \tag{2.6}$$

where $y_p(t)$ is the particular component and $y_h(t)$ is the homogeneous component of the homogeneous differential equation as:

$$y^{(n)} + a_{n-1}y^{(n-1)} + \cdots + a_1 y^{(1)}(t) + a_0 y(t) = 0 . \tag{2.7}$$

The functional form of the homogeneous solution is determined by the characteristic roots of Equation (2.7). The detailed forms can be seen in Table 2.1.

Tab. 2.1: Forms of the homogeneous component

Root λ	Homogeneous component $y_h(t)$
Single	$e^{\lambda t}$
Repeated r times	$(C_{r-1}t^{r-1} + C_{r-2}t^{r-2} + \cdots + C_1 t + C_0)e^{\lambda t}$
Complex $\lambda_{1,2} = \alpha \pm j\beta$	$e^{\alpha t}[C\cos(\beta t) + D\sin(\beta t)]$ or $A\cos(\beta t - \theta)$, where $Ae^{j\theta} = C + jD$

The functional form of the particular solution $y_p(t)$ is determined by the form of the input function. The detailed forms can be seen in Table 2.2.

Tab. 2.2: Functional forms of particular solutions to different inputs

Input $f(t)$	Particular component $y_p(t)$	
t^m	$P_0 + P_1 t + \cdots + P_m t^m$,	if none of the roots is 0;
	$t^r[P_0 + P_1 t + \cdots + P_m t^m]$,	if the root 0 repeats r times.
	$Pe^{\alpha t}$,	if α is not equal to the root;
$e^{\alpha t}$	$(P_1 t + P_0)e^{\alpha t}$,	if α is equal to a single root;
	$(P_r t^r + P_{r-1}t^{r-1} + \cdots + P_1 t + P_0)e^{\alpha t}$,	if α is equal to the r-times root.
$\cos\beta t$ or $\sin\beta t$	$P\cos\beta t + Q\sin\beta t$,	if none of the roots is equal to $\pm j\beta$;
	or $A\cos(\beta t - \theta)$, where $Ae^{j\theta} = P + jQ$.	

Example 2.2.1. The differential equation of a LTIC system is given by:

$$y''(t) + 5y'(t) + 6y(t) = f(t) . \tag{2.8}$$

(1) Determine the output signal when the input is given by $f(t) = 2e^{-t}$, $t \geq 0$, and the initial value is $y(0_+) = 2$, $y'(0_+) = -1$.
(2) Determine the output signal when the input is given by $f(t) = e^{-2t}$, $t \geq 0$, and the initial value is $y(0_+) = 1$, $y'(0_+) = 0$.

Solution:

(1) As the characteristic roots of characteristic equation $\lambda^2 + 5\lambda + 6 = 0$ are $\lambda_1 = -2$, $\lambda_2 = -3$, the homogeneous component is given by:

$$y_h(t) = C_1 e^{-2t} + C_2 e^{-3t}, \tag{2.9}$$

where C_1, C_2 are constants. From Table 2.2, for the input signal $f(t) = 2e^{-t}$, $t \geq 0$, the particular component is of the following form:

$$y_p(t) = Ke^{-t}, \tag{2.10}$$

where K is a constant. To calculate the value of the constant, we substitute the particular component $y_p(t)$ in Equation (2.8):

$$Ke^{-t} + 5(-Ke^{-t}) + 6Ke^{-t} = 2e^{-t} \quad \Rightarrow \quad K = 1.$$

The particular component is $y_p(t) = e^{-t}$. The overall response is given by:

$$y(t) = y_h(t) + y_p(t) = C_1 e^{-2t} + C_2 e^{-3t} + e^{-t} \tag{2.11}$$

where C_1, C_2 can be obtained by inserting initial values. This leads to the following simultaneous equations:

$$y(0_+) = C_1 + C_2 + 1 = 2, \quad y'(0_+) = -2C_1 - 3C_2 - 1 = -1,$$

with solutions $C_1 = 3$, $C_2 = -2$. The whole response is:

$$y(t) = 3e^{-2t} - 2e^{-3t} + e^{-t}, \quad t \geq 0. \tag{2.12}$$

(2) The homogeneous component is still $y_h(t) = C_1 e^{-2t} + C_2 e^{-3t}$. From Table 2.2, for the input signal $f(t) = e^{-2t}$, $t \geq 0$, the particular component is of the following form:

$$y_p(t) = (Q_0 + Q_1 t)e^{-2t}. \tag{2.13}$$

! **Note:** Given the differential equation, the form of homogeneous component is the same.

Substituting the particular component in Equation (2.8) yields:

$$Q_1 e^{-2t} = e^{-2t} \quad \Rightarrow \quad Q = 1. \tag{2.14}$$

The total response is, therefore, given by:

$$y(t) = C_1 e^{-2t} + C_2 e^{-3t} + te^{-2t} + Q_0 e^{-2t} = (C_1 + Q_0)e^{-2t} + C_2 e^{-3t} + te^{-2t}. \tag{2.15}$$

Substituting the initial values into Equation (2.15) leads to the following equations:

$$y(0_+) = (C_1 + Q_0) + C_2 = 1$$
$$y'(0_+) = -2(C_1 + Q_0) - 3C_2 + 1 - 0, \tag{2.16}$$

with solutions $C_1 + Q_0 = 2$, $C_2 = -1$. The overall response is:

$$y(t) = 2e^{-2t} - e^{-3t} + te^{-2t}, \quad t \geq 0. \tag{2.17}$$

2.2.2 Initial value of the system

Supposing the input $f(t)$ is inserted into the n-order system at $t = 0$, the *initial values* are defined as $y^{(j)}(0_+)$ $(j = 0, 1, 2, \ldots, n - 1)$ at $t = 0_+$. At $t = 0_-$; the input is not accessed. Therefore, the initial condition $y^{(j)}(0_-)$ reflects the historical situation of the system, which is independent of the input. Generally, the initial values $y^{(j)}(0_+)$ can be obtained from the initial condition $y^{(j)}(0_-)$ by the coefficient matching method, which is illustrated as the following.

Example 2.2.2. The differential equation of a LTIC system is given by:

$$y''(t) + 3y'(t) + 2y(t) = 2f'(t) + 6f(t) . \tag{2.18}$$

Determine the initial values $y(0_+)$ and $y'(0_+)$ when the input is given by $f(t) = \varepsilon(t)$, and the initial condition is $y(0_-) = 2, y'(0_-) = 1$.

Solution: Substituting $f(t) = \varepsilon(t)$ in Equation (2.18) yields:

$$y''(t) + 3y'(t) + 2y(t) = 2\delta(t) + 6\varepsilon(t) . \tag{2.19}$$

Note: The coefficient-matching method means that the coefficients of $\delta(t)$ on both sides of the equation are equal at intervals of $0_- < t < 0_+$. ❗

To match $2\delta(t)$ on the right-hand side of the equation, the impulse function must be included only in $y''(t)$, which is caused by the discontinuity of $y'(t)$ at $t = 0$. So, we have $y'(0_+) \neq y'(0_-)$. Moreover, $y'(t)$ cannot include $\delta(t)$, which leads to $y(0_+) = y(0_-) = 2$.

The integral calculation is made on two sides of Equation (2.19) in the infinitesimal interval $[0_-, 0_+]$:

$$\int_{0_-}^{0_+} y''(t)\,dt + 3\int_{0_-}^{0_+} y'(t)\,dt + 2\int_{0_-}^{0_+} y(t)\,dt = 2\int_{0_-}^{0_+} \delta(t)\,dt + 6\int_{0_-}^{0_+} \varepsilon(t)\,dt \tag{2.20}$$

This can be simplified as:

$$[y'(0_+) - y'(0_-)] + 3[y(0_+) - y(0_-)] = 2 \tag{2.21}$$

with the solution $y'(0_+) = y'(0_-) + 2 = 3$.

From the above coefficient-matching method, the following conclusions are drawn. (1) When the right-hand side of the n-order differential equation contains the impulse function (or the derivative of the impulse function), the response $y(t)$ and its derivatives will possibly jump at $t = 0$. (2) If the right-hand side does not contain the impulse function, $y(t) \rightarrow y^{(n-1)}(t)$ are continuous at $t = 0$.

Note: In Section 2.3, we give an alternative method of computing response for the case when the right-hand side of the equation contains the derivative of the impulse function. ❗

2.2.3 Zero-input response and zero-state response

Definition. *The zero-input response* $y_{zi}(t)$ is the output of the system when the external input is zero. It is produced by the system because of the initial conditions.

Definition. *The zero-state response* $y_{zs}(t)$ arises due to the input signal and does not depend on the initial conditions of the system. In calculating the zero-state response, the initial conditions of the system are assumed to be zero.

For LTIC systems, the complete response is the sum of the zero-input response and the zero-state response:

$$y(t) = y_{zi}(t) + y_{zs}(t) \tag{2.22}$$

Basic relationship: For the n-th order LTIC system, the initial conditions satisfy the following relationship:

(1)

$$y^{(j)}(0_-) = y_{zi}^{(j)}(0_-) + y_{zs}^{(j)}(0_-) = y_{zi}^{(j)}(0_-)$$

$$y^{(j)}(0_+) = y_{zi}^{(j)}(0_+) + y_{zs}^{(j)}(0_+), \quad j = 0, 1, \ldots, n-1 \tag{2.23}$$

(2) For the zero-state response, $y_{zs}^{(j)}(0_+)$ can be calculated by the coefficient -matching method.

(3) For the zero input response, $y_{zi}^{(j)}(t)$ are continuous at $t = 0$, which yields:

$$y_{zi}^{(j)}(0_+) = y_{zi}^{(j)}(0_-) = y^{(j)}(0_-). \tag{2.24}$$

Example 2.2.3. The differential equation of a LTIC system is given by:

$$y''(t) + 3y'(t) + 2y(t) = 2f'(t) + 6f(t). \tag{2.25}$$

Determine the zero-input response and zero-state response when the input is given by $f(t) = \varepsilon(t)$, and the initial condition is $y(0_-) = 2, y'(0_-) = 0$.

Solution:
(1) Determine the zero-input response
The homogeneous equation of $y_{zi}(t)$ is:

$$y_{zi}''(t) + 3y_{zi}'(t) + 2y_{zi}(t) = 0. \tag{2.26}$$

The initial values satisfy:

$$y_{zi}(0_+) = y_{zi}(0_-) = y(0_-) = 2$$

$$y_{zi}'(0_+) = y_{zi}'(0_-) = y'(0_-) = 0. \tag{2.27}$$

The characteristic equation is $\lambda^2 + 3\lambda + 2 = 0$, which has the characteristic roots at $\lambda = -1$ and $\lambda = -2$. The zero-input response is given by:

$$y_{zi}(t) = C_1 e^{-t} + C_2 e^{-2t} .$$

Substituting the initial values in the above equation yields $C_1 = 4$, $C_2 = -2$. The zero-input response is, therefore, given by:

$$y_{zi}(t) = 4e^{-t} - 2e^{-2t} , \quad t \geq 0 \tag{2.28}$$

(2) Determine the zero-state response
The zero-state response $y_{zs}(t)$ satisfies:

$$y_{zs}''(t) + 3y_{zs}'(t) + 2y_{zs}(t) = 2\delta(t) + 6\varepsilon(t)$$
$$y_{zs}(0_-) = y_{zs}'(0_-) = 0 . \tag{2.29}$$

According to the coefficient-matching method in Section 2.2.2, $y_{zs}(t)$ is continuous with $y_{zs}(0_+) = y_{zs}(0_-) = 0$. From the integral calculation of two sides of the equation,

$$[y_{zs}'(0_+) - y_{zs}'(0_-)] + 3[y_{zs}(0_+) - y_{zs}(0_-)] + 2 \int_{0-}^{0+} y_{zs}(t)\,dt = 2 + 6 \int_{0-}^{0+} \varepsilon(t)\,dt$$

we obtain $y_{zs}'(0_+) = 2 + y_{zs}'(0_-) = 2$.
The homogeneous component is $y_{zsh}(t) = D_1 e^{-t} + D_2 e^{-2t}$, which is similar to the zero-input response of the system:
when $t > 0$,

$$y_{zs}''(t) + 3y_{zs}'(t) + 2y_{zs}(t) = 6 \tag{2.30}$$

The particular component is $y_{zsp}(t) = K$. Substituting the particular component in Equation (2.30) yields $y_{zsp}(t) = 3$.

Note: How can we compute the particular component? !

The overall zero-state response is given by $y_{zs}(t) = D_1 e^{-t} + D_2 e^{-2t} + 3$. The initial values $y_{zs}(0_+) = 0$, $y_{zs}'(0_+) = 2$ are substituted to compute the zero-state response:

$$y_{zs}(t) = -4e^{-t} + e^{-2t} + 3 , \quad t \geq 0 \tag{2.31}$$

Example 2.2.4. The differential equation of a LTIC system is given by:

$$y''(t) + 5y'(t) + 6y(t) = f(t) . \tag{2.32}$$

Determine the response when the input is given by $f(t) = 10 \cos t$, $t \geq 0$, and the initial condition is $y(0_-) = 2$, $y'(0_-) = 0$.

Solution:

(1) The homogeneous component is $y_h(t) = C_1 e^{-2t} + C_2 e^{-3t}$.

(2) According to the input function $f(t) = 10\cos t, t \geq 0$, the particular component is represented as:

$$y_p(t) = P\cos t + Q\sin t \tag{2.33}$$

Substituting $y_p(t)$ and $f(t)$ in Equation (2.32) yields:

$$(-P + 5Q + 6P)\cos t + (-Q - 5P + 6Q)\sin t = 10\cos t \tag{2.34}$$

Equating the cosine and sine terms on the left- and right-hand sides of the equation, we obtain the following simultaneous equations:

$$5P + 5Q = 10,$$
$$-5P + 5Q = 0,$$

with the solution $P = Q = 1$. The particular component is given by:

$$y_p(t) = \cos t + \sin t = \sqrt{2}\cos\left(t - \frac{\pi}{4}\right) \tag{2.35}$$

(3) The overall response is the sum of homogeneous and particular components:

$$y(t) = y_h(t) + y_p(t) = C_1 e^{-2t} + C_2 e^{-3t} + \sqrt{2}\cos\left(t - \frac{\pi}{4}\right) \tag{2.36}$$

where C_1, C_2 are constants. Substituting initial values in the above equation yields the following simultaneous equations:

$$y(0_+) = C_1 + C_2 + 1 = 2, \quad y'(0_+) = -2C_1 - 3C_2 + 1 = 0,$$

with solutions $C_1 = 2$ and $C_2 = -1$. The overall response is given by:

$$y(t) = 2e^{-2t} - e^{-3t} + \sqrt{2}\cos\left(t - \frac{\pi}{4}\right), \quad t \geq 0 \tag{2.37}$$

In the above process of calculation, the homogeneous component does not depend on the external input, and hence the homogeneous component is also known as the *natural response*. The particular component forced by the input signal can be defined as the *forced response*.

In the overall response, the first two components decay to zero as $t \to \infty$, which are referred to as the *transient responses*. The last component shows oscillation with equal amplitude, which is known as the *steady-state response*. Generally, the stable response is composed of step functions or periodic cosine functions.

According to Equation (2.37), the components of the overall response can be classified as:

$$y(t) = \underbrace{\underbrace{2e^{-2t} - e^{-3t}}_{\text{transient response}}}_{\text{natural response}} + \underbrace{\sqrt{2}\cos\left(t - \frac{\pi}{4}\right)}_{\substack{\text{forced response} \\ \text{stable response}}}, \quad t \geq 0. \tag{2.38}$$

Note: Determine whether the natural response is a transient response. Determine whether the zero-input response is a natural response.

2.2.4 Response calculation with MATLAB

MATLAB provides the function for solving the zero-state response of the LTIC system. The function is *lsim*, and its method is as follows:

```
y=lsim(sys,f,t)
```

where *t* indicates the sampling point vector of the system response, *f* is the input signal of system, and *sys* establishes the model of the given LTIC system. In solving the differential equation, the *sys* model is obtained by function of *tf()*. For example, the differential equation of an LTIC system is given as:

$$a_3 y'''(t) + a_2 y''(t) + a_1 y'(t) + a_0 y(t) = b_3 f'''(t) + b_2 f''(t) + b_1 f'(t) + b_0 f(t) .$$

The *sys* model can be obtained by:

```
a=[a3,a2,a1,a0];
b=[b3,b2,b1,b0];
sys=tf(b,a)
```

where, *a* and *b* are the coefficients of the left and right-hand sides of the differential equation, respectively.

Example 2.2.5. The differential equation of an LTIC system is given by:

$$\frac{d^2 y(t)}{dt^2} + 2\frac{dy(t)}{dt} + 77y(t) = f(t) . \tag{2.39}$$

Determine the zero-state response when the input is given by $f(t) = 10\sin 2\pi t, t \geq 0$.

Solution:

```
ts=0; te=5; dt=0.01;
sys=tf([1],[1, 2, 77]);        % the model
t=ts:dt:te;
f=10*sin(2*pi*t);              % the input signal
y=lsim(sys,f,t);               % the zero-state response
plot (t,y);                    % waveform plotting
```

The waveform of the zero-state response is shown in Figure 2.6.

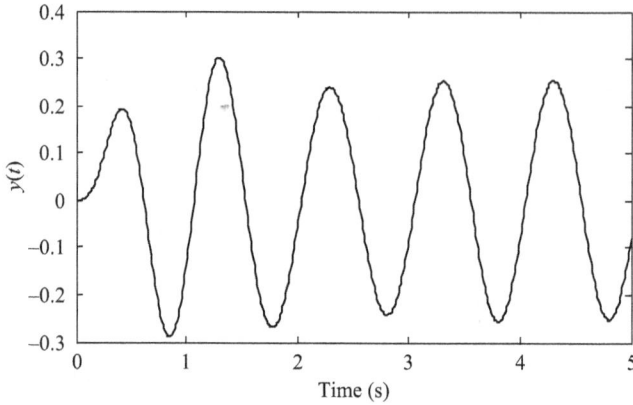

Fig. 2.6: The waveform of the response in Example 2.2.5

2.3 The impulse response and step response

2.3.1 CT impulse response

Definition. The *impulse response* $h(t)$ of an LTIC system is the zero-state output of the system when a unit impulse $\delta(t)$ is applied at the input, i.e., $h(t) = T[\{0\}, \delta(t)]$.

Example 2.3.1. The differential equation of an LTIC system is given by:

$$y''(t) + 5y'(t) + 6y(t) = f(t) . \qquad (2.40)$$

Determine the impulse response.

Solution: According to the definition, the impulse response can be obtained by solving the following differential equation:

$$h''(t) + 5h'(t) + 6h(t) = \delta(t)$$
$$h'(0_-) = h(0_-) = 0 . \qquad (2.41)$$

Note: The zero-state condition is included in the definition of the impulse response.

(i) Using the coefficient matching analysis, the initial value can be obtained as

$$h(0_+) = h(0_-)$$
$$h'(0_+) - h'(0_-) = 1 \quad \Rightarrow \quad h'(0_+) = 1 . \qquad (2.42)$$

(ii) For $t > 0$, the differential equation is given by the following homogeneous equation:

$$h''(t) + 5h'(t) + 6h(t) = 0 . \qquad (2.43)$$

As the characteristic roots of the homogeneous equation are -1 and -2, the impulse response is

$$h(t) = (C_1 e^{-2t} + C_2 e^{-3t})\varepsilon(t) .$$

(iii) Substituting the initial values yields $C_1 = 1$, $C_2 = -1$. So, the impulse response is

$$h(t) = (e^{-2t} - e^{-3t})\varepsilon(t) \,. \tag{2.44}$$

Example 2.3.2. The differential equation of an LTIC system is given by:

$$y''(t) + 5y'(t) + 6y'(t) = f''(t) + 2f'(t) + 3f(t) \,. \tag{2.45}$$

Determine the impulse response.

Solution: Suppose $h_1(t)$ satisfies

$$h_1''(t) + 5h'_1(t) + 6h_1(t) = \delta(t) \,. \tag{2.46}$$

Due to the differentiation property and linearity of the LTIC system, the impulse response satisfies

$$h(t) = h_1''(t) + 2h'_1(t) + 3h_1(t) \,. \tag{2.47}$$

(i) In Example 2.3.1, the response of Equation (2.46) is $h_1(t) = (e^{-2t} - e^{-3t})\varepsilon(t)$.
(ii) Calculate the first and second order derivatives of $h_1(t)$:

$$h'_1(t) = (e^{-2t} - e^{-3t})\delta(t) + (-2e^{-2t} + 3e^{-3t})\varepsilon(t) = (-2e^{-2t} + 3e^{-3t})\varepsilon(t)$$
$$h''_1(t) = (-2e^{-2t} + 3e^{-3t})\delta(t) + (4e^{-2t} - 9e^{-3t})\varepsilon(t) = \delta(t) + (4e^{-2t} - 9e^{-3t})\varepsilon(t)$$

According to Equation (2.47), the impulse response of the system is given by:

$$h(t) = \delta(t) + (3e^{-2t} - 6e^{-3t})\varepsilon(t) \,.$$

This example combines the linearity and differential property of the zero-state response to simplify the solution. If the input $\delta(t)$ is directly substituted into the equation, the derivatives of $\delta(t)$ will occur at the right-hand side of the equation. The process of calculating initial values with the coefficient matching method is a bit difficult.

Note: When the right-hand side of the equation contains the derivative of the impulse function, this is an alternative method of computing impulse response. !

2.3.2 CT step response

Definition. The *step response* $g(t)$ of an LTIC system is the zero-state output of the system when a unit step $\varepsilon(t)$ is applied at the input, i.e., $g(t) = T[\{0\}, \varepsilon(t)]$.

The relationship between the unit step function $\varepsilon(t)$ and the unit impulse function $\delta(t)$ is given by:

$$\delta(t) = \frac{d\varepsilon(t)}{dt} \,, \quad \varepsilon(t) = \int_{-\infty}^{t} \delta(x)\,dx \,.$$

According to the differential (or integral) properties of the LTIC system, the relationship between the step and the impulse response of the same system can be expressed as:

$$h(t) = \frac{dg(t)}{dt}, \quad g(t) = \int_{-\infty}^{t} h(x)\, dx \tag{2.48}$$

Example 2.3.3. Determine the input–output representations of the circuit in Figure 2.7, and compute its unit step response.

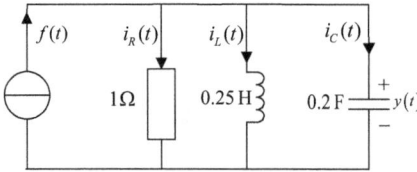

Fig. 2.7: Electrical circuit of Example 2.3.3

Solution:
(1) According to Kirchhoff's current law, the relationship between the input current $f(t)$ and the branch current is given by:

$$f(t) = i_R(t) + i_L(t) + i_C(t). \tag{2.49}$$

The output voltage $y(t)$ is measured across the inductor L. Expressed in terms of the branch current $i_L(t)$, the voltage is given by:

$$y(t) = L\frac{d}{dt}i_L(t).$$

It can be rewritten as:

$$i_L(t) = \frac{1}{L}\int_{0}^{t} y(\tau)\, d\tau + i_L(0-).$$

The voltage–current relation of capacitor C and resistor R is given by, respectively:

$$i_C(t) = C\frac{d}{dt}y(t)$$

$$i_R(t) = y(t)/1.$$

into Equation (2.49), we obtain:

$$f(t) = y(t) + \frac{1}{L}\int_{0}^{t} y(\tau)\, d\tau + i_L(0-) + C\frac{d}{dt}y(t). \tag{2.50}$$

Differentiating Equation (2.50) with respect to t yields:

$$\frac{d}{dt}f(t) = C\frac{d^2}{dt^2}y(t) + \frac{d}{dt}y(t) + \frac{1}{L}y(t) \rightarrow y''(t) + 5y'(t) + 20y(t) = 5f'(t), \quad (2.51)$$

which is a linear, second-order, constant-coefficient differential equation.

(2) The step response $g(t)$ is the zero-state output when $f(t) = \varepsilon(t)$ is applied at the input:

$$g''(t) + 5g'(t) + 20g(t) = 5\delta(t) \quad (2.52)$$

The initial value is $g(0_+) = g(0_-) = 0$, $g'(0_+) = i_C(0_+)/C = 1/C = 5$.
The characteristic roots of characteristic equation $\lambda^2 + 5\lambda + 20 = 0$ are $\lambda_{1,2} = (-5 \pm j\sqrt{55})/2$, and the homogeneous component is given by:

$$g_h(t) = c_1 e^{-\frac{5}{2}t} \cos\left(\frac{\sqrt{55}}{2}t\right) + c_2 e^{-\frac{5}{2}t} \sin\left(\frac{\sqrt{55}}{2}t\right). \quad (2.53)$$

The particular component is $g_p(t) = 0$. The overall response is given by $g(t) = g_h(t)$. The initial values are brought into Equation (2.53) to yield:

$$g(0_+) = 0 = c_1$$

$$g'(0_+) = 5 = c_2\frac{\sqrt{55}}{2} \quad \Rightarrow \quad c_2 = \frac{10}{\sqrt{55}} \cdot$$

Note: Think about why the particular component is 0. !

The unit step response is:

$$g(t) = \frac{10}{\sqrt{55}} e^{-\frac{5}{2}t} \sin\left(\frac{\sqrt{55}}{2}t\right), \quad t \geq 0.$$

2.3.3 Solution by MATLAB

MATLAB provides the functions *impulse(b, a)* and *step(b, a)* for solving the impulse and step response of the LTIC system. The vectors a and b are used to model the coefficients of the two sides in the differential equation as follows:

$$\sum_{i=0}^{n} a_i y^{(i)}(t) = \sum_{i=0}^{m} b_i f^{(i)}(t)$$

Example 2.3.4. The differential equation of an LTIC system is given by:

$$7y''(t) + 4y'(t) + 6y(t) = f'(t) + f(t).$$

Determine the impulse response and step response with MATLAB.

Solution:

```
a=[7 4 6]; b=[1 1];
subplot(2,1,1)
impulse(b,a)
subplot(2,1,2)
step(b,a)
```

The waveforms of the impulse response and step response are shown in Figure 2.8.

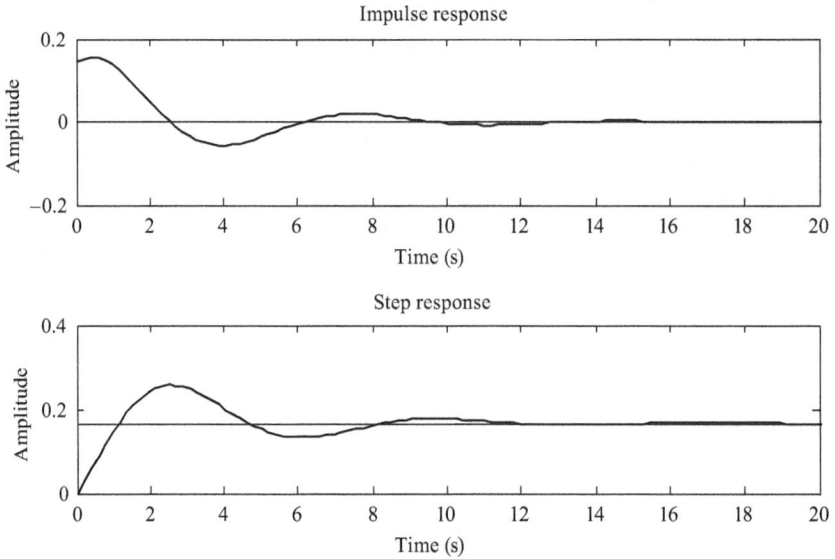

Fig. 2.8: The waveform of the response in Example 2.3.4

2.4 Convolution integral

2.4.1 Signal decomposition in the time domain

In this section, we will show that any arbitrary signal can be represented as a linear combination of time-shifted impulse functions [15]. First, a rectangular pulse with width Δ and height $1/\Delta$ is defined as in Figure 2.9 (a). The rectangular function in Figure 2.9 (b) can be described as $f_1(t) = A\Delta \cdot p(t)$.

To approximate $f(t)$ in Figure 2.10, the time axis is divided into uniform intervals of duration Δ. In each duration, the impulse is labeled as "$\ldots 0, 1, 2, -1, \ldots$". The "0" pulse with height $f(0)$ is denoted by $f(0)\Delta p(t)$; the "1" pulse with height $f(\Delta)$ is denoted by $f(\Delta)\Delta p(t-\Delta)$ and; the "-1" pulse with height $f(-\Delta)$ is denoted by $f(-\Delta)\Delta p(t+\Delta)$. The

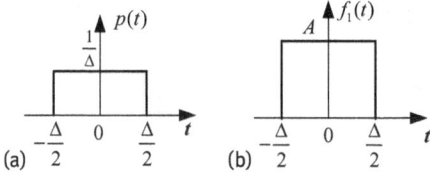

Fig. 2.9: Relationship between two rectangular pulses, (a) Rectangular pulse, (b) Rectangular function

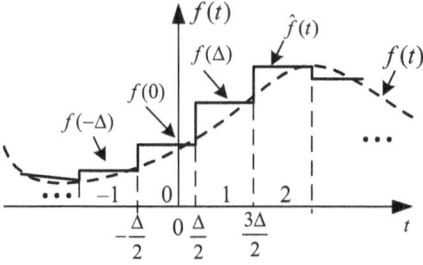

Fig. 2.10: Signal decomposition and approximation

staircase approximation $\hat{f}(t)$ can be represented as the linear combination of delayed pulses.

$$\hat{f}(t) = \sum_{n=-\infty}^{\infty} f(n\Delta)\Delta p(t - n\Delta) \qquad (2.54)$$

Applying the limit $\Delta \to 0$, the approximated function $p(t - n\Delta)$ approaches $\delta(t - \tau)$. The summation becomes integration. Substituting Δ by $d\tau$ and $n\Delta$ by τ, we obtain the following relationship:

$$\lim_{\Delta \to 0} \hat{f}(t) = f(t) = \int_{-\infty}^{\infty} f(\tau)\delta(t - \tau)\, d\tau$$

where τ is the dummy variable that disappears as the integration with limits is computed.

Note: The rectangular pulse changes into impulse function. ❗

2.4.2 Definition of the convolution integral

Following the above procedure of signal decomposition, an arbitrary CT signal $f(t)$ can be represented in terms of Equation (2.55):

$$f(t) = \int_{-\infty}^{\infty} f(\tau)\delta(t - \tau)\, d\tau \qquad (2.55)$$

The integral is referred to as the *convolution integral* and is denoted by $f(t) * \delta(t)$. It suggests that a CT signal can be represented as a weighted superposition of time-shifted impulse functions. We will use it to calculate the output of an LTIC system.

According to the linear time-invariant property of LTIC system, the zero-state response can be obtained as follows.

When the input signal is $\delta(t)$, the zero-state response is $h(t)$:

$$\delta(t) \rightarrow h(t)$$

According to the time-invariant property, we have:

$$\delta(t - \tau) \rightarrow h(t - \tau) .$$

According to the homogeneous property, we have:

$$f(\tau)\delta(t - \tau) \rightarrow f(\tau)h(t - \tau) .$$

According to the additive property, we have:

$$f(t) = \int_{-\infty}^{\infty} f(\tau)\delta(t - \tau)\, d\tau \rightarrow \int_{-\infty}^{\infty} f(\tau)h(t - \tau)\, d\tau . \tag{2.56}$$

The integral on the right-hand side of Equation (2.56) is denoted by $f(t) * \delta(t)$. Mathematically, the convolution of $f(t)$ and $h(t)$ is defined as follows:

$$y_{zs}(t) = f(t) * h(t) = \int_{-\infty}^{\infty} f(\tau)h(t - \tau)\, d\tau \tag{2.57}$$

! **Note:** This is the most important expression in this chapter.

When an input signal $f(t)$ passes through an LTIC system with impulse response $h(t)$, the resulting zero-state output can be calculated by convolving the input signal and the impulse response.

Similarly, the convolution of $f_1(t)$ and $f_2(t)$ can be defined as follows:

$$f_1(t) * f_2(t) = \int_{-\infty}^{\infty} f_1(\tau)f_2(t - \tau)\, d\tau \tag{2.58}$$

We now consider several examples of computing the convolution integral.

Example 2.4.1. Determine the zero-state response of an LTIC system when the input signal is given by $f(t) = e^t$, $(-\infty < t < \infty)$ and the impulse response is $h(t) = (6e^{-2t} - 1)\varepsilon(t)$.

Solution: According to Equation (2.57), $y_{zs}(t)$ is computed as $f(t) * h(t)$:

$$y_{zs}(t) = f(t) * h(t) = \int_{-\infty}^{\infty} e^\tau [6e^{-2(t-\tau)} - 1]\varepsilon(t - \tau)\, d\tau$$

Expressed as a function of the independent variable τ, the unit step function is given by:

$$\varepsilon(t - \tau) = \begin{cases} 1, & \tau < t \\ 0, & \tau > t \end{cases}.$$

The zero-state output can be expressed as:

$$y_{zs}(t) = \int_{-\infty}^{t} e^{\tau}[6e^{-2(t-\tau)} - 1]\,d\tau.$$

Therefore:

$$y_{zs}(t) = \int_{-\infty}^{t} (6e^{-2t}e^{3\tau} - e^{\tau})\,d\tau$$

$$= e^{-2t}\int_{-\infty}^{t}(6e^{3\tau})\,d\tau - \int_{-\infty}^{t} e^{\tau}\,d\tau$$

$$= e^{-2t} \cdot 2e^{3\tau}\big|_{-\infty}^{t} - e^{\tau}\big|_{-\infty}^{t} = 2e^{-2t}\cdot e^{3t} - e^{t} = e^{t}.$$

Example 2.4.2. Determine the zero-state response of an LTIC system when the input signal is given by $f(t) = e^{-t}\varepsilon(t)$ and the impulse response is $h(t) = e^{-2t}\varepsilon(t)$.

Solution: According to Equation (2.57), $y_{zs}(t)$ is computed as $f(t) * h(t)$:

$$y_{zs}(t) = f(t) * h(t) = \int_{-\infty}^{\infty} e^{-\tau}\varepsilon(\tau)e^{-2(t-\tau)}\varepsilon(t - \tau)\,d\tau$$

It can be expressed as:

$$y_{zs}(t) = e^{-2t}\int_{0}^{\infty} e^{\tau}\varepsilon(t - \tau)\,d\tau.$$

Expressed as a function of the independent variable τ, the unit step function is given by:

$$\varepsilon(t - \tau) = \begin{cases} 0, & \tau > t \\ 1, & \tau < t. \end{cases}$$

Based on the value of t, we have the following two cases for the output $y_{zs}(t)$.
(i) For $t < 0$, the shifted unit step function $\varepsilon(t-\tau) = 0$ within the limits of integration $[0, \infty]$. Therefore, $y_{zs}(t) = 0$, $t < 0$.
(ii) For $t > 0$, the shifted unit step function $\varepsilon(t-\tau)$ has two different values within the limits of integration $[0, \infty]$. For the range $[0, t]$, the unit step function $\varepsilon(t-\tau) = 1$. Otherwise, for the range $[t, \infty]$, the unit step function is zero.
Combining the two cases, the output is, therefore, given by:

$$y_{zs}(t) = \left(e^{-2t}\int_{0}^{t} e^{\tau}\,d\tau\right)\varepsilon(t) = e^{-2t}(e^{t} - 1)\varepsilon(t) = (e^{-t} - e^{-2t})\varepsilon(t).$$

2.4.3 Graphical method for evaluating the convolution integral

Given $f_1(t) * f_2(t) = \int_{-\infty}^{\infty} f_1(\tau) f_2(t-\tau) \, d\tau$, the convolution can be evaluated graphically by following Steps 1 to 4.

Step 1: Change the independent variable from t to τ to obtain $f_1(\tau)$ and $f_2(\tau)$.

Step 2: Reflect $f_2(\tau)$ about the vertical axis to obtain $f_2(-\tau)$ and shift it to get $f_2(t-\tau)$.

Step 3: Multiply function $f_1(\tau)$ by $f_2(t-\tau)$.

Step 4: Calculate the total area under the product function $f_1(\tau) f_2(t-\tau)$ by integrating it over $\tau = [-\infty, \infty]$.

> **!** **Note:** The integral variable is τ, and the convolution result is a function of the independent variable t.

Example 2.4.3. Using the graphical method, compute the convolution of the following two signals:
$$h(t) = 0.5[\varepsilon(t) - \varepsilon(t-1)]$$
$$f(t) = t \cdot [\varepsilon(t) - \varepsilon(t-2)] .$$

Solution: The functions $f(\tau)$, $h(\tau)$ and $h(-\tau)$ are plotted as a function of the variable τ in the subplots of Figure 2.11 (a) and (b).

(i) For $t < 0$, $h(-\tau)$ shifts to the left-hand side along the time axis. The nonzero parts of $h(t-\tau)$ and $f(\tau)$ do not overlap. In other words, the output $y_{zs}(t) = 0$.

From Figure 2.11 (b), we will observe the overlapping region in different cases.

(ii) For $0 \leq t \leq 1$, $h(-\tau)$ shifts to the right-hand side along the time axis. The non-zero parts of $h(t-\tau)$ and $f(\tau)$ overlap over duration $\tau = [0, t]$. Therefore,
$$y_{zs}(t) = \int_0^t \tau \cdot \frac{1}{2} \, d\tau = \frac{1}{4} t^2 .$$

(iii) For $1 \leq t \leq 2$, the non-zero parts of $h(t-\tau)$ and $f(\tau)$ overlap over duration $\tau = [t-1, t]$. Therefore,
$$y_{zs}(t) = \int_{t-1}^t \tau \cdot \frac{1}{2} \, d\tau = \frac{1}{2} t - \frac{1}{4} .$$

(iv) For $2 \leq t \leq 3$, the non-zero parts of $h(t-\tau)$ and $f(\tau)$ do overlap over duration $\tau = [t-1, 2]$. Therefore,
$$y_{zs}(t) = \int_{t-1}^2 \tau \cdot \frac{1}{2} \, d\tau = -\frac{1}{4} t^2 + \frac{1}{2} t + \frac{3}{4} .$$

(v) For $3 < t$, the nonzero parts of $h(t-\tau)$ and $f(\tau)$ do not overlap. Therefore, we obtain $y_{zs}(t) = 0$.

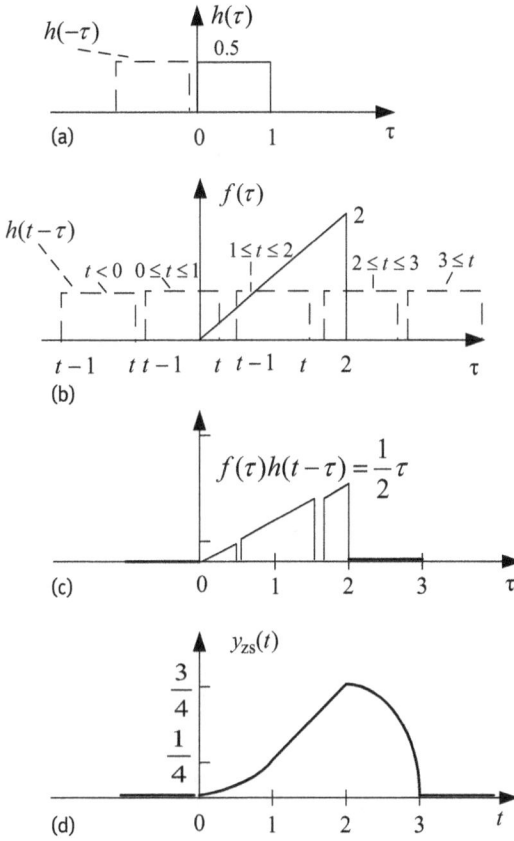

Fig. 2.11: Process of the graphic method of the convolution integral

Combining the above five cases, we obtain:

$$y_{zs}(t) = \begin{cases} 0, & t < 0 \\ \frac{1}{4}t^2, & 0 \le t \le 1 \\ \frac{1}{2}t - \frac{1}{4}, & 1 \le t \le 2 \\ -\frac{1}{4}t^2 + \frac{1}{2}t + \frac{3}{4} & 2 \le t \le 3 \\ 0, & t > 3. \end{cases}$$

The waveform for the output response is sketched in Figure 2.11 (d).

Note: Pay attention to the determination of the limits of integration.

Example 2.4.4. The functions $f_1(t)$ and $f_2(t)$ are plotted in Figure 2.12 (a) and (b) and $f(t) = f_2(t) * f_1(t)$. Compute the convolution value at $f(2)$.

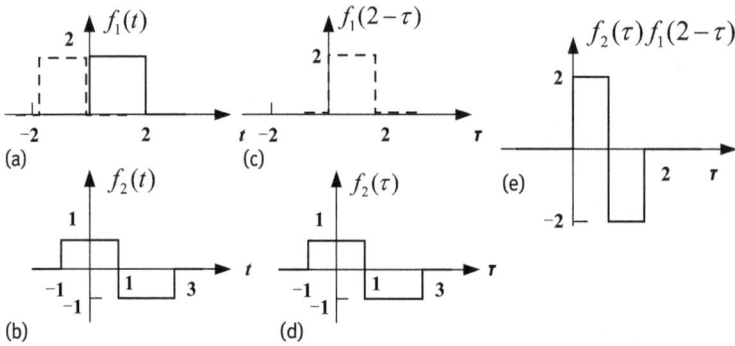

Fig. 2.12: Graphical method for the convolution integral in Example 2.4.4

Solution: According to the definition, $f(2)$ is given by:

$$f(2) = \int_{-\infty}^{\infty} f_2(\tau)f_1(2-\tau)\,d\tau .$$

Step 1: Change the independent variable from t to τ to obtain $f_1(\tau)$ and $f_2(\tau)$.

Step 2: Reflect $f_1(\tau)$ about the vertical axis to obtain $f_1(-\tau)$ and shift it to get $f_1(2-\tau)$, as shown in Figure 2.12 (c).

Step 3: Multiply $f_2(\tau)$ by $f_1(2-\tau)$; the result is plotted in Figure 2.12 (e).

Step 4: Calculate the total area under the product function $f_2(\tau)f_1(2-\tau)$ by integrating it over $\tau = [-\infty, \infty]$ to obtain $f(2) = 0$ (the net area is 0).

! **Note:** The value of $f(2)$ represents the net area under the wave.

In the graphical method for calculating convolution integral, it is critical to determine the upper and lower limits of the integral. The process of finding the overall convolution is generally complicated. It is more convenient to compute the convolution value at a certain moment.

2.4.4 Properties of the convolution integral

The convolution integral is a mathematical operation, which has several interesting properties. It can be used flexibly to simplify convolution. In the following discussion, convolution integrals are convergent (or existent).

1. Three properties of multiplication

(a) The commutative property

$$f_1(t) * f_2(t) = f_2(t) * f_1(t) \tag{2.59}$$

(b) The distributive property

$$f_1(t) * [f_2(t) + f_3(t)] = f_1(t) * f_2(t) + f_1(t) * f_3(t) \tag{2.60}$$

(c) The associative property

$$f_1(t) * [f_2(t) * f_3(t)] = [f_1(t) * f_2(t)] * f_3(t) \tag{2.61}$$

Note: Readers can prove it. ❗

The commutative property states that the order of the convolution operands does not affect the result of the convolution. The distributive property states that convolution is a linear operation. The associative property states that changing the order of the convolution operands does not affect the result of the convolution.

2. Impulse response of the composite system

As is shown in Figure 2.13 (a), the impulse response of the parallel system is the addition of that of the two subsystems $h(t) = h_1(t) + h_2(t)$.

As shown in Figure 2.13 (b), the impulse response of the cascade system is the convolution of that of the two subsystems $h(t) = h_1(t) * h_2(t) = h_2(t) * h_1(t)$.

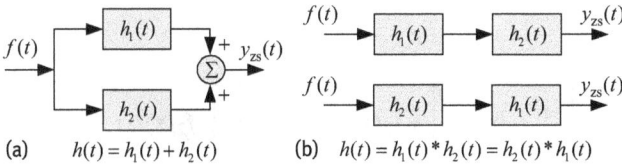

(a) $h(t) = h_1(t) + h_2(t)$

(b) $h(t) = h_1(t) * h_2(t) = h_2(t) * h_1(t)$

Fig. 2.13: Composite system

3. Convolution with an impulse or step function

(a) Convolving a signal with a unit impulse function whose origin is at $t = t_0$ shifts the signal to the origin of the unit impulse function:

$$f(t) * \delta(t) = f(t)$$
$$f(t) * \delta(t - t_0) = f(t - t_0) \tag{2.62}$$

(b) Convolving a signal with the derivative of impulse function is the derivative of the signal:

$$f(t) * \delta'(t) = f'(t)$$
$$f(t) * \delta^{(n)}(t) = f^{(n)}(t) \tag{2.63}$$

Proof.

$$f(t) * \delta'(t) = \delta'(t) * f(t) = \int_{-\infty}^{\infty} \delta'(\tau) f(t - \tau) \, d\tau = f'(t) \qquad \square$$

(c) Convolving a signal with a unit step function produces the running integral of the original signal $f(t)$ as a function of time t:

$$f(t) * \varepsilon(t) = \int_{-\infty}^{\infty} f(\tau) \varepsilon(t - \tau) \, d\tau = \int_{-\infty}^{t} f(\tau) \, d\tau = f^{(-1)}(t) \qquad (2.64)$$

Special case

$$\varepsilon(t) * \varepsilon(t) = t\varepsilon(t) \qquad (2.65)$$

! **Note:** Readers can prove it.

4. Differential and integral properties of convolution

(a)

$$\frac{d^n}{dt^n} [f_1(t) * f_2(t)] = \frac{d^n f_1(t)}{dt^n} * f_2(t) = f_1(t) * \frac{d^n f_2(t)}{dt^n} \qquad (2.66)$$

Proof..

$$\frac{d^n}{dt^n} [f_1(t) * f_2(t)] = \delta^{(n)}(t) * [f_1(t) * f_2(t)] = [\delta^{(n)}(t) * f_1(t)] * f_2(t)$$

$$= f_1^{(n)}(t) * f_2(t) \qquad \square$$

(b)

$$\int_{-\infty}^{t} [f_1(\tau) * f_2(\tau)] \, d\tau = \left[\int_{-\infty}^{t} f_1(\tau) \, d\tau \right] * f_2(\tau) = f_1(\tau) * \left[\int_{-\infty}^{t} f_2(\tau) \, d\tau \right] \qquad (2.67)$$

Proof..

$$\int_{-\infty}^{t} [f_1(\tau) * f_2(\tau)] \, d\tau = \varepsilon(t) * [f_1(t) * f_2(t)] = [\varepsilon(t) * f_1(t)] * f_2(t) = f_1^{(-1)}(t) * f_2(t) \quad \square$$

(c)

$$f_1(t) * f_2(t) = f'_1(t) * f_2^{(-1)}(t) + f_1(-\infty) \cdot \int_{-\infty}^{\infty} f_2(\tau) \, d\tau \qquad (2.68)$$

Under the condition of:

$$f_1(-\infty) = 0 \text{ or } f_2^{(-1)}(\infty) = 0, \qquad (2.69)$$

$$f_1(t) * f_2(t) = f'_1(t) * f_2^{(-1)}(t) \qquad (2.70)$$

Example 2.4.5. Given $f_1(t) = 1$ and $f_2(t) = e^{-t}\varepsilon(t)$, determine $f_1(t) * f_2(t)$.

Solution: According to the commutative property, the complex function is placed ahead to compute the integral:

$$f_1(t) * f_2(t) = f_2(t) * f_1(t) = \int_{-\infty}^{\infty} e^{-\tau}\varepsilon(\tau)\,d\tau = \int_{0}^{\infty} e^{-\tau}\,d\tau = -e^{-\tau}\big|_{0}^{\infty} = 1$$

Note: Applying Equation (2.70) is wrong because it does not satisfy the conditions of Equation (2.69). !

Example 2.4.6. Given $f_1(t)$ in Figure 2.14 and $f_2(t) = e^{-t}\varepsilon(t)$, determine $f_1(t) * f_2(t)$.

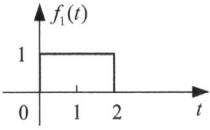

Fig. 2.14: The square wave signal

Solution 1.

$$f_1(t) * f_2(t) = f_1'(t) * f_2^{(-1)}(t)$$
$$f'_1(t) = \delta(t) - \delta(t-2)$$

$$f_2^{(-1)}(t) = \int_{-\infty}^{t} e^{-\tau}\varepsilon(\tau)\,d\tau = \left[\int_{0}^{t} e^{-\tau}\,d\tau\right]\varepsilon(t) = -e^{-\tau}\big|_{0}^{t}\cdot\varepsilon(t) = (1 - e^{-\tau})\varepsilon(t)$$

$$f_1(t) * f_2(t) = (1 - e^{-t})\varepsilon(t) - [1 - e^{-(t-2)}]\varepsilon(t-2)$$

5. Time-shift property
If $f(t) = f_1(t) * f_2(t)$, then

$$f_1(t - T_1) * f_2(t - T_2) = f(t - T_1 - T_2),\qquad(2.71)$$

for any arbitrary real constants T_1 and T_2. In other words, if the two operands of the convolution integral are shifted, then the result of the convolution integral is shifted in time by a duration that is the sum of the individual time shifts introduced in the operands.

Example 2.4.7. Repeat Example 2.4.6 and determine $f_1(t) * f_2(t)$ using the time-shift property.

Solution 2. According to Figure 2.14, the function of $f_1(t)$ can be expressed as:

$$f_1(t) = \varepsilon(t) - \varepsilon(t-2)$$

Then:

$$f_1(t) * f_2(t) = \varepsilon(t) * f_2(t) - \varepsilon(t-2) * f_2(t)$$
$$\varepsilon(t) * f_2(t) = f_2^{(-1)}(t) = (1 - e^{-t})\varepsilon(t)$$

Using the time-shift property:

$$\varepsilon(t-2) * f_2(t) = f_2^{(-1)}(t-2)$$

Therefore:

$$f_1(t) * f_2(t) = (1 - e^{-t})\varepsilon(t) - [1 - e^{-(t-2)}]\varepsilon(t-2)$$

! **Note:** Usually, if $f_1(-\infty) = 0$ holds, we can use Equation (2.70) to simplify computation.

Example 2.4.8. Given $f_1(t)$ and $f_2(t)$ in Figure 2.15, determine $f_1(t) * f_2(t)$.

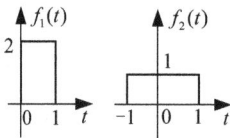

Fig. 2.15: Waveforms of two signals

Solution:

$$f_1(t) = 2\varepsilon(t) - 2\varepsilon(t-1)$$
$$f_2(t) = \varepsilon(t+1) - \varepsilon(t-1)$$
$$f_1(t) * f_2(t) = 2\varepsilon(t) * \varepsilon(t+1) - 2\varepsilon(t) * \varepsilon(t-1)$$
$$- 2\varepsilon(t-1) * \varepsilon(t+1) + 2\varepsilon(t-1) * \varepsilon(t-1)$$

By applying $\varepsilon(t) * \varepsilon(t) = t\varepsilon(t)$ and the time-shift property, the above expression is computed as:

$$f_1(t) * f_2(t) = 2(t+1)\varepsilon(t+1) - 2(t-1)\varepsilon(t-1) - 2t\varepsilon(t) + 2(t-2)\varepsilon(t-2)$$

6. Duration of convolution

Let the nonzero durations (or widths) of the convolution operands $f_1(t)$ and $f_2(t)$ be denoted by T_1 and T_2 time units, respectively. It can be shown that the nonzero duration (or width) of the convolution $f_1(t) * f_2(t)$ is $T_1 + T_2$ time units [16].

7. Important formulas of convolution

Table 2.3 shows some important formulas commonly used to compute the convolution integral.

! **Note:** Readers can prove it.

Tab. 2.3: Commonly used formulas to compute convolution

$K * f(t) = K \cdot [\text{net area of } f(t)]$

$f(t) * \delta(t) = f(t), \quad f(t) * \delta'(t) = f'(t), \quad f(t) * \delta^{(n)}(t) = f^{(n)}(t)$

$f(t) * \varepsilon(t) = f^{(-1)}(t), \quad \varepsilon(t) * \varepsilon(t) = t\varepsilon(t)$

$e^{-at}\varepsilon(t) * e^{-at}\varepsilon(t) = te^{-at}\varepsilon(t)$

$e^{-a_1 t}\varepsilon(t) * e^{-a_2 t}\varepsilon(t) = (1/(a_2 - a_1))(e^{-a_1 t} - e^{-a_2 t})\varepsilon(t) \qquad (a_1 \neq a_2)$

$\varepsilon(t) * e^{-at}\varepsilon(t) = (1/a)(1 - e^{-at})\varepsilon(t)$

$f(t) * \delta_T(t) = f(t) * \sum_{m=-\infty}^{\infty} \delta(t - mT) = \sum_{m=-\infty}^{\infty} f(t - mT)$

2.4.5 Comprehensive application instances

The computation of the convolution integral is one of the greatest difficulties in this chapter. The methods of computing convolution are summarized as follows.
(1) The definition is directly used to compute the convolution integral, which is effective for exponential functions or polynomial functions.
(2) The graphical method is suitable for convolution of simple waveforms, especially for computing convolution value at a certain time point.
(3) The properties and commonly used formulas are combined flexibly to simplify calculation.

Example 2.4.9. Given that $f_1(t) = e^{-2t}\varepsilon(t), f_2(t) = \varepsilon(t)$, determine the convolution $f_1(t) * f_2(t)$.

Solution 1. Definition

$$f_1(t) * f_2(t) = \int_{-\infty}^{\infty} e^{-2\tau}\varepsilon(\tau) \cdot \varepsilon(t - \tau)\,d\tau = \int_{0}^{t} e^{-2\tau}\,d\tau \cdot \varepsilon(t) = \frac{1}{2}(1 - e^{-2t})\varepsilon(t)$$

Solution 2. Graphical method (as shown in Figure 2.16)

$$f_1(t) * f_2(t) = \begin{cases} 0, & t < 0 \\ \int_0^t e^{-2\tau}\,d\tau = \frac{1}{2}(1 - e^{-2t}), & t \geq 0 \end{cases}$$

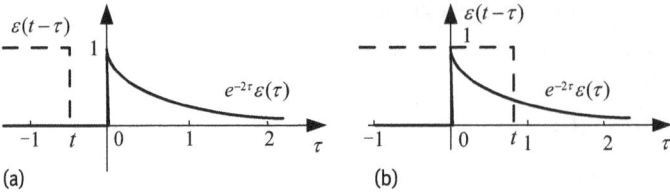

Fig. 2.16: Graphical method for convolution; (a) $t < 0$, (b) $t > 0$

Solution 3. Property

$$f_1(t) * f_2(t) = \varepsilon(t) * e^{-2t}\varepsilon(t) = \delta(t) * [e^{-2t}\varepsilon(t)]^{(-1)}$$

$$= [e^{-2t}\varepsilon(t)]^{(-1)} = \int_{-\infty}^{t} e^{-2\tau}\varepsilon(\tau)\,d\tau = \frac{1}{2}(1 - e^{-2t})\varepsilon(t)$$

Solution 4. Commonly used formulas

$$f_1(t) * f_2(t) = \varepsilon(t) * e^{-2t}\varepsilon(t) = \frac{1}{2}(1 - e^{-2t})\varepsilon(t)$$

Example 2.4.10. The periodic signal is constructed by convolution integral of the time-limited signal and the impulse train.

Solution: According to the time-shift property, the time-limited signal $f(t)$ convolved with time-shifted impulse $\delta(t - t_0)$ is $f(t - t_0)$. This result shows that the convolution of signal $f(t)$ with the unit impulse signal at $t = t_0$ results in the time shift of $f(t)$, as shown in Figure 2.17.

$$f(t) * \delta(t - t_0) = f(t - t_0)$$

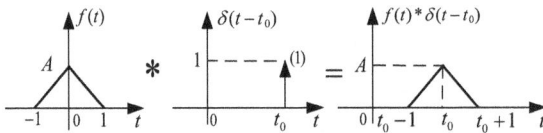

Fig. 2.17: Result of convolution with time-shifted impulse

Using this time-shift property, a periodic signal can be constructed by convolution. A time-limited signal $f_1(t)$ is shown in Figure 2.18 (a), and the periodic impulse train [17] $\delta_T(t)$ is shown in Figure 2.18 (b). The *impulse train* with period T is defined as:

$$\delta_T(t) \overset{\Delta}{=} \sum_{m=-\infty}^{\infty} \delta(t - mT), \tag{2.72}$$

where m is an integer. Then, $f_1(t)$ is convolved with the impulse train to obtain $f_T(t)$, as is shown in Figure 2.16 (c).

$$f_T(t) = f_1(t) * \delta_T(t) = f_1(t) * \left[\sum_{m=-\infty}^{\infty} \delta(t - mT) \right]$$

$$= \left[\sum_{m=-\infty}^{\infty} f_1(t) * \delta(t - mT) \right] = \sum_{m=-\infty}^{\infty} f_1(t - mT) \qquad (2.73)$$

The convolution integral of $f_1(t)$ and $\delta_T(t)$ is a periodic signal whose period is the same with that of the impulse train. Obviously, the time duration of $f_1(t)$ should satisfy $\tau < T$; otherwise, the waveforms between adjacent pulses at kT and $(k \pm 1)T$ will overlap with each other.

Note: Pay attention to the condition of $\tau < T$.

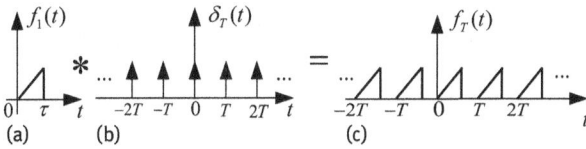

Fig. 2.18: Convolution with impulse train

Example 2.4.11. Given the rectangular pulse $g_\tau(t)$ (gate function) in Figure 2.19 (a) with width τ and height 1, determine the convolution integral $y(t) = g_\tau(t) * g_\tau(t)$.

Solution: Graphical method

$$y(t) = g_\tau(t) * g_\tau(t) = \int_{-\infty}^{\infty} g_\tau(x)g_\tau(t - x)\,dx \qquad (2.74)$$

The functions $g_\tau(x)$ and $g_\tau(-x)$ are plotted as a function of the variable x in the top subplot of Figure 2.19 (b).

(i) For $t < -\tau$ and $t > \tau$, the nonzero parts of $g_\tau(t-x)$ and $g_\tau(x)$ do not overlap, which is shown in Figure 2.19 (c). In other words, the output $y(t) = 0$.

(ii) For $-\tau \le t \le 0$, $g_\tau(-x)$ shifts to the left-hand side along the time axis. The nonzero parts of $g_\tau(t - x)$ and $g_\tau(x)$ overlap over duration $x = [-\tau/2, t + \tau/2]$, which is shown in Figure 2.19 (d). Therefore:

$$g_\tau(t) * g_\tau(t) = \int_{-\frac{\tau}{2}}^{t+\frac{\tau}{2}} (1 \times 1)\,dx = t + \tau .$$

(iii) For $0 \leq t \leq \tau$, the nonzero parts of $g_\tau(t - x)$ and $g_\tau(x)$ overlap over duration $x = [t - \tau/2, \tau/2]$, which is shown in Figure 2.19 (e). Therefore:

$$g_\tau(t) * g_\tau(t) = \int_{t-\frac{\tau}{2}}^{\frac{\tau}{2}} (1 \times 1)\, dx = \tau - t.$$

Combining the above three cases, we obtain:

$$g_\tau(t) * g_\tau(t) = \begin{cases} 0, & t < -\tau,\ t > \tau \\ t + \tau & ,-\tau \leq t < 0 \\ \tau - t & ,0 \leq t \leq \tau. \end{cases}$$

The waveform for the result is sketched in Figure 2.19 (f).

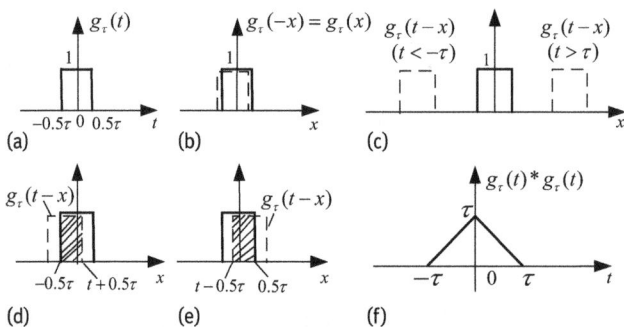

Fig. 2.19: Graphical method for convolution of two gate functions; (a) $g_\tau(t)$, (b) $t = 0$, (c) $t < -\tau, t > 0$, (d) $-\tau \leq t < 0$, (e) $0 \leq t < \tau$, (f) $g_\tau(t) * g_\tau(t)$

Example 2.4.12. In wireless communication system, the received signal is composed of the expected transmitting signal and other echo signal. An echo signal can be formed by a reflected signal of a building. In a system of indoor recording, the microphone receives the normal signal and the reflected signal from the wall. In order to eliminate the interference of these echoes, an inverse system should be designed to compensate the echo signal and restore the original signal.

Figure 2.20 shows an echo cancellation system composed of two blocks. The echo subsystem is to produce echoes, and the inverse subsystem is to counteract echo. Determine the relationship between two impulse responses $h(t)$ and $h_i(t)$.

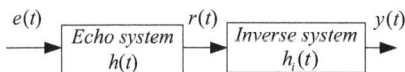

Fig. 2.20: Block diagram of an echo cancellation system

Solution: A mathematical model is established for this multipath transmission phenomenon. The received signal $r(t)$ composed of the normal signal $e(t)$ and its echo component of each path is given by:

$$r(t) = \sum_{m=0}^{N} a_m \cdot e(t - T_m) . \tag{2.75}$$

where T_m and a_m represent the time delayed in each path and the corresponding attenuation coefficient ($0 \leq a_m < 1$), respectively. These signals from $N + 1$ paths are combined together to form an echo signal. The impulse response of the echo system is given by:

$$h(t) = \sum_{m=0}^{N} a_m \cdot \delta(t - T_m) . \tag{2.76}$$

According to the convolution integral, the output $r(t)$ of the echo system can be obtained as follows:

$$r(t) = h(t) * e(t) \tag{2.77}$$

In order to extract the normal signal $e(t)$ from the echo signal, the inverse subsystem is designed to obtain the final recovered signal. The output $y(t)$ of the inverse system is expected to be equal to $e(t)$:

$$y(t) = e(t) = r(t) * h_i(t) = [e(t) * h(t)] * h_i(t) = e(t) * [h(t) * h_i(t)] \tag{2.78}$$

Obviously, the convolution integral of $h(t)$ and $h_i(t)$ equals $\delta(t)$.

$$h(t) * h_i(t) = \delta(t) \tag{2.79}$$

In Equation (2.79), the first operand of the convolution $h(t)$ is given by Equation (2.76), and the convolution result is known as $\delta(t)$. The $h_i(t)$ can be solved by deconvolution. We do not explain the calculation of deconvolution in detail.

Note: Readers can refer to other information about deconvolution.

2.4.6 Convolution computation with MATLAB

The convolution can be implemented using MATLAB. The tool function *conv(f1, f2)* is convenient for discrete-time domain convolution. In the continuous-time domain, the signal is uniformly-spaced sampled to compute convolution, and the result should multiply with the sampling interval.

Example 2.4.13. Two continuous-time signals are given:

$$f_1(t) = \begin{cases} 2, & 0 < t < 1 \\ 0, & \text{else} \end{cases} , \quad f_2(t) = \begin{cases} t, & 0 < t < 2 \\ 0, & \text{else} . \end{cases}$$

Use MATLAB to draw the time-domain waveform of $f(t) = f_1(t) * f_2(t)$.

Solution:

```
t1=0:0.001:1                    % the sampling interval is 0.001
ft1=2*rectpuls(t1-0.5,1)        % generate f_1(t)
t2=0:0.001:2
ft2=t2                          % generate f_2(t)
t3=0:0.001:3
ft3=conv(ft1,ft2)               % compute the convolution
ft3=ft3*0.001                   % multiply with the sampling interval
```

The waveform for the convolution result is sketched in Figure 2.21.

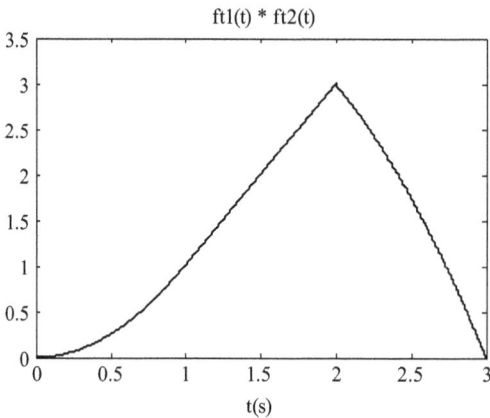

Fig. 2.21: Waveform of the convolution result

2.5 Summary

This chapter introduced the basic time-domain analysis methods of LTIC systems. The system can be represented by a differential equation and solved by the classical methods in Section 2.2 to compute the zero-input response and zero-state response. In Section 2.3, the impulse response $h(t)$ was defined as the output when the elementary signal $\delta(t)$ was applied at the input. Section 2.4 gave an alternative method of convolution to compute the zero-state response $y_{zs}(t) = f(t) * h(t)$.

Chapter 2 problems

2.1 The differential equation of an LTIC system is given as follows. Determine the zero-input response $y_{zi}(t)$ of each system:

(1) $y''(t) + 5y'(t) + 6y(t) = f(t)$

$y(0-) = 1$, $y'(0-) = -1$

(2) $y''(t) + 2y'(t) + 5y(t) = f'(t) + f(t)$

$y(0-) = 2$, $y'(0-) = -2$

(3) $y''(t) + 2y'(t) + y(t) = f(t)$

$y(0-) = 1$, $y'(0-) = 1$

(4) $y''(t) + y(t) = f(t)$

$y(0-) = 2$, $y'(0-) = 0$

2.2 Determine the initial values of $y(0_+)$, $y'(0_+)$ of the given LTIC systems:

(1) $y''(t) + 3y'(t) + 2y(t) = f(t)$

$y(0-) = 0$, $y'(0-) = 1$, $f(t) = \varepsilon(t)$

(2) $y''(t) + 6y'(t) + 8y(t) = f'(t)$

$y(0-) = 0$, $y'(0-) = 1$, $f(t) = \varepsilon(t)$

(3) $y''(t) + 4y'(t) + 3y(t) = f'(t) + f(t)$

$y(0-) = 0$, $y'(0-) = 1$, $f(t) = \varepsilon(t)$

(4) $y''(t) + 4y'(t) + 5y(t) = f'(t)$

$y(0-) = 1$, $y'(0-) = 2$, $f(t) = e^{-2t}\varepsilon(t)$

2.3 Determine the zero-input response, the zero-state response and overall response of the following LTIC systems:

(1) $y''(t) + 4y'(t) + 3y(t) = f(t)$

$f(t) = \varepsilon(t)$, $y(0_-) = 1$, $y'(0_-) = 1$

(2) $y''(t) + 4y'(t) + 4y(t) = f'(t) + 3f(t)$

$f(t) = e^{-t}\varepsilon(t)$, $y(0_-) = 1$, $y'(0_-) = 2$

(3) $y''(t) + 2y'(t) + 2y(t) = f'(t)$

$f(t) = \varepsilon(t)$, $y(0_-) = 0$, $y'(0_-) = 1$

2.4 Calculate the unit impulse response $h(t)$ of the LTIC system shown in Figure P2.1:

$$h_1(t) = \varepsilon(t), \quad h_2(t) = e^{-2t}\varepsilon(t), \quad h_3(t) = \delta(t-1), \quad h_4(t) = 2e^{-3t}\varepsilon(t)$$

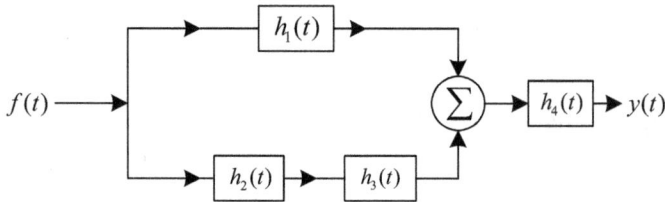

Fig. P2.1: Block diagram of system in Problem 2.4

2.5 The RC circuit is shown in Figure P2.2 and the output is the capacitor voltage. Given that $R = 1\,\Omega$, $C = 0.5\,\text{F}$, and $u_c(0_-) = -1\,\text{V}$, determine the overall response of the following inputs:

(1) $u_s(t) = \varepsilon(t)$

(2) $u_s(t) = e^{-t}\varepsilon(t)$

(3) $u_s(t) = e^{-2t}\varepsilon(t)$

(4) $u_s(t) = t\varepsilon(t)$

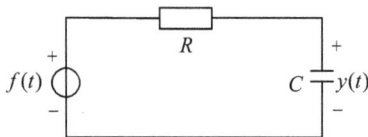

Fig. P2.2: RC circuit used in Problem 2.5

2.6 The RL circuit is shown in Figure P2.3 and the output is the inductor current. Determine the unit impulse response and step response for the output.

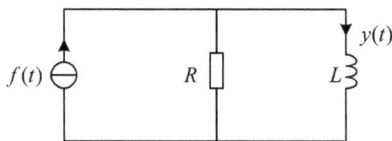

Fig. P2.3: RL circuit used in Problem 2.6

2.7 Calculate the impulse response of the following systems:

(1) $y(t) = f(t - 1) + 2f(t - 3)$

(2) $y'(t) + 4y(t) = 2f(t)$

(3) $y'(t) + 2y(t) = f'(t) - f(t)$

2.8 Determine the unit impulse response and step response of an LTIC system described by the following differential equation:

$$y'(t) + 2y(t) = f''(t)$$

2.9 The input $f(t)$ and the unit impulse response $h(t)$ of an LTIC system are plotted in Figure P2.4. Determine the zero-state response of the system by graphical method for convolution.

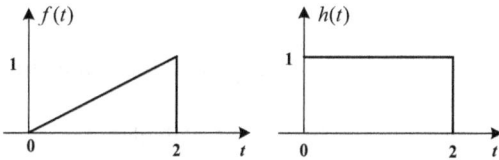

Fig. P2.4: Waveforms in Problem 2.9

2.10 Calculate the output for the following input signal and impulse response:

$$f(t) = \begin{cases} 1.5, & -2 \le t \le 3 \\ 0, & \text{otherwise} \end{cases} \quad \text{and} \quad h(t) = \begin{cases} 2, & -1 \le t \le 2 \\ 0, & \text{otherwise} \end{cases}$$

2.11 Determine the convolution integral $f(t) = f_1(t) * f_2(t)$ of each group of signals:

(1) $f_1(t) = \varepsilon(t)$, $f_2(t) = \varepsilon(t - 3)$

(2) $f_1(t) = \varepsilon(t)$, $f_2(t) = e^{-2t}\varepsilon(t)$

(3) $f_1(t) = \sin \pi t\varepsilon(t)$, $f_2(t) = \varepsilon(t) - \varepsilon(t - 4)$

(4) $f_1(t) = t\varepsilon(t)$, $f_2(t) = e^{-2t}\varepsilon(t)$

(5) $f_1(t) = \varepsilon(t - 3)$, $f_2(t) = e^{-2t}\varepsilon(t + 1)$

2.12 The function waveforms are shown in Figure P2.5. Plot the waveforms of the following convolutions:

(1) $f_1(t) * f_2(t)$ (2) $f_1(t) * f_3(t)$

(3) $f_1(t) * f_4(t)$ (4) $f_1(t) * f_2(t) * f_2(t)$

(5) $f_1(t) * [2f_4(t) - f_3(t - 3)]$

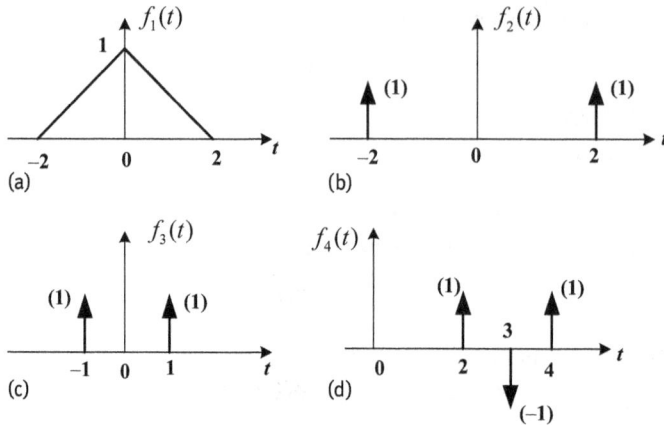

Fig. P2.5: Waveforms in Problem 2.12

2.13 Given that $f_1(t) = t\varepsilon(t)$, $f_2(t) = \varepsilon(t) - \varepsilon(t-2)$, determine the convolution integral: $y(t) = f_1(t) * f_2(t-1) * \delta'(t-2)$.

2.14 Diagrams of an LTIC system are shown in Figure P2.6. Calculate the zero-state response with the input $f(t) = e^{-t}\varepsilon(t)$.

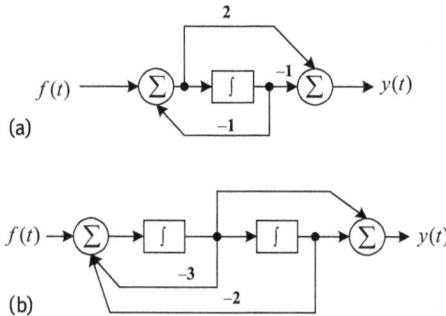

(a)

(b)

Fig. P2.6: Block diagrams in Problem 2.14

2.15 The differential equation of an LTIC system is given as follows:

$$y''(t) + 4y'(t) + 3y(t) = 2f'(t) + f(t)$$

(1) Determine the unit impulse response of system with MATLAB.
(2) Determine the unit step response of system with MATLAB.

2.16 Determine the convolution of the two functions and plot the results with MAT-LAB:

(1) $f_1(t) = \varepsilon(t) - \varepsilon(t-1)$; $f_2(t) = 2t[\varepsilon(t) - \varepsilon(t-1)]$
(2) $f_1(t) = \cos\omega t\varepsilon(t)$; $f_2(t) = \varepsilon(t) - \varepsilon(t-4)$.

3 Time-domain analysis of LTID systems

Please focus on the following key questions.

1. How can we represent the calculation of interest rates on a bank deposit as a problem of discrete-time system analysis?
2. What is the classical solution to the difference equations of LTID systems?
3. What is the elementary signal and its response in discrete-time domain analysis?
4. How can we derive a new operation of the convolution sum from discrete-time signal decomposition in the time domain? What is the equation of solving zero-state response in the discrete-time domain?

3.0 Introduction

In this chapter, the linear, time-invariant, discrete-time (LTID) system is analyzed to solve the response in the time domain. The LTID system is modeled with a linear, constant-coefficient difference equation representing the relationships of the response and input. As was the case for the LTIC systems discussed in Chapter 2, we are primarily interested in calculating the output response of an LTID system to any arbitrary input signal. The basic idea is to decompose the signal by the unit impulse function for calculating the output based on convolving the applied input with the impulse response.

This chapter first introduces the linear constant-coefficient difference equation to model an LTID system. Classical time-domain analysis is used to solve the difference equation to obtain the zero-input response and the zero-state response. In Section 3.3, the unit impulse response is defined as the output of an LTID system to the unit impulse function applied at the input. In addition, the unit step response is defined as the output of the unit step function as the input. In Section 3.4, the signal decomposition in the time domain and the convolution sum are discussed in detail. This development leads to a second approach to calculating the output based on convolving the applied input sequence with the impulse response in the DT domain. The chapter is concluded in Section 3.5 with a summary of the important concepts covered in the chapter.

https://doi.org/10.1515/9783110593907-003

3.1 Representation of an LTID system

3.1.1 Analytical description based on mathematical models

As discussed in Chapter 2, an LTIC system can be modeled using a linear constant-coefficient differential equation. Likewise, the input–output relationship of a linear DT system can be described using a difference equation.

Example 3.1.1. A person deposits a certain amount $f(k)$ of money in the bank in the beginning of each month. The monthly interest rate is β. Determine the model of the bank deposit $y(k)$ of the k-th month.

Solution: The deposit in the beginning of the k-th month is $y(k)$, which includes three parts. They are the new deposit of each month $f(k)$, the deposit of the previous month $y(k-1)$, and its interest $\beta \cdot y(k-1)$:

$$y(k) = y(k-1) + \beta \cdot y(k-1) + f(k) \tag{3.1}$$

$$y(k) - (1+\beta) \cdot y(k-1) = f(k) \tag{3.2}$$

If the deposit starts at $k = 0$, then the initial state is $y(0) = f(0)$.

Equation (3.2) is called the first-order difference equation between $y(k)$ and $f(k)$.

! **Note:** This is the forward-difference equation.

3.1.2 Description based on the block diagram

For a linear, constant-coefficient difference equation, this includes three basic operations: multiplication, difference and addition. These basic operations can be expressed by the ideal parts to be connected with each other, which is drawn as a block diagram. As is shown in Figure 3.1, the basic operation units are the unit delayer, adder and multiplier.

! **Note:** The integrator in the CT system is replaced by a delayer in the DT system.

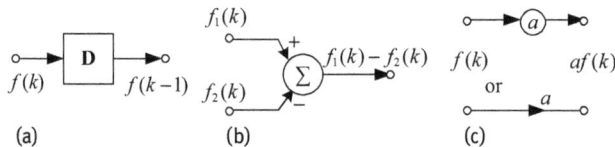

Fig. 3.1: Basic components of the DT system; (a) Unit delayer, (b) Adder, (c) Multiplier

Example 3.1.2. The block diagram of an LTID system is shown in Figure 3.2; determine the difference equation of the input $f(k)$ and the output $y(k)$.

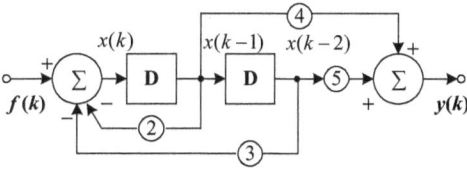

Fig. 3.2: The block diagram of Example 3.1.2

Solution:
(i) Set the auxiliary variable $x(k)$ at the input of the first delayer.
(ii) For the left adder, the input–output representation is:

$$x(k) = f(k) - 2x(k-1) - 3x(k-2),\qquad(3.3)$$

which is rewritten as:

$$x(k) + 2x(k-1) + 3x(k-2) = f(k).\qquad(3.4)$$

(iii) For the right adder, the input–output equation is:

$$y(k) = 4x(k-1) + 5x(k-2).\qquad(3.5)$$

(iv) According to the characteristics of the LTID system, Equations (3.4) and (3.5) are combined to obtain the difference equation:

$$y(k) + 2y(k-1) + 3y(k-2) = 4f(k-1) + 5f(k-2)$$

Note: Readers can prove this.

3.1.3 General form of the difference equation

The general form of the difference equation is as follows:

$$y(k) + a_{n-1}y(k-1) + \cdots + a_0 y(k-n) = b_m f(k) + \cdots + b_0 f(k-m)\qquad(3.6)$$

where $f(k)$ denotes the input sequence and $y(k)$ denotes the resulting output sequence, and coefficients a_i (for $0 \le i \le n-1$) and b_i (for $0 \le i \le m$) are parameters that characterize the DT system.

We now consider an iterative procedure for solving linear, constant-coefficient difference equations.

Example 3.1.3. The linear difference equation of an LTID system is

$$y(k) + 3y(k-1) + 2y(k-2) = f(k) .$$

Determine the output signal $y(k)$ when the input is given by $f(k) = 2^k \varepsilon(k)$ and the initial value is $y(0) = 0$, $y(1) = 2$.

Solution: The difference equation is expressed as follows:

$$y(k) = -3y(k-1) - 2y(k-2) + f(k) .$$

By iterating from $k = 2$, the output response is computed as follows:

$$y(2) = -3y(1) - 2y(0) + f(2) = -2 ,$$
$$y(3) = -3y(2) - 2y(1) + f(3) = 10 ,\ldots$$

Additional values of the output sequence for $k > 3$ can be similarly evaluated from further iteration with respect to k.

Note: This recursive iterative calculation method is generally not applicable to the analytic form of the solution.

3.2 Classical solution of the difference equation

3.2.1 Classical solution of the direct method

Consider the difference equation of the input $f(k)$ and the output $y(k)$ for the LTID system as follows:

$$y(k) + a_{n-1}y(k-1) + \ldots a_0 y(k-n) = b_m f(k) + \ldots b_0 f(k-m)$$

In Chapter 2, we showed that the output response of a CT system can be decomposed into two components: the homogeneous solution and a special solution. This is also valid for DT systems. The classical solution of difference equation is:

$$y(k) = y_h(k) + y_p(k) , \tag{3.7}$$

where $y_p(k)$ is the particular component and $y_h(k)$ is the homogeneous component.

1. The homogeneous component $y_h(k)$
The homogeneous component is the solution of the corresponding homogeneous difference equation as follows:

$$y(k) + a_{n-1}y(k-1) + \ldots a_0 y(k-n) = 0 . \tag{3.8}$$

The functional form of $y_h(k)$ is determined by the characteristic roots λ_k of the characteristic function as follows:

$$\lambda^n + a_{n-1}\lambda^{n-1} + \ldots + a_0 = 0.$$

The form of homogeneous component is determined by different cases of characteristic roots, as shown in Table 3.1.

Tab. 3.1: The forms of the homogeneous component

Root λ	The homogeneous component $y_h(k)$
Single	$C\lambda^k$
Repeated r times	$(C_{r-1}k^{r-1} + C_{r-2}k^{r-2} + \cdots + C_1 k + C_0)\lambda^k$
Complex $\lambda_{1,2} = \alpha \pm j\beta$	$\rho^k[C\cos(\beta k) + D\sin(\beta k)]$ or $A\rho^k\cos(\beta k - \theta)$, where $Ae^{j\beta} = C + jD$

2. The particular component $y_p(k)$

The functional form of the particular solution $y_p(k)$ is determined by the form of the input function. The detailed forms are shown in Table 3.2.

Tab. 3.2: The functional forms of particular solutions to different inputs

Input $f(t)$	Particular component $y_p(k)$	
k^m	$P_m k^m + P_{m-1}k^{m-1} + \cdots + P_1 k + P_0,$	if none of the roots is 1;
	$k^r[P_m k^m + P_{m-1}k^{m-1} + \cdots + P_1 k + P_0],$	if the root 1 repeats r times.
a^k	$Pa^k,$	if a is not equal to the root;
	$(P_1 k + P_0)a^k,$	if a is equal to a single root;
	$(P_r k^r + P_{r-1}k^{r-1} + \cdots + P_1 k + P_0)a^k,$	if a is equal to the r-times root.
$\cos\beta k$ or $\sin\beta k$	$P\cos(\beta k) + Q\sin(\beta k),$	if none of the roots is equal to $e^{\pm j\beta}$;
	or $A\cos(\beta k - \theta)$, where $Ae^{j\theta} = P + jQ.$	

Note: Pay attention to the distinction compared to the differential equations. ❗

Example 3.2.1. The linear difference equation of an LTID system is:

$$y(k) + 4y(k-1) + 4y(k-2) = f(k).$$

Determine the output signal $y(k)$ when the input is given by $f(k) = 2^k \varepsilon(k)$ and the initial value is $y(0) = 0$, $y(1) = -1$.

Solution: As the characteristic roots of characteristic equation $\lambda^2 + 4\lambda + 4 = 0$ are $\lambda_1 = \lambda_2 = -2$, the homogeneous component is given by:

$$y_h(k) = (C_1 k + C_2)(-2)^k, \quad k \geq 0. \tag{3.9}$$

From Table 3.2, for the input signal $f(k) = 2^k \varepsilon(k)$, the particular component is of the following form:

$$y_p(k) = P(2)^k, \quad k \geq 0. \tag{3.10}$$

Substituting the particular component in the difference equation yields:

$$P(2)^k + 4P(2)^{k-1} + 4P(2)^{k-2} = 2k \quad \Rightarrow \quad P = 1/4.$$

The particular component is $y_p(k) = 2^{k-2}, k \geq 0$.

The overall response is given by:

$$y(k) = y_h + y_p = (C_1 k + C_2)(-2)^k + 2^{k-2}, \quad k \geq 0, \tag{3.11}$$

where C_1, C_2 are obtained by inserting initial values. This leads to the following simultaneous equations:

$$y(0) = C_2 + \frac{1}{4} = 0, \quad y(1) = (C_1 + C_2)(-2) + \frac{1}{2} = -1,$$

with solutions $C_1 = 1, C_2 = -1/4$. The overall response is:

$$y(k) = \left(k - \frac{1}{4}\right)(-2)^k + 2^{k-2}, \quad k \geq 0. \tag{3.12}$$

The homogeneous component does not depend on the external input, and hence the homogeneous component is also called the *natural response*. The particular component forced by the input signal can be defined as the *forced response*.

! **Note:** Determine the natural and forced response in Equation (3.12).

Example 3.2.2. The difference equation of an LTID system is given by:

$$6y(k) - 5y(k-1) + y(k-2) = f(k) \tag{3.13}$$

Determine the response when the input is given by $f(k) = 10\cos(0.5\pi k), k \geq 0$, and the initial value is $y(0) = 0, y(1) = 1$.

Solution:

(1) The homogeneous component is $y_h(k) = C_1(1/2)^k + C_2(1/3)^k$.
(2) According to the input function $f(k) = 10\cos(0.5\pi k), k \geq 0$, the particular component is represented as:

$$y_p(k) = P\cos(0.5\pi k) + Q\sin(0.5\pi k) \tag{3.14}$$

Substituting $y_p(k)$ and $f(k)$ into Equation (3.13) yields:

$$(6P + 5Q - P)\cos(0.5\pi k) + (6Q - 5P - Q)\sin(0.5\pi k) = 10\cos(0.5\pi k) \tag{3.15}$$

Equating the cosine and sine terms on the left-hand and right-hand sides of the equation, we obtain the following simultaneous equations:

$$6P + 5Q - P = 10$$
$$6Q - 5P - Q = 0,$$

With the solution $P = Q = 1$. The particular component is given by:

$$y_P(k) = \cos(0.5\pi k) + \sin(0.5\pi k) = \sqrt{2}\cos\left(0.5\pi k - \frac{\pi}{4}\right), \quad k \geq 0 \qquad (3.16)$$

(3) The overall response is the sum of the homogeneous and the particular components:

$$y(k) = y_h(k) + y_p(k)$$

$$= C_1\left(\frac{1}{2}\right)^k + C_2\left(\frac{1}{3}\right)^k + \cos(0.5\pi k) + \sin(0.5\pi k), \quad k \geq 0 \qquad (3.17)$$

where C_1, C_2 are obtained by inserting initial values. This leads to the following simultaneous equations:

$$y(0) = C_1 + C_2 + 1 = 0, \quad y(1) = 0.5C_1 + \frac{1}{3}C_2 + 1 = 1$$

with the solutions $C_1 = 2$, $C_2 = -3$. The overall response is given by:

$$y(k) = 2\left(\frac{1}{2}\right)^k - 3\left(\frac{1}{3}\right)^k + \cos(0.5\pi k) + \sin(0.5\pi k)$$

$$= 2\left(\frac{1}{2}\right)^k - 3\left(\frac{1}{3}\right)^k + \sqrt{2}\cos\left(0.5\pi k - \frac{\pi}{4}\right), \quad k \geq 0 \qquad (3.18)$$

In the overall response, the first two components decay to zero as $k \to \infty$, which is referred to as the *transient response*. The last component shows equal amplitude oscillation, which is known as the *steady-state response*. Generally, if each of the characteristic roots $|\lambda| < 1$, the natural response is a transient response.

According to Equation (3.18), the components of the overall response can be classified as:

$$y(k) = \underbrace{\overbrace{2\left(\frac{1}{2}\right)^k - 3\left(\frac{1}{3}\right)^k}^{\text{natural response}}}_{\text{transient response}} + \underbrace{\overbrace{\sqrt{2}\cos\left(0.5\pi k - \frac{\pi}{4}\right)}^{\text{forced response}}}_{\text{steady-state response}}, \quad k \geq 0. \qquad (3.19)$$

Note: Determine whether the zero-state response is the forced response. !

3.2.2 Zero-input response and zero-state response

In Chapter 2, we showed that the output response of a CT system can be decomposed into two components: the zero-state response and the zero-input response. This is also valid for DT systems. The output response can be expressed as:

$$y(k) = y_{zi}(k) + y_{zs}(k), \qquad (3.20)$$

where $y_{zi}(k)$ denotes the *zero-input response* of the system and $y_{zs}(k)$ denotes the *zero-state response* of the DT system.

The zero-input response $y_{zi}(k)$ is the response produced by the initial conditions of the system without any external input. To calculate the zero-input component, we assume that the applied input sequence is $f(k) = 0$. On the other hand, the zero-state response $y_{zs}(k)$ arises due to the input sequence and does not depend on the initial conditions of the system. To calculate the zero-state component, the initial conditions are assumed to be zero.

Supposing that the input sequence is inserted into the n-order system at $k = 0$, the initial conditions are denoted by the ancillary condition $y(-1), y(-2), \ldots y(-n)$. The initial values should be calculated by iterating with respect to $k = 0, 1 \ldots, n$.

Example 3.2.3. The difference equation of an LTID system is given by:

$$y(k) + 3y(k - 1) + 2y(k - 2) = f(k) . \tag{3.21}$$

Determine the zero-input response and zero-state response when the input is given by $f(k) = 2^k$, $k \geq 0$ and the initial condition is $y(-1) = 0, y(-2) = 1/2$.

Solution:
(1) Determine the zero-input response
The homogeneous equation of $y_{zi}(k)$ is:

$$y_{zi}(k) + 3y_{zi}(k - 1) + 2y_{zi}(k - 2) = 0 , \tag{3.22}$$

with ancillary condition

$$y_{zi}(-1) = y(-1) = 0 , \quad y_{zi}(-2) = y(-2) = 1/2 .$$

Note: Why is the initial condition of zero-input response the same as that of the overall response?

The initial value is obtained by iterating Equation (3.22) for $k = 0, 1$:

$$y_{zi}(0) = -3y_{zi}(-1) - 2y_{zi}(-2) = -1$$
$$y_{zi}(1) = -3y_{zi}(0) - 2y_{zi}(-1) = 3 .$$

Note: The ancillary condition $y_{zi}(-1), \ldots, y_{zi}(-n)$ can be directly used to determine the coefficients of the zero-input response.

As the characteristic roots of the homogeneous equation are $\lambda_1 = -1, \lambda_2 = -2$, the zero-input response is $y_{zi}(k) = C_1(-1)^k + C_2(-2)^k$. Substituting the initial value yields:

$$C_1 = 1 , \quad C_2 = -2$$

The zero-input response is, therefore, given by:

$$y_{zi}(k) = (-1)^k - 2(-2)^k , \quad k \geq 0$$

(2) Determine the zero-state response
The zero-state response $y_{zs}(k)$ satisfies the equation:

$$y_{zs}(k) + 3y_{zs}(k-1) + 2y_{zs}(k-2) = 2^k ,\qquad (3.23)$$

with the ancillary condition:

$$y_{zs}(-1) = y_{zs}(-2) = 0$$

Iterating Equation (3.23) for $k = 0, 1$ yields

$$y_{zs}(0) = -3y_{zs}(-1) - 2y_{zs}(-2) + 1 = 1$$
$$y_{zs}(1) = -3y_{zs}(0) - 2y_{zs}(-1) + 2 = -1$$

Note: The initial value of the zero-state response must be calculated iteratively. !

The homogeneous component is of the same form as the zero-input response:

$$y_{zsh}(k) = D_1(-1)^k + D_2(-2)^k .$$

The particular component is $y_{zsp}(k) = P \cdot 2^k$. Substituting the particular component in Equation (3.23) yields $y_{zsp}(k) = (1/3)2^k$. The overall zero-state response is obtained as follows:

$$y_{zs}(k) = D_1(-1)^k + D_2(-2)^k + (1/3)2^k .\qquad (3.24)$$

Substituting the initial values $y_{zs}(0)$ and $y_{zs}(1)$ in Equation (3.24) yields $D_1 = -1/3, D_2 = 1$. Therefore, the zero-state response is given by:

$$y_{zs}(k) = -(-1)^k/3 + (-2)^k + (1/3)2^k ,\quad k \geq 0$$

Example 3.2.4. The difference equation of an LTID system is given by:

$$y(k) - 2y(k-1) + 2y(k-2) = f(k)\qquad (3.25)$$

Determine the zero-input response, the zero-state response and the overall response when the input is given by $f(k) = k, k \geq 0$ and the initial condition is $y(-1) = 1$, $y(-2) = 0.5$.

Solution:
(1) Determine the zero-input response
The zero-input response $y_{zi}(k)$ satisfies the following equation:

$$\left.\begin{array}{l} y_{zi}(k) - 2y_{zi}(k-1) + 2y_{zi}(k-2) = 0 \\ y_{zi}(-1) = y(-1) = 1 , \ \ y_{zi}(-2) = y(-2) = 0.5 \end{array}\right\}\qquad (3.26)$$

The characteristic function is $\lambda^2 - 2\lambda + 2 = 0$ with the characteristic roots $\lambda_{1,2} = 1 \pm j1 = \sqrt{2}e^{\pm j(\pi/4)}$:

$$y_{zi}(k) = (\sqrt{2})^k \left[C_1 \cos\left(\frac{k\pi}{4}\right) + D_1 \sin\left(\frac{k\pi}{4}\right) \right] \tag{3.27}$$

The initial values $y_{zi}(0)$ and $y_{zi}(1)$ are calculated by iterating Equation (3.26):

$$y_{zi}(0) = 2y_{zi}(-1) - 2y_{zi}(-2) = 1$$
$$y_{zi}(1) = 2y_{zi}(0) - 2y_{zi}(-1) = 0$$

Substituting the initial values into Equation (3.27) yields:

$$y_{zi}(0) = C_1 = 1$$
$$y_{zi}(1) = \sqrt{2}\left(C_1 \frac{\sqrt{2}}{2} + D_1 \frac{\sqrt{2}}{2} \right) = 0 \quad \Rightarrow \quad D_1 = -1.$$

The zero-input response is expressed as:

$$y_{zi}(k) = (\sqrt{2})^k \left[\cos\left(\frac{k\pi}{4}\right) - \sin\left(\frac{k\pi}{4}\right) \right], \quad k \geq 0.$$

(2) Determine the zero-state response

The zero-state response $y_{zs}(k)$ satisfies the following equation:

$$\left. \begin{array}{r} y_{zs}(k) - 2y_{zs}(k-1) + 2y_{zs}(k-2) = k \\ y_{zs}(-1) = y_{zs}(-2) = 0 \end{array} \right\} \tag{3.28}$$

The initial values $y_{zs}(0)$, $y_{zs}(1)$ are obtained by:

$$y_{zs}(0) = 2y_{zs}(-1) - 2y_{zs}(-2) = 0$$
$$y_{zs}(1) = 2y_{zs}(0) - 2y_{zs}(-1) + 1 = 1.$$

According to the input sequence, the particular component is $y_{zsp}(k) = P_1 k + P_0$. Substituting $y_{zsp}(k)$ into Equation (3.28) yields:

!

Note: The particular component is based on the form of the input sequence.

$$P_1 k + P_0 - 2[P_1(k-1) + P_0] + 2[P_1(k-2) + P_0] = k$$

Equating the coefficients of the two sides of the equations, we obtain:

$$P_1 = 1, \quad P_0 = 2.$$

The particular component is $y_{zsp}(k) = k + 2, k \geq 0$.

The homogeneous component is:

$$y_{zsh}(k) = (\sqrt{2})^k \left[C_2 \cos\left(\frac{k\pi}{4}\right) + D_2 \sin\left(\frac{k\pi}{4}\right) \right],$$

and the overall zero-state response is:

$$y_{zs}(k) = (\sqrt{2})^k \left[C_2 \cos\left(\frac{k\pi}{4}\right) + D_2 \sin\left(\frac{k\pi}{4}\right) \right] + k + 2 .$$

Substituting the initial values $y_{zs}(0)$, $y_{zs}(1)$ into the above equation, we obtain:

$$y_{zs}(0) = C_2 + 2 = 0$$

$$y_{zs}(1) = \sqrt{2}\left(C_2\frac{\sqrt{2}}{2} + D_2\frac{\sqrt{2}}{2} \right) + 3 = 1$$

with solutions $C_2 = -2$, $D_2 = 0$. Therefore, the zero-state response is:

$$y_{zs}(k) = -2\left(\sqrt{2}\right)^k \cos\left(\frac{k\pi}{4}\right) + k + 2, \quad k \geq 0 .$$

(3) Determine the overall response
   The overall response is given by:

$$y(k) = y_{zi}(k) + y_{zs}(k)$$

$$= \left(\sqrt{2}\right)^k \left[\cos\left(\frac{k\pi}{4}\right) - \sin\left(\frac{k\pi}{4}\right) \right] - 2\left(\sqrt{2}\right)^k \cos\left(\frac{k\pi}{4}\right) + k + 2$$

$$= -\left(\sqrt{2}\right)^{k+1} \cos\left(\frac{k\pi}{4} - \frac{\pi}{4}\right) + k + 2, \quad k \geq 0$$

3.2.3 Response calculation with MATLAB

MATLAB provides the function for solving the zero-state response of the LTID system. The function is a *filter*, and its format is as follows:

```
y=filter( b, a, f )
```

where, f is the input signal of system; $b=[b0, b1, b2, \ldots, bm]$, $a=[a0, a1, a2, \ldots, an]$ are the coefficients of the right- and left-hand sides of the difference equation, respectively.

Example 3.2.5. Consider the LTID system of noise smoothing with the following input–output relationship:

$$y(k) = \frac{1}{M} \sum_{n=0}^{M-1} f(k-n) . \qquad (3.29)$$

The input sequence is given by $f(k) = s(k) + d(k)$, where $s(k) = (2k)0.9^k$, and $d(k)$ is a random noise sequence. Calculate the zero-state response with MATLAB.

Solution:

```
R=51;                  % the length of the input sequence
d=rand(1,R)-0.5;       % produce a random sequence in (-0.5,0.5)
k=0:R-1;
s=2*k.*(0.9.^k);
f=s+d;                 % the input sequence
M=5;
b=ones(M,1)/M;         % the coefficients in right-hand side of the
                       % difference equation
a=1;                   % the coefficient on left-hand side of the
                       % difference equation
y=filter(b,a,f);       % calculate the zero-state response by function
```

The input and output of the system are plotted in Figure 3.3.

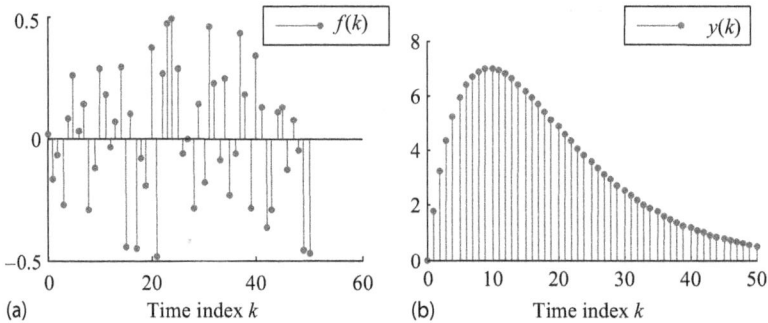

(a) Time index k (b) Time index k

Fig. 3.3: Waveforms of Example 3.2.5

3.3 Impulse response and step response

3.3.1 Basic discrete-time sequence

1. DT unit impulse sequence

Recall that a DT impulse function is defined in Chapter 1 as follows:

$$\delta(k) = \begin{cases} 1, & k = 0 \\ 0, & k \neq 0. \end{cases}$$

The shifted impulse function $\delta(k - k_0)$ is defined as follows:

$$\delta(k - k_0) = \begin{cases} 1, & k = k_0 \\ 0, & k \neq k_0. \end{cases}$$

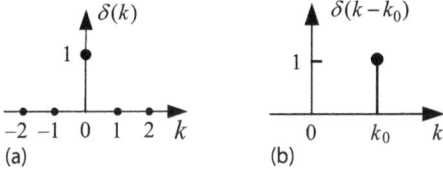

Fig. 3.4: The DT unit impulse function and shifted impulse function; (a) DT unit impulse sequence, (b) Time-shifted impulse sequence

The waveform for a DT unit impulse function is shown in Figure 3.4 (a) and the time-shifted DT unit impulse function is shown in Figure 3.4 (b).

(1) Addition: The addition of two unit impulse functions is $\delta(k) + 2\delta(k) = 3\delta(k)$. The sequence shown in Figure 3.5 can be represented as $f(k) = \delta(k-1) + \delta(k-2)$.

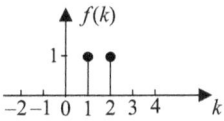

Fig. 3.5: The DT function

(2) Multiplication:

$$\delta(k) \cdot \delta(k) = \delta(k), \quad \delta(k-1) \cdot \delta(k-2) = 0.$$

Note: Multiplication cannot be made on two CT impulses. ❗

(3) The sampling properties:

$$f(k)\delta(k) = f(0)\delta(k)$$
$$f(k)\delta(k-k_0) = f(k_0)\delta(k-k_0)$$

(3.30)

$$\sum_{k=-\infty}^{\infty} f(k)\delta(k) = f(0)$$

(3.31)

$$\sum_{k=-\infty}^{\infty} f(k)\delta(k-k_0) = f(k_0)$$

(4) The parity:

$$\delta(k) = \delta(-k)$$

2. DT unit step sequence
Recall that a DT step function is defined in Chapter 1 as follows:

$$\varepsilon(k) = \begin{cases} 0, & k < 0 \\ 1, & k \geq 0 \end{cases}.$$

The waveform for a DT unit step function is shown in Figure 3.6.

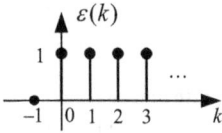

Fig. 3.6: The DT unit step function

(1) Addition: The addition of two unit step functions is $\varepsilon(k) + 2\varepsilon(k) = 3\varepsilon(k)$. The sequence shown in Figure 3.6 can be represented as $f(k) = \varepsilon(k-1) - \varepsilon(k-3)$.

(2) Multiplication:

$$\varepsilon(k) \cdot \varepsilon(k) = \varepsilon(k) , \quad \varepsilon(k-2) \cdot \varepsilon(k-5) = \varepsilon(k-5) .$$

(3) Integration:

$$\sum_{i=-\infty}^{k} \varepsilon(i) = \begin{cases} 0 , & k < 0 \\ k+1 , & k \geq 0 \end{cases}$$

$$= (k+1)\varepsilon(k) \tag{3.32}$$

Note: $\int_{-\infty}^{t} \varepsilon(\tau)\,d\tau = t\varepsilon(t).$

(4) Relationship with the DT unit impulse sequence

The relationship between $\delta(k)$ and $\varepsilon(k)$ is as follows:

$$\delta(k) = \varepsilon(k) - \varepsilon(k-1) , \quad \varepsilon(k) = \sum_{i=-\infty}^{k} \delta(i)$$

3. DT sinusoidal sequence

The DT sinusoid sequence is defined as follows:

$$f(k) = A \cos(\Omega_0 k + \varphi) ,$$

where A is the amplitude, and Ω_0 is the DT radian frequency. As discussed in Section 1.1.3, a DT sinusoidal sequence is not necessarily periodic. The DT sinusoidal signal is periodic only if the fraction $2\pi/\Omega_0$ is a rational number.

The DT sinusoid sequence $f(k) = \cos(\Omega_0 k)$ can be obtained by sampling of the CT sinusoidal signal $\cos \omega_0 t$, which takes the following form:

$$f(k) = \cos(\omega_0 t)\big|_{t=kT_s} = \cos\left(\frac{2\pi}{T_0} \times kT_s\right) = \cos(\Omega_0 k)$$

where $\Omega_0 = 2\pi T_s/T_0$. The signal $f(k)$ is periodic only if the fraction T_0/T_s is a rational number.

4. DT exponential sequence

The DT complex exponential sequence with radian frequency Ω_0 is defined as follows:

$$f(k) = e^{\beta k} = e^{(\sigma + j\Omega_0)k}$$
$$= e^{\sigma k}[\cos(\Omega_0 k) + j\sin(\Omega_0 k)]$$
$$= r^k[\cos(\Omega_0 k) + j\sin(\Omega_0 k)]$$

It can be seen that the real and imaginary parts of the complex exponential sequence are sinusoidal sequences whose amplitudes r^k vary exponentially. It can be classified as rising $(r > 1)$, decaying $(0 < r < 1)$ and constant-valued $(r = 1)$ exponentials depending upon the value of $r = e^\sigma$.

5. Correspondence between CT and DT elementary signals

Table 3.3 summarizes the correspondence between CT and DT elementary signals.

Tab. 3.3: Correspondence between CT and DT elementary signals

CT	Elementary signal	DT
Unit impulse signal	$\delta(t) \leftrightarrow \delta(k)$	Unit impulse sequence
Unit step signal	$\varepsilon(t) \leftrightarrow \varepsilon(k)$	Unit step sequence
Sinusoidal signal	$A\cos(\omega t + \varphi) \leftrightarrow A\cos(\Omega_0 k + \varphi)$	Sinusoidal sequence
Exponential signal	$Ae^{j\omega t} \leftrightarrow Ae^{j\Omega_0 k}$	Exponential sequence
Complex exponential function	$e^{st} \leftrightarrow e^{\beta k}$ (or z^k)	Complex exponential sequence

3.3.2 Unit impulse response and step response of an LTID system

1. DT unit impulse response

Definition. The *impulse response* $h(k)$ of an LTID system is the zero-state response of the system when a unit impulse $\delta(k)$ is applied at the input. The effect of $h(k)$ is similar to the impulse response $h(t)$ in an LTIC system.

Since the unit impulse $\delta(k)$ is zero when $k > 0$, it represents that the impulse response has the same form with the zero-input response. The initial values can be determined by iterating with respect to $k = 0, 1 \ldots, n$ under the condition of zero state.

2. DT unit step response

Definition. The *step response* $g(k)$ of an LTID system is the zero-state response of the system when a unit step $\varepsilon(k)$ is applied at the input. The effect of $g(k)$ is similar to the impulse response $g(t)$ in an LTIC system. Given the difference equation of LTID system, the unit step response can be obtained by the classical method in Section 3.2.1.

Applying the relation between $\varepsilon(k)$ and $\delta(k)$, the relation between $g(k)$ and $h(k)$ satisfies the following expression:

$$\varepsilon(k) = \sum_{i=-\infty}^{k} \delta(i) \qquad \Rightarrow \quad g(k) = \sum_{i=-\infty}^{k} h(i) \tag{3.33}$$

$$\delta(k) = \varepsilon(k) - \varepsilon(k-1) \quad \Rightarrow \quad h(k) = g(k) - g(k-1) \tag{3.34}$$

Example 3.3.1. Determine the unit impulse response $h(k)$ and the step response $g(k)$ for the LTID system shown in Figure 3.7.

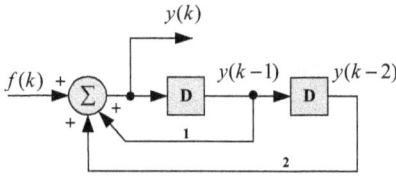

Fig. 3.7: Block diagram of an LTID system

Solution:
(1) Determine the difference equation
The output of the adder can be obtained as follows:

$$y(k) = f(k) + y(k-1) + 2y(k-2) . \tag{3.35}$$

The difference equation is represented by:

$$y(k) - y(k-1) - 2y(k-2) = f(k) . \tag{3.36}$$

(2) Determine the unit impulse response
According to the definition of the unit impulse response, $h(k)$ can be obtained by substituting $y(k)$ by $h(k)$ and $f(k)$ by $\delta(k)$ in Equation (3.36):

$$\left. \begin{array}{r} h(k) - h(k-1) - 2h(k-2) = \delta(k) \\ h(-1) = h(-2) = 0 \end{array} \right\} \tag{3.37}$$

> **Note:** The zero-state condition is included in the definition of the impulse response.

The initial values can be determined by iterating with respect to $k = 0, 1$:

$$\left. \begin{array}{l} h(0) = h(-1) + 2h(-2) + \delta(0) = 1 \\ h(1) = h(0) + 2h(-1) + \delta(1) = 1 \end{array} \right\}$$

When $k > 0$, $h(k)$ satisfies the homogeneous equation:

$$h(k) - h(k-1) - 2h(k-2) = 0 .$$

The characteristic equation is as follows:

$$\lambda^2 - \lambda - 2 = (\lambda + 1)(\lambda - 2) = 0 ,$$

with characteristic roots $\lambda_1 = -1$ and $\lambda_2 = 2$. Therefore, the homogeneous solution is given by:

$$h(k) = C_1(-1)^k + C_2(2)^k , \quad k > 0 ,$$

where C_1, C_2 are obtained by inserting initial values. This leads to the following simultaneous equations:

$$h(0) = C_1 + C_2 = 1 , \quad h(1) = -C_1 + 2C_2 = 1 ,$$

with solutions $C_1 = 1/3$ and $C_2 = 2/3$. The unit impulse response is:

$$h(k) = \frac{1}{3}(-1)^k + \frac{2}{3}(2)^k , \quad k \geq 0 . \tag{3.38}$$

(3) Determine the unit step response
According to the definition of the unit step response, $g(k)$ can be obtained by substituting $y(k)$ by $g(k)$ and $f(k)$ by $\varepsilon(k)$ in Equation (3.36).

$$\left. \begin{array}{l} g(k) - g(k-1) - 2g(k-2) = \varepsilon(k) \\ g(-1) = g(-2) = 0 \end{array} \right\}$$

The initial values can be determined by iterating with respect to $k = 0, 1$.

$$\left. \begin{array}{l} g(0) = g(-1) - 2g(-2) + \varepsilon(0) = 1 \\ g(1) = g(0) - gh(-1) + \varepsilon(1) = 2 \end{array} \right\}$$

When $k > 0$, $g(k)$ satisfies the following equation:

$$g(k) - g(k-1) - 2g(k-2) = 1 .$$

The particular solution is $g_p(k) = -1/2, k \geq 0$, and the homogeneous solution is given by $g_h(k) = D_1(-1)^k + D_2(2)^k$. So, the overall response is expressed as follows:

$$g_h(k) = D_1(-1)^k + D_2(2)^k - \frac{1}{2} , \quad k \geq 0$$

Substituting the initial values yields $C_1 = 1/6, C_2 = 4/3$. The unit step response is given by:

$$g(k) = \frac{1}{6}(-1)^k + \frac{4}{3}(2)^k - \frac{1}{2} , \quad k \geq 0 .$$

Solution 2. According to Equation (3.33), $g(k)$ can be obtained by integration of $h(k)$:

$$h(k) = \frac{1}{3}(-1)^k + \frac{2}{3}(2)^k , \quad k \geq 0$$

$$g(k) = \sum_{i=-\infty}^{k} h(i) = \frac{1}{3}\sum_{i=0}^{k}(-1)^i + \frac{2}{3}\sum_{i=0}^{k}(2)^i , \quad k \geq 0 \tag{3.39}$$

According to summation formula of the geometric sequence:

$$\sum_{i=0}^{k}(-1)^i = \frac{1-(-1)^{k+1}}{1-(-1)} = \frac{1}{2}[1+(-1)^k], \quad \sum_{i=0}^{k}(2)^i = \frac{1-(2)^{k+1}}{1-2} = 2(2)^k - 1$$

The unit step response is given by:

$$g(k) = \frac{1}{3} \cdot \frac{1}{2}[1+(-1)^k] + \frac{2}{3}[2(2)^k - 1] = \frac{1}{6}(-1)^k + \frac{4}{3}(2)^k - \frac{1}{2}, \quad k \geq 0.$$

Example 3.3.2. Determine the unit impulse response $h(k)$ for the LTID system shown in Figure 3.8.

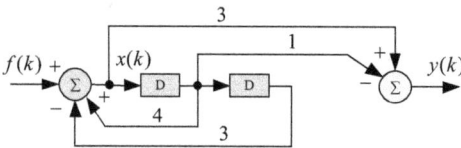

Fig. 3.8: Block diagram of an LTID system

Solution:
(1) Determine the difference equation
Set the auxiliary variable $x(k)$ at the input of the first delayer. For the left adder, the input–output representation is:

$$x(k) = f(k) + 4x(k-1) - 3x(k-2), \tag{3.40}$$

which is rewritten as:

$$x(k) - 4x(k-1) + 3x(k-2) = f(k). \tag{3.41}$$

For the right adder, the input–output equation is:

$$y(k) = 3x(k) - x(k-1). \tag{3.42}$$

According to the characteristics of the LTID system, Equations (3.41) and (3.42) are combined to obtain the difference equation:

$$y(k) - 4y(k-1) + 3y(k-2) = 3f(k) - f(k-1) \tag{3.43}$$

! **Note:** The combination of two equations was mentioned in Section 3.1.2.

(2) Determine the unit impulse response
When $k \geq 2$, Equation (3.43) becomes a homogeneous equation:

$$h(k) - 4h(k-1) + 3h(k-2) = 0, \quad k \geq 2, \tag{3.44}$$

with zero ancillary conditions: $h(-1) = h(-2) = 0$.

The initial values can be obtained as follows:

$$h(0) = 4h(-1) - 3h(-2) + 3 = 3$$
$$h(1) = 4h(0) - 3h(-1) - 1 = 11$$

Based on the characteristic roots $\lambda_1 = 1$ and $\lambda_2 = 3$, the unit impulse response is expressed as follows:

$$h(k) = c_1(1)^k + c_2(3)^k, \quad k \geq 2 \tag{3.45}$$

Substituting $h(0)$, $h(1)$ into Equation (3.45) yields $C_1 = -1$ and $C_2 = 4$. The unit impulse response is given by:

$$h(k) = -1 + 4(3)^k, \quad k \geq 0.$$

3.3.3 Calculation with MATLAB

Consider the following linear, constant-coefficient difference equation:

$$a_n y(k) + a_{n-1} y(k - 1) + \ldots + a_0 y(k - n) = b_m f(k) + \ldots + b_0 f(k - m),$$

which models the relationship between the input sequence $f(k)$ and the output response $y(k)$ of an LTID system.

MATLAB provides a built-in function for calculating the unit impulse response of an LTID system. The function is *impz* with the syntax

```
h=impz(b,a,k).
```

where, $b=[b0, b1, b2, \ldots, bm]$ and $a=[a0, a1, a2, \ldots, an]$ is the coefficient vector of the right- and left-hand side of the difference equation, respectively; k is the time index.

Example 3.3.3. The difference equation of an LTID system is described as follows:

$$y(k) + 3y(k) + 2y(k) = f(k).$$

Compute the unit impulse response for $0 \leq k \leq 10$ using MATLAB.

Solution:

```
k=0:10;              % time index
a=[1 3 2];           % coefficient vector in the left-hand side of
                     % the equation
b=[1];               % coefficient vector in the right-hand side of
                     % the equation
h=impz(b,a,k);       % calculate the response
stem(k,h,'.')
```

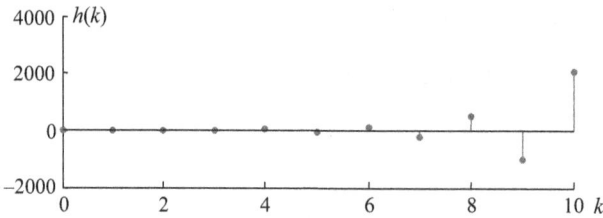

Fig. 3.9: The waveform of the response in Example 3.3.3

3.4 Convolution sum

3.4.1 Representation of sequences using Dirac delta functions

Any discrete sequence $f(k)$ can be represented as a linear combination of time-shifted, DT impulse functions. As in Figure 3.10, $f(k)$ is represented by:

$$f(k) = \cdots + f(-2)\delta(k + 2) + f(-1)\delta(k + 1) + f(0)\delta(k)$$
$$+ f(1)\delta(k - 1) + \cdots + f(i)\delta(k - i) + \cdots ,$$

which reduces to:

$$f(k) = \sum_{i=-\infty}^{\infty} f(i)\delta(k - i) . \tag{3.46}$$

Equation (3.46) provides an alternative representation of an arbitrary DT function using a linear combination of time-shifted DT impulses. In Equation (3.46), variable m denotes the dummy variable for the summation that disappears as the summation is computed. Recall that a similar representation exists for the CT functions and is given by $f(t) = \int_{-\infty}^{\infty} f(\tau)\delta(t - \tau)\,d\tau$.

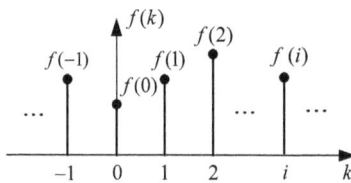

Fig. 3.10: Sequence decomposition

3.4.2 Convolution sum

In this section, we need to compute the zero-state response of any complex input sequences. The input sequence is represented as a linear combination of time-shifted impulse functions:

$$f(k) = \sum_{i=-\infty}^{\infty} f(i)\delta(k - i)$$

When the input signal is $\delta(k)$, the zero-state response is $h(k)$:

$$\delta(k) \rightarrow h(k)$$

According to the properties of time-invariance and homogeneity, the output of the input $f(i)\delta(k - i)$ is given by:

$$f(i)\delta(k - i) \rightarrow f(i)h(k - i) .$$

Applying the principle of superposition, the overall output $y_{zs}(k)$ resulting from the input sequence $f(k)$ is given by:

$$f(k) = \sum_{i=-\infty}^{\infty} f(i)\delta(k - i) \rightarrow \sum_{i=-\infty}^{\infty} f(i)h(k - i) = y_{zs}(k) , \qquad (3.47)$$

where the summation on the right-hand side used to compute the output response is referred to as the *convolution sum*. The zero-state response is the convolution of the input sequence $f(k)$ with the impulse response $h(k)$ of the LTID system. Mathematically, Equation (3.47) is expressed as follows:

$$y_{zs}(k) = f(k) * h(k) = \sum_{i=-\infty}^{\infty} f(i)h(k - i) , \qquad (3.48)$$

where $*$ denotes the convolution sum.

Note: This is the most important expression in this chapter. ❗

Similarly, the convolution sum of $f_1(k)$ and $f_2(k)$ is given by:

$$f(k) = f_1(k) * f_2(k) = \sum_{i=-\infty}^{\infty} f_1(i)f_2(k - i) \qquad (3.49)$$

We now consider several examples of computing the convolution sum.

Example 3.4.1. Assuming that the impulse response of an LTID system is given by $h(k) = b^k \varepsilon(k)$. Determine the zero-state response to the input sequence $f(k) = a^k \varepsilon(k)$.

Solution: Using the convolution sum, the response is given by:

$$y_{zs}(k) = f(k) * h(k) = \sum_{i=-\infty}^{\infty} f(i)h(k - i) = \sum_{i=-\infty}^{\infty} a^i \varepsilon(i) b^{k-i} \varepsilon(k - i) .$$

Using the definition of the unit step function $\varepsilon(i)$, the above summation simplifies as follows:

$$y_{zs}(k) = \sum_{i=0}^{\infty} a^i b^{k-i} \varepsilon(k - i) .$$

Depending on the value of k, the output response may take two different forms for $k \geq 0$ or $k < 0$.

Case 1: When $k < 0$, the unit step function $\varepsilon(k - i) = 0$ within the limits of summation $0 \leq i < \infty$. Therefore, the output sequence $y_{zs}(k) = 0$ for $k < 0$.

Case 2: When $k \geq 0$, the unit step function $\varepsilon(k - i)$ has the following values:

$$\varepsilon(k - i) = \begin{cases} 1, & i \leq k \\ 0, & i > k . \end{cases}$$

Combining the above cases, the output sequence is therefore given by:

$$y_{zs}(k) = \left[\sum_{i=0}^{k} a^i b^{k-i}\right] \varepsilon(k) = b^k \left[\sum_{i=0}^{k} \left(\frac{a}{b}\right)^i\right] \varepsilon(k) = \begin{cases} b^k \frac{1-\left(\frac{a}{b}\right)^{k+1}}{1-\frac{a}{b}} \varepsilon(k), & a \neq b \\ b^k(k+1)\varepsilon(k), & a = b . \end{cases} \quad (3.50)$$

Example 3.4.2. Determine the following convolution sum:

(i) $\varepsilon(k) * \varepsilon(k)$

(ii) $a^k \varepsilon(k) * \varepsilon(k - 4)$

(iii) $\varepsilon(k - 3) * \varepsilon(k - 4)$

(iv) $(0.5)^k \varepsilon(k) * 1$

Solution:

(i)

$$\varepsilon(k) * \varepsilon(k) = \sum_{i=-\infty}^{\infty} \varepsilon(i)\varepsilon(k - i)$$

$$= \left(\sum_{i=0}^{k} 1\right) \varepsilon(k) = (k + 1)\varepsilon(k)$$

(ii)

$$a^k \varepsilon(k) * \varepsilon(k - 4) = \sum_{i=-\infty}^{\infty} a^i \varepsilon(i) \cdot \varepsilon(k - 4 - i) = \left(\sum_{i=0}^{k-4} a^i\right) \varepsilon(k - 4)$$

$$= \begin{cases} \frac{1-a^{k-3}}{1-a} \varepsilon(k - 4), & a \neq 1 \\ (k - 3)\varepsilon(k - 4), & a = 1 \end{cases}$$

(iii)

$$\varepsilon(k - 3) * \varepsilon(k - 4) = \sum_{i=-\infty}^{\infty} \varepsilon(i - 3)\varepsilon(k - 4 - i)$$

$$= \left(\sum_{i=3}^{k-4} 1\right) \varepsilon(k - 4 - 3) = (k - 6)\varepsilon(k - 7)$$

(iv)

$$(0.5)^k \varepsilon(k) * 1 = \sum_{i=-\infty}^{\infty} (0.5)^i \varepsilon(i) \times 1$$

$$= \sum_{i=0}^{\infty} (0.5)^i = \frac{1}{1 - 0.5} = 2$$

3.4.3 Graphical method for evaluating the convolution sum

The graphical approach for calculating the convolution sun is similar to the graphical procedure for calculating the convolution integral for the LTIC system, discussed in Chapter 2. The main steps of the convolution sum $f(k) = \sum_{i=-\infty}^{\infty} f_1(i)f_2(k - i)$ are the following.

Step 1: Change the independent variable from k to i to obtain $f_1(i), f_2(i)$.

Step 2: Reflect $f_2(i)$ about the vertical axis to obtain $f_2(-i)$ and shift it to obtain $f_2(k-i)$.

Step 3: Multiply the sequence $f_1(i)$ by $f_2(k - i)$ and plot the product function $f_1(i)f_2(k-)$.

Step 4: Calculate the summation to obtain the output response
$$f(k) = \sum_{i=-\infty}^{\infty} f_1(i)f_2(k - i).$$

Note: The summing variable is i, and the convolution result is a function of the independent variable k. **!**

Example 3.4.3. For the two DT sequences shown in Figure 3.11, using the graphical convolution approach, determine $f(k) = f_1(k) * f_2(k)$.

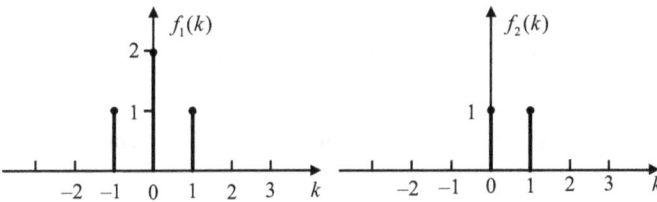

Fig. 3.11: Discrete sequences

Solution:

(1) Change the independent variable to obtain $f_1(i)$ and $f_2(i)$. The reflection $f_2(-i)$ is plotted in Figure 3.12.

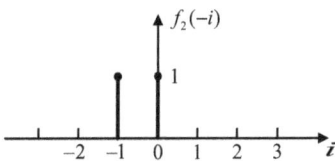

Fig. 3.12: Change variable and reverse to obtain $f_2(-i)$

(2) The DT sequence $f_2(k - i)$ is obtained by shifting the time-reflected function $f_2(-i)$ by k samples. Depending on the value of k, the process of multiplication in the overlapping area is given in Figure 3.13.

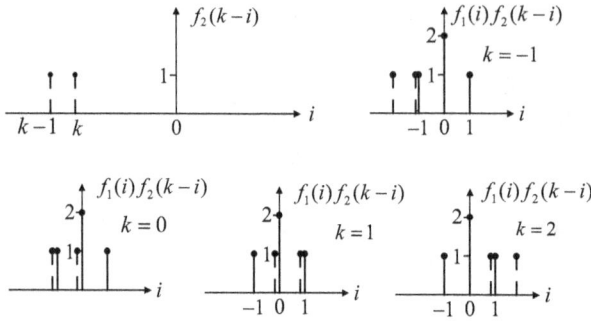

Fig. 3.13: Graphical method for the convolution sum

(3) The nonzero parts of $f_1(i)$ and $f_2(k-i)$ overlap over the duration $k = [-1, 0, 1, 2]$. Therefore, the convolution sum is given by:

$$f_1(k) * f_2(k) = \begin{cases} 0, & k \le -2 \\ \sum_{i=-\infty}^{\infty} f_1(i)f_2(-1-i) = 1, & k = -1 \\ \sum_{i=-\infty}^{\infty} f_1(i)f_2(0-i) = 3, & k = 0 \\ \sum_{i=-\infty}^{\infty} f_1(i)f_2(1-i) = 3, & k = 1 \\ \sum_{i=-\infty}^{\infty} f_1(i)f_2(2-i) = 1, & k = 2 \\ 0, & k \ge 3 \end{cases} \qquad (3.51)$$

Example 3.4.4. The sequences $f_1(k)$, $f_2(k)$ are shown in Figure 3.14. Determine $f(2)$ of their convolution $f(k) = f_1(k) * f_2(k)$.

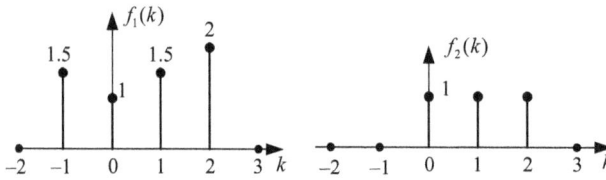

Fig. 3.14: Two discrete sequences

Solution:

$$f(2) = \sum_{i=-\infty}^{\infty} f_1(i)f_2(2-i) \qquad (3.52)$$

(1) Change the independent variable to obtain $f_1(i)$ and $f_2(i)$.
(2) Reflect $f_2(i)$ about the vertical axis to obtain $f_2(-i)$ and shift it to the right-hand side by a selected value of 2 to obtain $f_2(2-i)$.
(3) As shown in Figure 3.15, multiply the sequence $f_1(i)$ by $f_2(2-i)$ to compute the sum as $f(2) = 1 \cdot 1 + 1.5 \cdot 1 + 2 \cdot 1 = 4.5$.

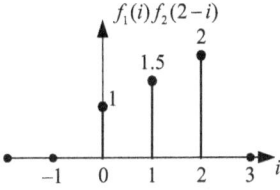

Fig. 3.15: Graphic method for computing the convolution sum

3.4.4 The carry-less multiplication method

When the convolved sequences are relatively short in length, carry-less multiplication is a convenient method to compute the convolution. From the definition of the convolution sum in Equation (3.53), the value of $f(k)$ equals the sum of the multiplication of $f_1(i)$ and $f_2(k-i)$, whose time indices meet $k = i + (k-i)$. This is the basic principle of the carry-less multiplication method of convolution sum:

$$f(k) = \sum_{i=-\infty}^{\infty} f_1(i)f_2(k-i)$$
$$= \cdots + f_1(-1)f_2(k+1) + f_1(0)f_2(k) + f_1(1)f_2(k-1)$$
$$+ f_1(2)f_2(k-2) + \cdots + f_1(i)f_2(k-i) + \cdots \tag{3.53}$$

Example 3.4.5. Given two sequences as follows:

$$f_1(k) = \{0,2,1,5,0\} \quad \text{and} \quad f_2(k) = \{0,3,4,0,6,0\} \quad,$$
$$\uparrow k = 1 \qquad\qquad\qquad \uparrow k = 0$$

determine their convolution sum $f(k) = f_1(k) * f_2(k)$.

Solution: Using the carry-less multiplication method, we obtain:

```
          3,   4,   0,  6
     ×    2,   1,  5
        15,  20,  0,  30
     3,   4,   0,   6
+ 6,  8,   0,  12
  6,  11,  19,  32,  6,  30  .
```

So, the convolution sum is

$$f(k) = \{0,6,11,19,32,6,30\} .$$
$$\uparrow k = 1$$

The carry-less multiplication method is more convenient for computing the convolution sum at any arbitrary position.

Note: For example, $f(2) = \cdots + f_1(-1)f_2(3) + f_1(0)f_2(2) + f_1(1)f_2(1) + f_1(2)f_2(0) + \cdots$.   **!**

3.4.5 Properties of the convolution sum

The properties of the convolution sum are similar to the properties of the convolution integral presented in Chapter 2. They both meet the commutative property, distributive property and the associative property.

(a) The commutative property:

$$f_1(k) * f_2(k) = f_2(k) * f_1(k) \tag{3.54}$$

(b) The distributive property:

$$f_1(k) * [f_2(k) + f_3(k)] = f_1(k) * f_2(k) + f_1(k) * f_3(k) \tag{3.55}$$

(c) The associative property:

$$f_1(k) * [f_2(k) * f_3(k)] = [f_1(k) * f_2(k)] * f_3(k) \tag{3.56}$$

As shown in Figure 3.16 (a), the impulse response of the parallel system is the addition of that of the two subsystems: $h(k) = h_1(k) + h_2(k)$

As shown in Figure 3.16 (b), the impulse response of the cascade system is the convolution of that of the two subsystems: $h(k) = h_1(k) * h_2(k) = h_2(k) * h_1(k)$

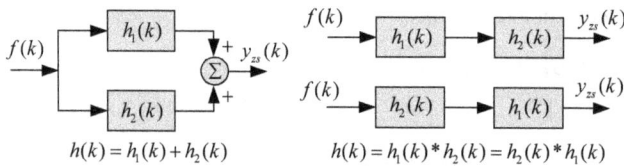

Fig. 3.16: Composite system; (a) Parallel, (b) Cascade

The commonly used properties of the convolution sum for DT sequences are listed as follows:

(i) $f(k) * \delta(k) = f(k)$

(ii) $f(k) * \delta(k - k_0) = f(k - k_0)$

(iii) $f(k) * \varepsilon(k) = \sum_{i=-\infty}^{k} f(i)$

(iv) $f_1(k - k_1) * f_2(k - k_2) = f_1(k - k_1 - k_2) * f_2(k) = f_1(k) * f_2(k - k_1 - k_2)$

! **Note:** Readers can prove it.

Example 3.4.6. Two DT sequences are given in Figure 3.17. Determine their convolution sum $f(k) = f_1(k) * f_2(k)$.

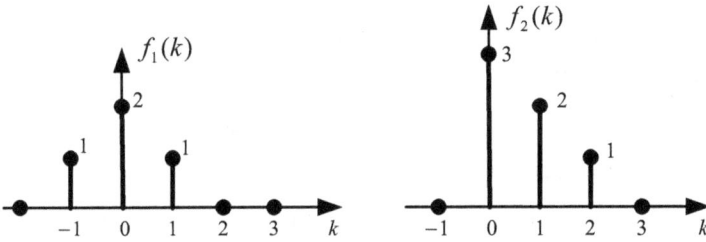

Fig. 3.17: Two discrete sequences

Solution 1. The carry-less multiplication is applied to compute the convolution sum:

```
          1    2    1
      ×   3    2    1
      ─────────────────
          1    2    1
     2    4    2
3    6    3
─────────────────────────
3    8    8    4    1
```

Fig. 3.18: Process of carry-less multiplication in Example 3.4.6

We note that the first non-zero value starts at the time indices $k = -1 + 0 = -1$. Therefore, the convolution sum is given by

$$f_1(k) * f_2(k) = \begin{cases} 0, & k < -1 \\ 3, & k = -1 \\ 8, & k = 0 \\ 8, & k = 1 \\ 4, & k = 2 \\ 1, & k = 3 \\ 0, & k > 3 \end{cases} \tag{3.57}$$

Solution 2. The functions of two DT sequences are written as:

$$f_1(k) = \delta(k+1) + 2\delta(k) + \delta(k-1)$$
$$f_2(k) = 3\delta(k) + 2\delta(k-1) + \delta(k-2). \tag{3.58}$$

According to the properties, we have:

$$f_1(k) * f_2(k)$$
$$= [\delta(k+1) + 2\delta(k) + \delta(k-1)] * [3\delta(k) + 2\delta(k-1) + \delta(k-2)]$$
$$= [3\delta(k+1) + 2\delta(k) + \delta(k-1)] + [6\delta(k) + 4\delta(k-1) + 2\delta(k-2)]$$
$$+ [3\delta(k-1) + 2\delta(k-2) + \delta(k-3)]$$
$$= 3\delta(k+1) + 8\delta(k) + 8\delta(k-1) + 4\delta(k-2) + \delta(k-3).$$

! **Note:** Which solution method is easier?

Example 3.4.7. The composite system shown in Figure 3.19 is composed of three subsystems. Their unit impulse responses are $h_1(k) = \delta(k)$, $h_2(k) = \delta(k-N)$, and $h_3(k) = \varepsilon(k)$, respectively. Calculate the unit impulse responses of the complex system.

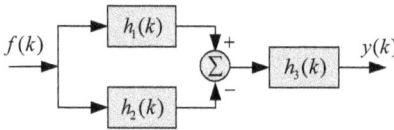

Fig. 3.19: The system diagram

Solution: According to the cascade and parallel structure of the system, the unit impulse response is computed as follows:

$$h(k) = [h_1(k) - h_2(k)] * h_3(k)$$
$$= [\delta(k) - \delta(k-N)] * \varepsilon(k)$$
$$= \delta(k) * \varepsilon(k) - \delta(k-N) * \varepsilon(k)$$
$$= \varepsilon(k) - \varepsilon(k-N).$$

3.4.6 Convolution calculation with MATLAB

To perform DT convolution, MATLAB provides a built-in function *conv*. To represent each DT signal, two vectors are required. The first vector contains the sample values, while the second vector stores the time indices corresponding to the sample values. We illustrate its usage by the following example.

Example 3.4.8. Given two sequences $x_1(k) = \sin(k)$, $0 \le k \le 10$ and $x_2(k) = 0.8^k$, $0 \le k \le 15$, calculate the convolution sum $y(k) = x_1(k) * x_2(k)$.

Solution:

```
k1=0:10;          % time indices of x_1
x1=sin(k1);       % values of sequence x_1
k2=0:15;          % time indices of x_2
x2=0.8.^k2;       % values of sequence x_2
y=conv(x1,x2);    % convolve x_1 with x_2
```

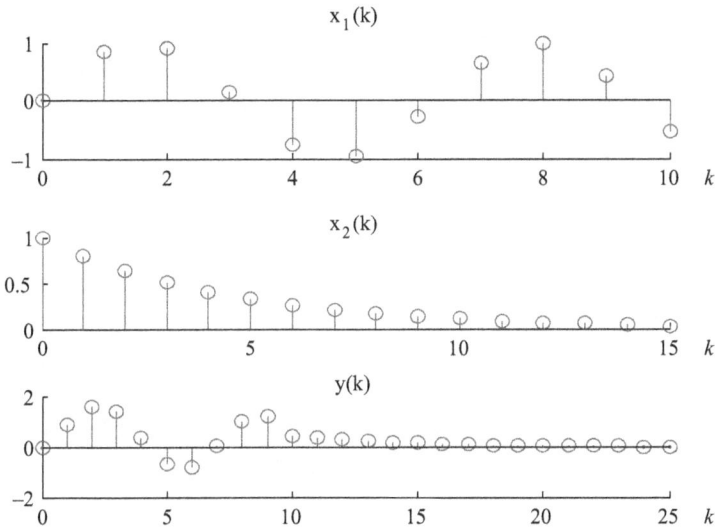

Fig. 3.20: Results of Example 3.4.8

The two DT sequences and the convolution result are plotted in Figure 3.20.

3.4.7 Application of the convolution sum

The convolution sum has many applications in different fields. In statistics, the weighted moving average on data is a convolution. In probability theory, the probability density function of the sum of two statistical independent variables X and Y is the convolution of the probability density functions of X and Y. In acoustics, echoes can be represented as a convolution of the source sound and a function that represents various reflection effects [18]. In electrical engineering and signal processing, the output of any linear system can be obtained by convolving the input signal and the unit impulse response.

For a moving average filter, the output sequence $y(k)$ is the average of the input sequence $x(k)$ within the filter window length [19, 20]. For example, when the filter length is 5, the output is given by:

$$y(k) = \frac{1}{5}[x(k) + x(k-1) + x(k-2) + x(k-3) + x(k-4)] \,.$$

Note: A larger filter length can obtain a smoother filtering result.

Its unit impulse response is defined as:

$$h(k) = \frac{1}{5}[\delta(k) + \delta(k-1) + \delta(k-2) + \delta(k-3) + \delta(k-4)] \,.$$

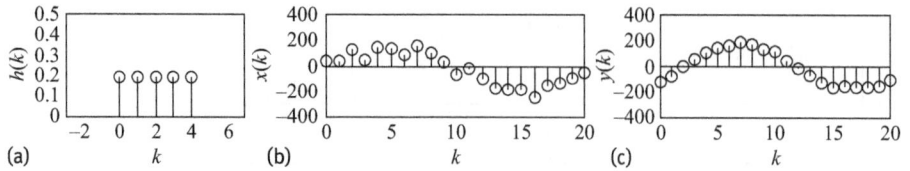

Fig. 3.21: Example of a convolution for data smoothing

The unit impulse response $h(k)$ is a rectangular pulse, which is plotted in Figure 3.21 (a). The input sequence $x(k)$, as shown in Figure 3.21 (b), is convolved with the unit impulse response to achieve the data smoothing. The result $y(k) = x(k) * h(k)$ is plotted in Figure 3.21 (c).

The DT convolution can be used to filter two-dimensional signals [21], such as digital images. The 2D unit impulse response H is called as the convolution kernel. For example, the 2D convolution kernel $H_{3\times3}$ for the average filter is given by:

$$H_{3\times3} = \begin{bmatrix} \frac{1}{9} & \frac{1}{9} & \frac{1}{9} \\ \frac{1}{9} & \frac{1}{9} & \frac{1}{9} \\ \frac{1}{9} & \frac{1}{9} & \frac{1}{9} \end{bmatrix},$$

which can be used to filter the pixel values of a 2D image [22]. Using the convolution kernel, the pixel value is filtered as the average value of the point and the neighboring eight points.

As Figure 3.22 shows, the noisy image in (a) is filtered by the convolution with the average kernel. The average kernel can eliminate the noise and achieve smooth image in (b); meanwhile the edges are blurred. More discussion can be found in Section 8.7 in Chapter 8

(a) (b)

Fig. 3.22: Example of image smoothing filtering; (a) Image with added noise, (b) Image after average filtering

3.5 Summary

This chapter introduced the basic time-domain analysis methods of LTID systems. The system can be represented by a difference equation and solved by the classical methods in Section 3.2 to compute the zero-input response and zero-state response. In Section 3.3, the impulse response $h(k)$ was defined as the output when the elementary signal $\delta(k)$ was applied at the input. Section 3.4 gave an alternative method of convolution to compute the zero-state response $y_{zs}(k) = f(k) * h(k)$.

Chapter 3 problems

3.1 Determine the response of the LTID system described by the following difference equation:

(1) $y(k) - 0.5y(k - 1) = 0$

$y(0) = 1$

(2) $y(k) + \frac{1}{3}y(k - 1) = 0$

$y(-1) = -1$

(3) $y(k) - 5y(k - 1) + 6y(k - 2) = \varepsilon(k - 2)$

$y(0) = 1, \ y(1) = 5$

3.2 Determine the zero input response of the LTID system described by the following difference equation:

(1) $y(k) + 3y(k - 1) + 2y(k - 2) = f(k)$

$y(-1) = 1, \ y(-2) = 1$

(2) $y(k) + 2y(k - 1) + y(k - 2) = f(k) - f(k - 1)$

$y(-1) = 1, \ y(-2) = -3$

(3) $y(k) + y(k - 2) = f(k - 2)$

$y(-1) = -2, \ y(-2) = -1$

3.3 Consider the LTID systems with the following input–output relationship:

(1) $y(k) = f(k - 1) + 2f(k - 3)$
(2) $y(k + 1) - 0.4y(k) = f(k)$

Calculate the impulse response for the two LTID systems. Also, determine the output responses of the LTID systems when the input is given by $f(k) = 2\delta(k) + 3\delta(k - 1)$.

3.4 The waveforms of three discrete time signals are plotted in Figure P3.1. Calculate the following convolution sum and plot the results:

(1) $f_1(k) * f_2(k)$

(2) $f_2(k) * f_3(k)$

(3) $[f_1(k) - f_2(k)] * f_3(k)$

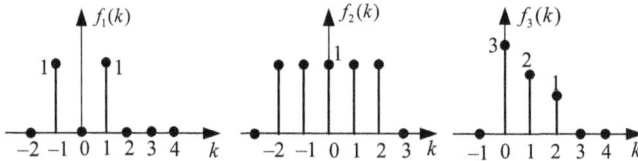

Fig. P3.1: Waveforms in Problem 3.4

3.5 For the following DT sequences:

$$f(k) = \begin{cases} 2, & 0 \le k \le 2 \\ 0, & \text{otherwise} \end{cases} \quad \text{and} \quad h(k) = \begin{cases} k+1, & 0 \le k \le 4 \\ 0, & \text{otherwise} \end{cases}$$

Calculate the convolution sum $y(k) = f(k) * h(k)$ using the graphical approach.

3.6 For the following DT sequences:

$$f(k) = \begin{cases} k, & 0 \le k \le 3 \\ 0, & \text{otherwise} \end{cases} \quad \text{and} \quad h(k) = \begin{cases} 2, & -1 \le k \le 2 \\ 0, & \text{otherwise} \end{cases}$$

Calculate the convolution sum $y(k) = f(k) * h(k)$ using the carry-less multiplication method.

3.7 The unit impulse responses of three subsystems in Figure P3.2 are as follows:

$$h_1(k) = \varepsilon(k) \quad h_2(k) = \delta(k-3) \quad h_3(k) = 0.8^k \varepsilon(k)$$

Calculate the unit impulse response $h(k)$ of the system.

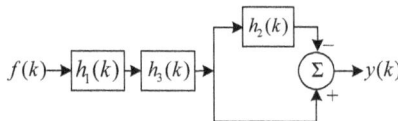

Fig. P3.2: Block diagram used in Problem 3.7

3.8 Calculate the zero-input, zero-state and overall responses of the LTID systems described by the following difference equations:

(1) $y(k) - 2y(k-1) = f(k)$

$y(-1) = -1, f(k) = 2\varepsilon(k)$

(2) $y(k) + 2y(k-1) = f(k)$

$y(-1) = 1, f(k) = 2^k \varepsilon(k)$

(3) $y(k) + 3y(k-1) + 2y(k-2) = f(k)$

$y(-1) = 1, y(-2) = 0, f(k) = \varepsilon(k)$

(4) $y(k) + 2y(k-1) + y(k-2) = f(k)$

$y(-1) = 3, y(-2) = -5, f(k) = 3(0.5)^k \varepsilon(k)$

3.9 Determine the unit impulse response of the LTID system described in the following difference equations:

(1) $y(k) + 2y(k-1) = f(k-1)$

(2) $y(k) - y(k-2) = f(k)$

(3) $y(k) + y(k-1) + \dfrac{1}{4}y(k-2) = f(k)$

(4) $y(k) + 4y(k-2) = f(k)$

(5) $y(k) - 4y(k-1) + 8y(k-2) = f(k)$

3.10 Determine the zero-state response of the LTID systems:

(1) $h(k) = f(k) = \varepsilon(k) - \varepsilon(k-4)$

(2) $h(k) = 2^k [\varepsilon(k) - \varepsilon(k-4)], f(k) = \delta(k) - \delta(k-2)$

(3) $h(k) = (0.5)^k \varepsilon(k), f(k) = \varepsilon(k) - \varepsilon(k-5)$

3.11 The difference equation for an LTID system is given:

$$y(k) - \frac{5}{6}y(k-1) + \frac{1}{6}y(k-2) = f(k) - f(k-1)$$

$$y(-1) = 0, \quad y(-2) = 1, \quad f(k) = \varepsilon(k)$$

(1) Calculate the unit impulse response of the system.

(2) Calculate the zero-input response, zero-state response and overall response.

(3) Find the transient component, the steady-state component, the natural response and the forced response.

3.12 Determine the input-output difference equation for each LTID system shown in Figure P3.3.

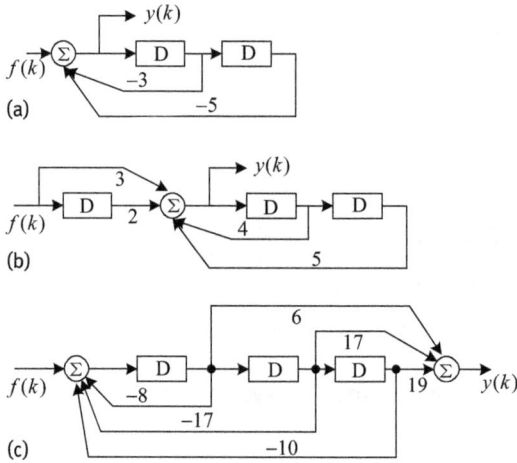

Fig. P3.3: Block diagrams used in Problem 3.12

3.13 The DT sequence $f(k) = 2k\varepsilon(k)$ is applied at the input of an LTID system described by the following difference equation:

$$y(k) - 0.4y(k-1) = f(k-1)$$

with the ancillary condition $y(-1) = 4$. Compute the output response $y(k)$ of the LTID system for $0 \le k \le 50$ using MATLAB.

3.14 Consider the following two DT sequences $f(k)$ and $h(k)$.

$$f(k) = \begin{cases} -1, & k = -1 \\ 1, & k = 0 \\ 2, & k = 1 \\ 0, & \text{otherwise} \end{cases} \quad \text{and} \quad h(k) = \begin{cases} 3, & k = -1, 2 \\ 1, & k = 0 \\ -2, & k = 1, 3 \\ 0, & \text{otherwise}. \end{cases}$$

Compute the convolution $y(k) = f(k) * h(k)$ using MATLAB.

4 Frequency-domain analysis of LTIC systems

Please focus on the following key questions.

1. How can we use the examples in life to explain the signal expressions in different domains?
2. What are two elementary signals for the decomposition of the periodic signals? How do we decompose periodic signals?
3. What is the elementary signal for decomposition of aperiodic signals?
4. What is the method in the frequency-domain to analyze LTIC systems?
5. What is the problem solved by the sampling theorem?

4.0 Introduction

In various applications, the process to change the magnitude of some frequency components of a signal, or to eliminate some specified frequency components, is referred to as signal filtering. For example, Figure 4.1 depicts the block diagram of a low-pass filtering system. In an audio recording system, noise with high frequency needs to be removed or weakened by a low-pass filter.

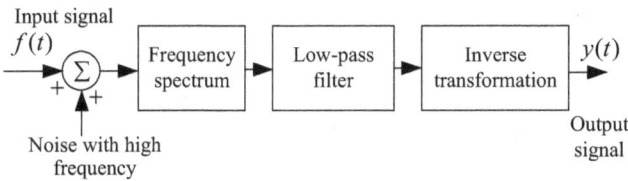

Fig. 4.1: Block diagram of a low-pass filtering system

The sampling and quantization in an audio recording system is also an important concept. The frequency of voice signals is usually in the range of 20 Hz~20 kHz. The sampling is to extract the CT voice signals at fixed intervals to obtain DT signals. The sampling frequency should be appropriate to recover the original CT signal. In such applications, frequency-domain analysis of signals and systems provides a convenient means of solving for the response of LTIC systems.

In this chapter, we focus on CT signals and introduce frequency-domain analysis methods. In Section 4.1, the CTFS (continuous-time Fourier series) is used to decompose periodic signals into their frequency components. Based on two basic functions, CTFS has the sinusoidal and complex exponential representations. The definition and properties of frequency spectrum are detailed in Section 4.2. Section 4.3 considers ape-

https://doi.org/10.1515/9783110593907-004

riodic CT signals and develops an equivalent Fourier representation, CTFT (continuous-time Fourier transform). The frequency-domain analysis of LTIC systems is mainly discussed in Section 4.4. Some application cases are briefly introduced in Section 4.5. Finally, the sampling theorem is discussed in Section 4.6. The chapter is concluded in Section 4.7 with a summary of the important concepts covered in the chapter.

4.1 CTFS of periodic signals

4.1.1 Trigonometric CTFS

Definition. An arbitrary periodic function $f(t)$ with a fundamental period T can be expressed as follows:

$$f(t) = \frac{a_0}{2} + \sum_{n=1}^{\infty} a_n \cos(n\Omega t) + \sum_{n=1}^{\infty} b_n \sin(n\Omega t), \qquad (4.1)$$

where $\Omega = 2\pi/T$ is the fundamental frequency of $f(t)$. Equation (4.1) is the trigonometric Fourier series expansion of continuous-time periodic signal, which is called *trigonometric CTFS*. The *trigonometric CTFS coefficients* $a_0/2$, a_n and b_n are calculated as follows:

$$\frac{a_0}{2} = \frac{1}{T} \int_{(T)} f(t)\, dt \qquad (4.2)$$

$$a_n = \frac{2}{T} \int_{(T)} f(t) \cos(n\Omega t)\, dt \qquad (4.3)$$

$$b_n = \frac{2}{T} \int_{(T)} f(t) \sin(n\Omega t)\, dt \qquad (4.4)$$

From the above equations, it is straightforward to verify that coefficients $a_0/2$ represent the average or mean value (also referred to as the DC component) of $f(t)$. Obviously, a_n is the even function of n (or $n\Omega$), and b_n is the odd function of n (or $n\Omega$):

$$\begin{cases} a_n = a_{-n} \\ b_n = -b_{-n} \end{cases} \qquad (4.5)$$

! **Note:** Here, n must be a positive integer.

It should be noted that not all periodic signals can have CTFS. The periodic signal $f(t)$ must satisfy the following sufficient "Dirichlet conditions":

(i) Absolutely integrable. The periodic signal $f(t)$ must be absolutely integrable over a period. The area under one period is finite:

$$\int_{(T)} |f(t)|\, dt < \infty$$

(ii) Bounded variation. The signal $f(t)$ must have a finite number of maxima or minima in one period.

(iii) Finite discontinuities. The signal $f(t)$ must have a finite number of discontinuities in one period. In addition, each of the discontinuity has a finite value.

Most practical signals satisfy these three conditions. Within the scope of this book, the periodic signals satisfy the Dirichlet conditions. We no longer consider verifying the condition other than in special circumstances.

Note: Try to draw a function curve that does not satisfy the Dirichlet conditions.   **!**

In Equation (4.1), the same frequency terms $a_n \cos(n\Omega t)$ and $b_n \sin(n\Omega t)$ are combined into a sinusoidal component:

$$f(t) = \frac{a_0}{2} + \sum_{n=1}^{\infty} a_n \cos(n\Omega t) + \sum_{n=1}^{\infty} b_n \sin(n\Omega t)$$

$$= \frac{A_0}{2} + \sum_{n=1}^{\infty} A_n \cos(n\Omega t + \varphi_n). \tag{4.6}$$

Equation (4.6) shows that the periodic signal $f(t)$ can be expressed as the sum of a DC component and a series of harmonic components. The DC component $A_0/2 (= a_0/2)$ is the average value of function $f(t)$ in one period.

When $n = 1$, $A_1 \cos(\Omega t + \varphi_1)$ is called the fundamental component or the first harmonic component, A_1 is the amplitude of the fundamental component, φ_1 is the phase of the fundamental wave, $A_n \cos(n\Omega t + \varphi_n)$ is called n-th harmonic component, and A_n and φ_n are the amplitude and phase of the n-th harmonic component, respectively.

The relationship between A_n, φ_n and a_n, b_n is given by:

$$\begin{cases} A_n = \sqrt{a_n^2 + b_n^2} \\ \varphi_n = -\arctan \frac{b_n}{a_n} \end{cases} \text{and} \begin{cases} a_n = A_n \cos(\varphi_n) \\ b_n = -A_n \sin(\varphi_n). \end{cases} \tag{4.7}$$

It is obvious that A_n is the even function of n (or $n\Omega$), and φ_n is the odd function of n (or $n\Omega$):

$$\begin{cases} A_n = A_{-n} \\ \varphi_n = -\varphi_{-n} \end{cases} \tag{4.8}$$

Note: Here, n must be a positive integer.   **!**

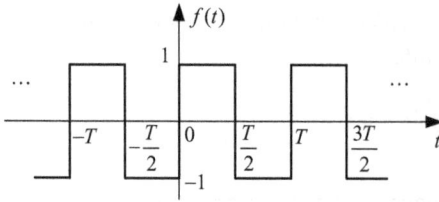

Fig. 4.2: Waveform of the periodic signal in Example 4.1.1

Example 4.1.1. Calculate the trigonometric CTFS of the periodic rectangular pulse shown in Figure 4.2.

Solution: Using Equation (4.3), the CTFS coefficient a_n is given by:

$$a_n = \frac{2}{T} \int_{-\frac{T}{2}}^{\frac{T}{2}} f(t)\cos(n\Omega t)\,dt = \frac{2}{T} \int_{-\frac{T}{2}}^{0} (-1) \times \cos(n\Omega t)\,dt + \frac{2}{T} \int_{0}^{\frac{T}{2}} 1 \times \cos(n\Omega t)\,dt$$

$$= \frac{2}{T}\frac{1}{n\Omega}[-\sin(n\Omega t)]\Big|_{-\frac{T}{2}}^{0} + \frac{2}{T}\frac{1}{n\Omega}[\sin(n\Omega t)]\Big|_{0}^{\frac{T}{2}}$$

$$= 0$$

Using Equation (4.4), the CTFS coefficient b_n is given by:

$$b_n = \frac{2}{T} \int_{-\frac{T}{2}}^{\frac{T}{2}} f(t)\sin(n\Omega t)\,dt = \frac{2}{T} \int_{-\frac{T}{2}}^{0} (-1) \times \sin(n\Omega t)\,dt + \frac{2}{T} \int_{0}^{\frac{T}{2}} 1 \times \sin(n\Omega t)\,dt$$

$$= \frac{2}{T}\frac{1}{n\Omega}[\cos(n\Omega t)]\Big|_{-\frac{T}{2}}^{0} + \frac{2}{T}\frac{1}{n\Omega}[-\cos(n\Omega t)]\Big|_{0}^{\frac{T}{2}}$$

$$= \frac{2}{T}\frac{T}{n2\pi}\{[1 - \cos(-n\pi)] + [1 - \cos(n\pi)]\}$$

$$= \frac{2}{n\pi}[1 - \cos(-n\pi)]$$

$$= \begin{cases} 0, & n = 2, 4, 6, \ldots \\ \frac{4}{n\pi}, & n = 1, 3, 5, \ldots \end{cases}$$

Therefore, the CTFS expansion of the periodic signal is expressed as follows:

$$f(t) = \frac{4}{\pi}\left[\sin(\Omega t) + \frac{1}{3}\sin(3\Omega t) + \cdots + \frac{1}{n}\sin(n\Omega t) + \cdots\right], \quad n = 1, 3, 5, \ldots$$

According to Equation (4.7), we can obtain:

$$A_0 = 0$$

$$A_n = \sqrt{a_n^2 + b_n^2} = \frac{4}{n\pi}, \quad n = 1, 3, 5, \ldots$$

$$\varphi_n = -\arctan\frac{b_n}{a_n} = -\frac{\pi}{2}$$

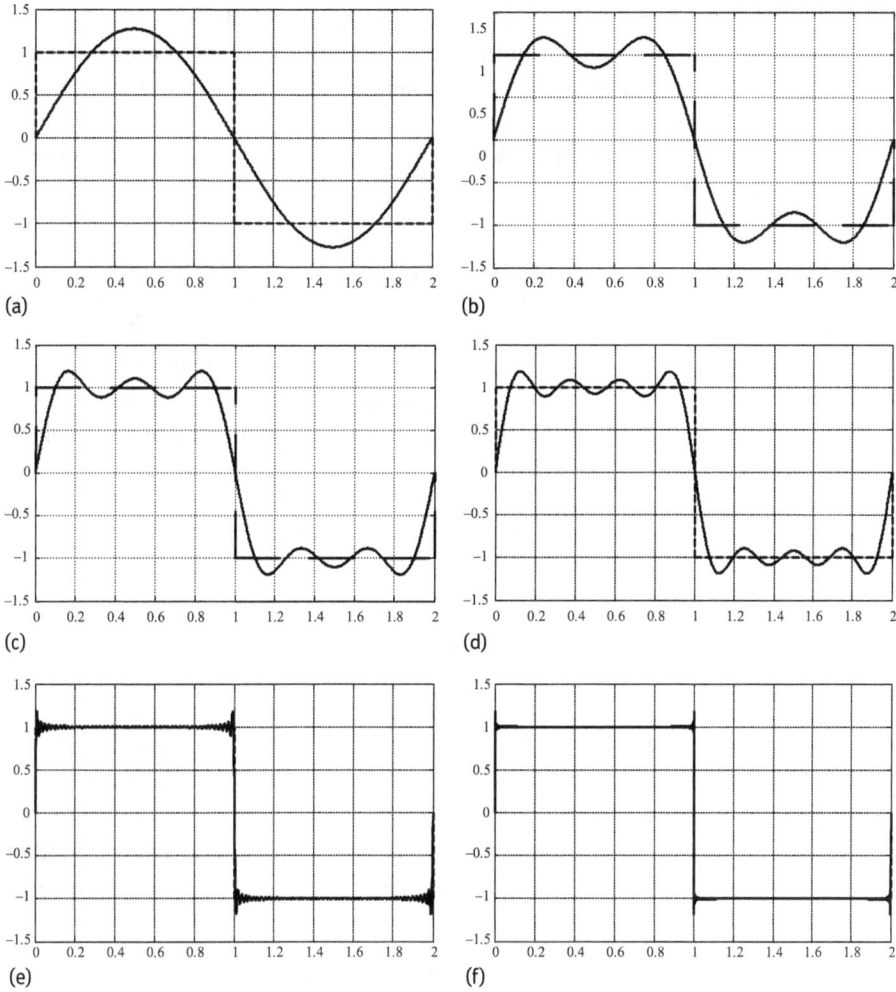

Fig. 4.3: Rectangular pulse reconstructed with a finite number of harmonics; (a) $N = 1$, (b) $N = 3$, (c) $N = 5$, (d) $N = 7$, (e) $N = 99$, (f) $N = 999$

The equivalent form of the CTFS expansion can be expressed as follows:

$$f(t) = \frac{A_0}{2} + \sum_{n=1}^{\infty} A_n \cos(n\Omega t + \varphi_n)$$

$$= \frac{4}{\pi} \left[\cos\left(\Omega t - \frac{\pi}{2}\right) + \frac{1}{3}\cos\left(3\Omega t - \frac{\pi}{2}\right) + \cdots \right.$$

$$\left. + \frac{1}{n}\cos\left(n\Omega t - \frac{\pi}{2}\right) + \cdots \right], \quad n = 1, 3, 5, \ldots$$

Figure 4.3 gives the convergence of the Fourier series representation of the rectangular pulse with period $T = 2$. The harmonics are added from $N = 1$ to $N = 999$ to reconstruct the approximation $f_N(t) \approx \sum_{n=1}^{N} A_n \cos(n\Omega t + \varphi_n)$.

From this example, some observations are concluded as follows:

(i) To obtain a more accurate waveform, more harmonic components are needed. As N increases, the error between the combination of harmonics and the original waveform decreases.

(ii) The low-frequency harmonics reflect the general shape of the waveform and the high-frequency harmonics supplement the discontinuity details.

(iii) As more terms are added to the CTFS, the separation between the ripples becomes narrower, and the approximated function is closer to the original function. The peak amplitude of the ripples, however, does not decrease with more CTFS terms. The presence of ripples near the discontinuity is a limitation of the CTFS representation of discontinuous signals and is known as the *Gibbs phenomenon*. Specifically, for a discontinuity of unit height 1, the partial sum exhibits the value of 1.09 (i.e., an overshoot of 9% of the height of the discontinuity), no matter how large N becomes.

! **Note:** The value 9% of the overshoot is computed in Section 4.5.3.

4.1.2 Symmetry of waveform and harmonic characteristics

In this section, we consider the symmetry properties if the real-valued periodic signal is even, odd, even harmonic, or odd harmonic function. Some Fourier coefficients will be zero, which will simplify the computation of Fourier series. The DC, cosine or sine components of various symmetric signals are listed in Table 4.1.

Some functions are neither odd nor odd harmonic functions but may be odd or odd harmonic functions plus DC components. Now, the DC component should be removed first to determine the harmonic components of the Fourier series.

! **Note:** Here, the signal is supposed to be a real-valued periodic signal.

4.1.3 Exponential Fourier series

In Section 4.1.1, we considered the trigonometric CTFS expansion using a set of sinusoidal terms as the basis functions. An alternative expression for the CTFS is obtained if complex exponentials $\{e^{jn\Omega t}, n \in Z\}$ are used as the basis functions to expand a CT periodic signal. The resulting CTFS representation is referred to as exponential CTFS, which is defined below.

Definition. An arbitrary periodic function $f(t)$ with a fundamental period T can be expressed as follows:

$$f(t) = \sum_{n=-\infty}^{\infty} F_n e^{jn\Omega t}, \tag{4.9}$$

Tab. 4.1: Harmonic characteristics of different symmetrical signals

Function type	Waveform example	DC	Sine term	Cosine term
Even function $f(t) = f(-t)$		Possibly included	Not included	Included
Odd function $f(t) = -f(-t)$		Not included	Included	Not included
Odd harmonic function $f(t \pm (T_0/2)) = -f(t)$		Possibly included	Odd sine included	Odd cosine included
Even harmonic function $f(t \pm (T_0/2)) = f(t)$		Possibly included	Even sine included	Even cosine included

where the exponential CTFS coefficients F_n are calculated as:

$$F_n = \frac{1}{T} \int_{(T)} f(t) e^{-jn\Omega t} \, dt \,, \tag{4.10}$$

Ω being the fundamental frequency given by $\Omega = 2\pi/T$.

Equation (4.9) is known as the *exponential CTFS* representation of $f(t)$.

Since the basic functions corresponding to the trigonometric and exponential CTFS are related by Euler's identity:

$$\cos(n\Omega t + \varphi_n) = \frac{1}{2} \left[e^{j(n\Omega t + \varphi_n)} + e^{-j(n\Omega t + \varphi_n)} \right] \,,$$

it is intuitively pleasing to believe that the exponential and trigonometric CTFS coefficients are also related to each other. The exact relationship is derived by expanding the trigonometric CTFS series as follows:

$$\begin{aligned}
f(t) &= \frac{A_0}{2} + \sum_{n=1}^{\infty} A_n \cos(n\Omega t + \varphi_n) \\
&= \frac{A_0}{2} + \frac{1}{2} \sum_{n=1}^{\infty} A_n e^{j(n\Omega t + \varphi_n)} + \frac{1}{2} \sum_{n=1}^{\infty} A_n e^{-j(n\Omega t + \varphi_n)} \\
&= \frac{A_0}{2} + \frac{1}{2} \sum_{n=1}^{\infty} A_n e^{j(n\Omega t + \varphi_n)} + \frac{1}{2} \sum_{n=-1}^{-\infty} A_{-n} e^{-j(-n\Omega t + \varphi_{-n})} \,.
\end{aligned} \tag{4.11}$$

Since A_n is the even function of n, and φ_n is the odd function of n, the second summation can be expressed as follows:

$$\frac{1}{2} \sum_{n=-1}^{-\infty} A_{-n} e^{-j(-n\Omega t + \varphi_{-n})} = \frac{1}{2} \sum_{n=-1}^{-\infty} A_n e^{j(n\Omega t + \varphi_n)} , \tag{4.12}$$

which leads to the following expression:

$$f(t) = \frac{A_0}{2} + \frac{1}{2} \sum_{n=1}^{\infty} A_n e^{j(n\Omega t + \varphi_n)} + \frac{1}{2} \sum_{n=-1}^{-\infty} A_n e^{j(n\Omega t + \varphi_n)}$$

$$= \frac{1}{2} \sum_{n=-\infty}^{\infty} A_n e^{j(n\Omega t + \varphi_n)} = \sum_{n=-\infty}^{\infty} \frac{1}{2} A_n e^{j\varphi_n} e^{jn\Omega t} \tag{4.13}$$

Comparing the above expansion with the definition of exponential CTFS in Equation (4.9) yields:

$$F_n = |F_n| e^{j\varphi_n} = \frac{1}{2} A_n e^{j\varphi_n} = \frac{1}{2}(a_n - jb_n) . \tag{4.14}$$

Obviously, $|F_n|$ is the even function of n (or $n\Omega$), and φ_n is the odd function of n (or $n\Omega$):

$$\begin{cases} |F_n| = |F_{-n}| \\ \varphi_n = -\varphi_{-n} \end{cases} \tag{4.15}$$

! **Note:** Here, n is an integer within $(-\infty, +\infty)$.

Substituting the definitions of a_n, b_n, we can verify the expression for the exponential CTFS coefficients F_n:

$$F_n = \frac{1}{2} \left[\frac{2}{T} \int_{(T)} f(t) \cos(n\Omega t) \, dt - j\frac{2}{T} \int_{(T)} f(t) \sin(n\Omega t) \, dt \right]$$

$$= \frac{1}{T} \int_{(T)} f(t) e^{-jn\Omega t} \, dt . \tag{4.16}$$

Any periodic signal $f(t)$ can be decomposed into the exponential functions of different frequencies; F_n is the coefficient of the exponential component with frequency $n\Omega$, and $F_0 = A_0/2$ is the DC component.

Example 4.1.2. Calculate the exponential CTFS expansion for the periodic function $f(t)$ shown in Figure 4.4.

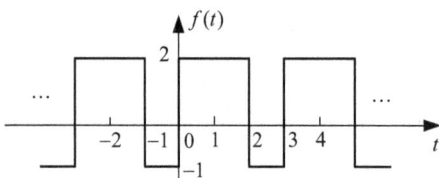

Fig. 4.4: Waveform of the periodic signal in Example 4.1.2

Solution: The fundamental period T is 3, which shows the fundamental frequency $\Omega = 2\pi/T = 2\pi/3$. The exponential CTFS coefficients F_n are given by:

$$F_n = \frac{1}{3} \int_0^T f(t) e^{-jn\Omega t}\, dt = \frac{1}{3} \left[\int_0^2 2e^{-jn\Omega t}\, dt - \int_2^3 e^{-jn\Omega t}\, dt \right]$$

$$= \frac{2}{3} \frac{1}{-jn\Omega} e^{-jn\Omega t}\Big|_0^2 - \frac{1}{3}\frac{1}{-jn\Omega} e^{-jn\Omega t}\Big|_2^3$$

$$= \frac{2}{j3n\Omega} \left[1 - e^{-j2n\Omega} \right] + \frac{1}{j3n\Omega} \left[e^{-j3n\Omega} - e^{-j2n\Omega} \right]$$

$$= \frac{2 - 3e^{-j2n\Omega} + e^{-j3n\Omega}}{j3n\Omega}$$

Substituting $\Omega = 2\pi/3$, we obtain the following expression for the exponential CTFS coefficients:

$$F_n = \frac{2 - 3e^{-j\frac{4\pi}{3}n} + e^{-j2\pi n}}{j2\pi n} = \frac{3}{j2\pi n}\left(1 - e^{-j\frac{4\pi}{3}n}\right)$$

The exponential Fourier series is:

$$f(t) = \sum_{n=-\infty}^{\infty} F_n e^{jn\Omega t} = \sum_{n=-\infty}^{\infty} \frac{3}{j2\pi n}\left(1 - e^{-j\frac{4\pi}{3}n}\right) e^{jn\Omega t}.$$

4.1.4 Parseval's power theorem

Definition. The average power of $f(t)$ is computed by *Parseval's power theorem*, which is shown in Equation (4.17). Thus, the total average power of a periodic signal equals the sum of the average powers of all of its harmonic components:

$$P = \left(\frac{A_0}{2}\right)^2 + \sum_{n=1}^{\infty} \frac{1}{2} A_n^2 = \sum_{n=-\infty}^{\infty} |F_n|^2 \tag{4.17}$$

Proof. The average power P in one period of a real function $f(t)$ is defined as the energy per unit time:

$$P = \frac{1}{T} \int_{-\frac{T}{2}}^{\frac{T}{2}} f^2(t)\, dt$$

Substituting Equation (4.6) into the above expression yields:

$$P = \frac{1}{T} \int_{-\frac{T}{2}}^{\frac{T}{2}} f^2(t)\, dt = \frac{1}{T} \int_{-\frac{T}{2}}^{\frac{T}{2}} \left[\frac{A_0}{2} + \sum_{n=1}^{\infty} A_n \cos(n\Omega t + \varphi_n) \right]^2 dt$$

$$= \left(\frac{A_0}{2}\right)^2 + \sum_{n=1}^{\infty} \frac{1}{2} A_n^2 \tag{4.18}$$

Based on $|F_n| = 1/2A_n$, the above expression simplifies as follows:

$$P = \left(\frac{A_0}{2}\right)^2 + \sum_{n=1}^{\infty} \frac{1}{2}A_n^2 = |F_0|^2 + 2\sum_{n=1}^{\infty} |F_n|^2 = \sum_{n=-\infty}^{\infty} |F_n|^2 \qquad \square$$

! **Note:** In Equation (4.18), the integrals of $\cos(n\Omega t + \varphi_n)$ and $\cos(m\Omega t + \varphi_m) \cdot \cos(n\Omega t + \varphi_n)$ in one period are zero; the integral of $\cos^2(n\Omega t + \varphi_n)$ is $T/2$.

4.2 Fourier spectrum of periodic signals

4.2.1 Definition of the Fourier spectrum

Definition. The plot of the magnitude of the CTFS coefficients versus $n\Omega$ is defined as the *magnitude (or amplitude) spectrum*, while the plot of the phase of the CTFS coefficients versus $n\Omega$ is defined as the *phase spectrum*.

A periodic signal can be represented by trigonometric CTFS and exponential CTFS:

$$f(t) = \frac{A_0}{2} + \sum_{n=1}^{\infty} A_n \cos(n\Omega t + \varphi_n) \qquad (4.19)$$

$$f(t) = \sum_{n=-\infty}^{\infty} F_n e^{jn\Omega t} \qquad (4.20)$$

In trigonometric CTFS, the amplitude A_n and phase φ_n plotted versus $n\Omega$ ($n > 0$) are referred to as the unilateral spectrum. In exponential CTFS, the amplitude $|F_n|$ and phase φ_n plotted versus $n\Omega$, $n \in (-\infty, \infty)$ are referred to as the bilateral spectrum.

The CTFS coefficients provide frequency information about the content of a signal. The spectrum can visually reflect the distribution of harmonic components, which can help us understand the nature of the signal by looking at the values of the coefficients.

Example 4.2.1. The periodic signal is given in Equation (4.21). Calculate the fundamental period T, the fundamental frequency Ω, and plot the unilateral and bilateral spectrum of magnitude and phase, respectively:

$$f(t) = 1 - \frac{1}{2}\cos\left(\frac{\pi}{4}t - \frac{2\pi}{3}\right) + \frac{1}{4}\sin\left(\frac{\pi}{3}t - \frac{\pi}{6}\right) \qquad (4.21)$$

Solution: Rewrite the expression of $f(t)$ as follows:

$$f(t) = 1 + \frac{1}{2}\cos\left(\frac{\pi}{4}t - \frac{2\pi}{3} + \pi\right) + \frac{1}{4}\cos\left(\frac{\pi}{3}t - \frac{\pi}{6} - \frac{\pi}{2}\right)$$

$$= 1 + \frac{1}{2}\cos\left(\frac{\pi}{4}t + \frac{\pi}{3}\right) + \frac{1}{4}\cos\left(\frac{\pi}{3}t - \frac{2\pi}{3}\right) \qquad (4.22)$$

Obviously, the period of $1/2 \cos(\pi/4t+\pi/3)$ is $T_1 = 8$, and the period of $1/4 \cos(\pi/3t-2\pi/3)$ is $T_2 = 6$. The fundamental period of $f(t)$ is $T = 24$. The fundamental frequency is $\Omega = 2\pi/T = \pi/12$ rad/s.

Note: The spectral lines locate at $n\Omega$.

(1) Plot the unilateral spectra
The signal has components of the DC, the third ($n = 3$) and fourth ($n = 4$) harmonics. The unilateral magnitude and phase spectra are plotted in Figure 4.5.

Note: Here, $A_0/2$ cis the DC component.

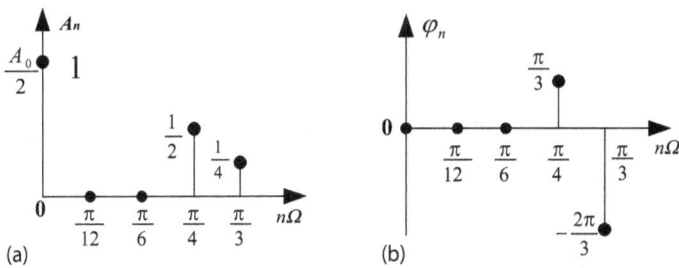

Fig. 4.5: Unilateral amplitude and phase spectra; (a) the amplitude spectrum, (b) the phase spectrum

(2) Plot the bilateral spectra
From the relationship between the trigonometric and exponential coefficients in Equation (4.14), the bilateral magnitude can be obtained as:

$$|F_n| = \frac{1}{2}A_n$$

In addition, from Equation (4.15), the bilateral spectrum of magnitude is of even symmetry, and the bilateral spectrum of phase is of odd symmetry [23]. The bilateral magnitude and phase spectra are plotted in Figure 4.6.

Note: Here, F_0 is the DC component.

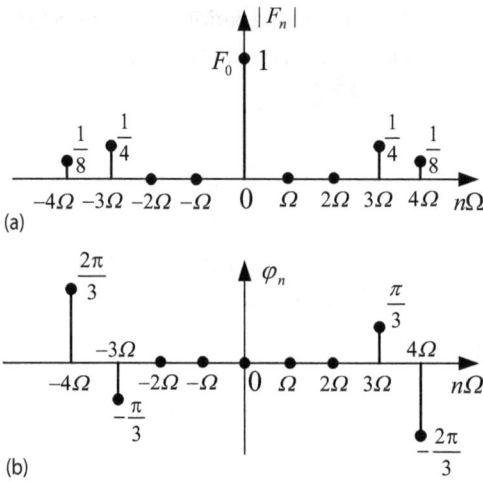

(a)

(b)

Fig. 4.6: The bilateral amplitude and phase spectra; (a) the amplitude spectrum, (b) the phase spectrum

4.2.2 Characteristics of the spectrum of periodic signals

The magnitude of the exponential CTFS coefficients indicates the strength of the frequency component in the signal. We now make an analysis on the properties of the spectrum. The square wave sketched in Figure 4.7 is a periodic signal with fundamental period T The pulse amplitude is 1, and the pulse width is τ.

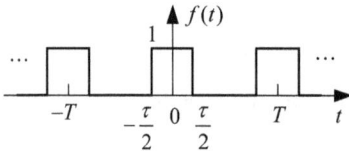

Fig. 4.7: Periodic square wave

Using Equation (4.10), the Fourier series coefficients of $f(t)$ are determined as follows:

$$F_n = \frac{1}{T} \int_{-\frac{\tau}{2}}^{\frac{\tau}{2}} f(t) e^{-jn\Omega t}\, dt = \frac{1}{T} \int_{-\frac{\tau}{2}}^{\frac{\tau}{2}} 1 \times e^{-jn\Omega t}\, dt = \frac{1}{T} \frac{e^{-jn\Omega t}}{-jn\Omega}\Big|_{-\frac{\tau}{2}}^{\frac{\tau}{2}}$$

$$= \frac{2}{T} \frac{\sin\left(\frac{n\Omega\tau}{2}\right)}{n\Omega} = \frac{\tau}{T} \frac{\sin\frac{n\Omega\tau}{2}}{\frac{n\Omega\tau}{2}}, \qquad n = 0, \pm1, \pm2, \ldots \tag{4.23}$$

Here, the function $\sin x/x$ is referred to as the sampling function $Sa(x) = \sin x/x$. Substituting $\Omega = 2\pi/T$, Equation (4.23) can be written as:

$$F_n = \frac{\tau}{T} Sa\left(\frac{n\Omega\tau}{2}\right) = \frac{\tau}{T} Sa\left(\frac{n\pi\tau}{T}\right), \qquad n = 0, \pm1, \pm2, \ldots (4.2-6) \tag{4.24}$$

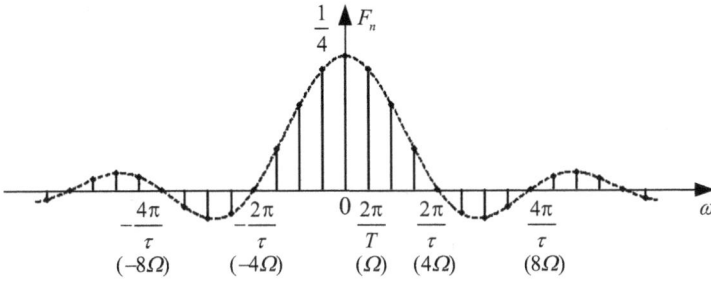

Fig. 4.8: Bilateral spectrum of the periodic square wave ($T = 4\tau$)

For $T = 4\tau$, from Equation (4.24), the coefficients are calculated as:

$$F_n = \frac{1}{4} \mathrm{Sa}\left(\frac{n\pi}{4}\right) = \frac{\sin\left(\frac{n\pi}{4}\right)}{n\pi}, \quad n \neq 0 \tag{4.25}$$

While:

$$F_0 = \frac{\tau}{T} = \frac{1}{4}$$

From Equation (4.25), $F_n = 0$ for $n = 4m$, $m \neq 0$. Figure 4.8 is a bar graph of the Fourier series coefficients. Note that the coefficients are all real valued but periodically vary between positive and negative values. Because the CTFS coefficients F_n do not have imaginary components, the phase corresponding to the CTFS coefficients is calculated from its sign as follows:

if $\quad F_n \geq 0$, $\quad$ then the associated phase $\quad \varphi_n = 0$;

if $\quad F_n < 0$, $\quad$ then the associated phase $\quad \varphi_n = \pm\pi$.

Note: Try to obtain the magnitude and phase spectra from Figure 4.8.

It can be concluded that the spectrum of the periodic signal is discrete and convergent. The detailed characteristics are analyzed as follows:

(1) Discreteness
As shown in Figure 4.8, the spectrum of the periodic signal is composed of a number of discrete spectral lines separated by a fundamental frequency Ω. The envelope of the discrete spectrum is an Sa function. The coefficients are regularly spaced samples of the envelope ($\tau/T \, \mathrm{Sa}(n\Omega\tau/2)$).

We now further investigate the impact of increasing the period on the frequency spectrum. Figure 4.9 depicts the signals of periodic square pulse with the same pulse width τ and different period T and their spectra. The spacing between spectral lines, $\Omega = 2\pi/T$, decreases as T increases. When $T \to \infty$, the spacing $\Omega \to 0$. That is,

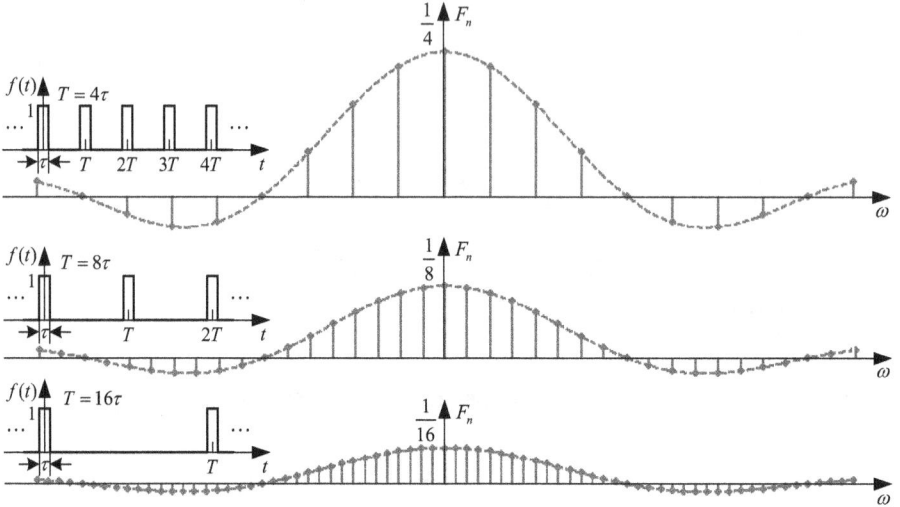

Fig. 4.9: Plots of the spectra with different periods

the periodic signal becomes aperiodic, and the spectral interval approaches becoming infinitely small. So, the aperiodic signal has a continuous spectrum. At the same time, the amplitude of each harmonic component tends to be infinitesimal.

(2) Convergence

The overall trend of the height of spectral lines in the spectrum of a periodic signal decreases with an increase of the harmonic number n. When $n \to \infty$, the amplitude of its harmonic component approaches zero. In a communication system, it is necessary to transmit harmonic components with low frequencies, which contain the main energy of the signal. The power of harmonic components between $\omega = 0$ and $\omega = 2\pi/\tau$ (the frequency of the first zero amplitude) takes up 90.3% of the total power. For a periodic square wave, the frequency bandwidth is defined as:

$$B_\omega = \frac{2\pi}{\tau} \quad \text{or} \quad B_f = \frac{1}{\tau} \tag{4.26}$$

where the bandwidth B is inversely proportional to the pulse width τ.

! **Note:** Readers can carry out an analysis of the relationship between B and τ.

Figure 4.10 depicts the signals of periodic square pulse with the same period T and different pulse width τ, and their spectra. The spectra have the same spectral interval $\Omega = 2\pi/T$. The signal bandwidth, $B_\omega = 2\pi/\tau$, increases as the pulse width τ decreases. In short, the width compression in the time domain broadens band width in the frequency domain.

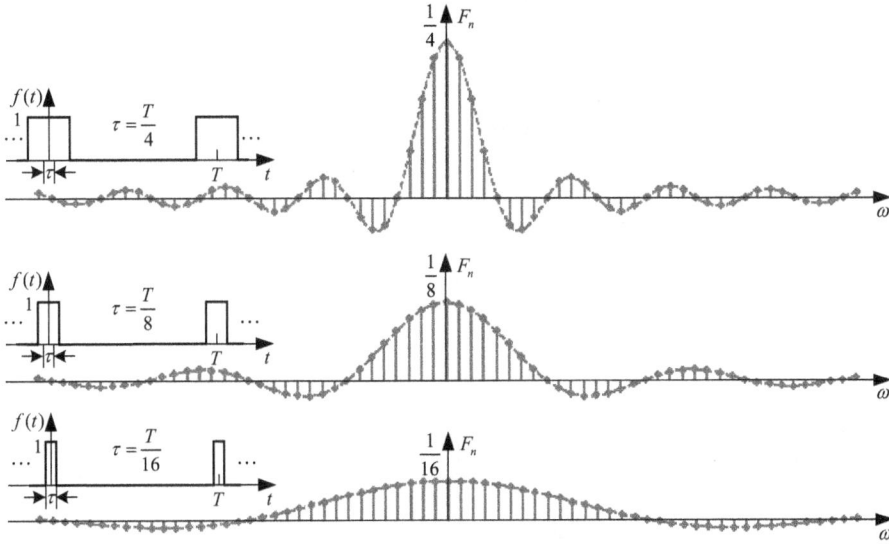

Fig. 4.10: Plots of the spectra with different pulse widths

The ratio τ/T is referred to as the duty cycle, which is defined as the ratio between the time τ that the waveform has a high value and the fundamental period T. Here, the duty cycle decreases by reducing the value of τ, while maintaining the fundamental period T at a constant value. As the duty cycle is decreased, the energy within one period of the waveform in the time domain is concentrated over a relatively narrower fraction of the time period. The energy in the corresponding CTFS representations is distributed over a larger number of the CTFS coefficients. In other words, the width of the main lobe and side lobes of the discrete Sa function increases with a reduction in the duty cycle.

4.2.3 Application of the Fourier series

Example 4.2.2. A simple DC-to-AC converter that based on periodic conversion is shown in Figure 4.11. The conversion period is $1/60$ s. We consider two cases: (a) the converter is turned on or off, and (b) the converter inverts the polarity. Figure 4.12 (a) and (b) depicts the output waveform for the two cases, respectively.

The conversion efficiency is defined as the ratio of the average power of the fundamental component to that of the original DC signal. Calculate the efficiency in the above two cases.

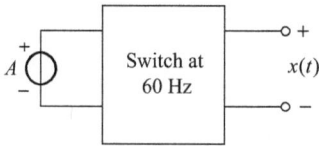

Fig. 4.11: DC-to-AC converter

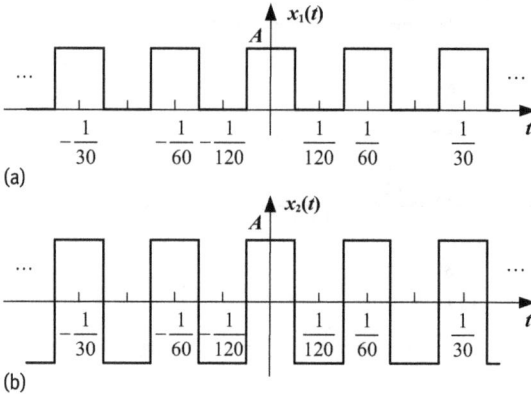

(a)

(b)

Fig. 4.12: Output waveform of the two cases

Solution: The period and fundamental frequency is $T = 1/60\,\text{s}$, $\Omega = 2\pi/T = 120\pi\,\text{rad/s}$ respectively

The average power of the original DC signal is A^2.

(1) From the square wave $x_1(t)$ in Figure 4.12(a), the coefficients of trigonometric CTFS are given by:

$$a_0 = \frac{A}{2}, \quad a_n = \frac{2A \sin\left(\frac{n\pi}{2}\right)}{n\pi}, \quad n = 1, 2, \ldots, \quad b_n = 0$$

The efficiency is computed as follows:

$$C_{\text{eff1}} = \frac{(a_1)^2/2}{A^2} = \frac{\frac{1}{2} \cdot \left(\frac{2A \sin\left(\frac{\pi}{2}\right)}{\pi}\right)^2}{A^2} = \frac{2}{\pi^2} \approx 0.2 .$$

(2) From the square wave $x_2(t)$ in Figure 4.12(b), the coefficients of the trigonometric CTFS are given by:

$$a_0 = 0, \quad a_n = \frac{4A \sin\left(\frac{n\pi}{2}\right)}{n\pi}, \quad n = 1, 3, 5, \ldots, \quad b_n = 0$$

The efficiency is computed as follows:

$$C_{\text{eff2}} = \frac{(a_1)^2/2}{A^2} = \frac{\frac{1}{2} \cdot \left(\frac{4A \sin\left(\frac{\pi}{2}\right)}{\pi}\right)^2}{A^2} = \frac{8}{\pi^2} \approx 0.8 .$$

! **Note:** Why is the second efficiency four times that of the first converter?

4.3 Continuous-time Fourier transforms

4.3.1 Definition of CTFT

In Section 4.1, we introduced the frequency representations for periodic signals based on the trigonometric and exponential continuous-time Fourier series (CTFS). In Section 4.2.2, applying the period $T \to \infty$ caused the spacing $\Omega = 2\pi/T$ in the spectrum to decrease to zero. The CTFS spectrum of the aperiodic signal are continuous along the frequency w-axis. At the same time, the amplitude of each harmonic component tends to be zero. In order to show the relative difference between these amplitudes of harmonic components, the exponential coefficients F_n are expanded T times to derive the mathematical definition of the CTFT:

$$T \cdot F_n = \int_{(T)} f(t)e^{-jn\Omega t}\, dt \tag{4.27}$$

Applying $T \to \infty$ to the above equation, we have $n\Omega \to w$. The resulting expression is as follows:

$$F(jw) = \lim_{T\to\infty} F_n T = \lim_{T\to\infty} \int_{(T)} f(t)e^{-jn\Omega t}\, dt = \int_{-\infty}^{\infty} f(t)e^{-jwt}\, dt \tag{4.28}$$

The term $F(jw)$ denotes the exponential CTFT coefficients of signal $f(t)$.

Using the exponential CTFS definition, $f(t)$ can be evaluated from the CTFS coefficients F_n as follows:

$$f(t) = \sum_{n=-\infty}^{\infty} F_n e^{jn\Omega t} = \sum_{n=-\infty}^{\infty} F_n T \cdot e^{jn\Omega t}\frac{1}{T} \tag{4.29}$$

When the period $T \to \infty$, the fundamental frequency Ω approaches a small value denoted by dw. The fundamental period T is, therefore, given by $1/T = \Omega/(2\pi) \to (dw)/(2\pi)$. Substituting all the above symbols into Equation (4.29) yields:

$$f(t) = \lim_{T\to\infty} \sum_{n=-\infty}^{\infty} F_n T \cdot e^{jn\Omega t}\frac{1}{T} = \frac{1}{2\pi}\int_{-\infty}^{\infty} F(jw)e^{jwt}\, dw . \tag{4.30}$$

Combining Equations (4.28) and (4.30), we have the CTFT synthesis equation and the CTFT analysis equation as follows:

CTFT analysis equation (CTFT):

$$F(jw) = \int_{-\infty}^{\infty} f(t)e^{-jwt}\, dt \tag{4.31}$$

CTFT synthesis equation (inverse CTFT):

$$f(t) = \frac{1}{2\pi}\int_{-\infty}^{\infty} F(jw)e^{jwt}\, dw \tag{4.32}$$

Note: The signal is expressed as a combination of the basic signal e^{jwt}.   !

Collectively, the CTFT pair is denoted by:

$$f(t) \leftrightarrow F(j\omega).$$

(4.33)

Alternatively, the CTFT pair may also be represented as:

$$\left.\begin{aligned} F(j\omega) &= \mathcal{F}[f(t)] \\ f(t) &= \mathcal{F}^{-1}[F(j\omega)] \end{aligned}\right\}$$

(4.34)

In general, the CTFT $F(j\omega)$ is a complex function of the angular frequency ω, which can be written as:

$$F(j\omega) = |F(j\omega)|\, e^{j\varphi(\omega)} = R(\omega) + jX(\omega)$$

(4.35)

The plots of magnitude $|F(j\omega)|$ and phase $\varphi(\omega)$ with respect to ω are, respectively, referred to as the magnitude and phase spectra of the aperiodic function.

Comparing CTFT with CTFS, we draw the following conclusions:

(1) The basis function of the CTFS is $\{e^{jn\Omega t}\}$, and the basis function of the CTFT is $e^{j\omega t}$. The variable ω is a continuous variable, and the CTFT $F(j\omega)$ is, therefore, defined for all frequencies ω within the range $-\infty < \omega < \infty$.

! **Note:** The spectrum of CTFS is discrete and the spectrum of DTFT is continuous.

(2) The sufficient condition for the existence of CTFT is the absolutely integrable condition $\int_{-\infty}^{\infty} |f(t)|\, dt < \infty$.

(3) It is also easy to calculate some integrals using the following equations:

$$F(0) = \int_{-\infty}^{\infty} f(t)\, dt \qquad f(0) = \frac{1}{2\pi} \int_{-\infty}^{\infty} F(j\omega)\, d\omega$$

! **Note:** Readers can prove it.

4.3.2 CTFT pairs for elementary CT signals

Table 4.2 lists the CTFTs for elementary CT signals, which are useful for frequency analysis [24]. Almost all of the listed CTFTs can be calculated by definition.

! **Note:** Readers can refer to other references and try to prove it.

Tab. 4.2: CTFT for elementary signals

$f(t)$	$F(j\omega)$	Time-domain waveform and spectra
Causal decaying exponential function $f(t) = e^{-\alpha t}\varepsilon(t)$, $\quad \alpha > 0$	$\dfrac{1}{\alpha + j\omega}$	
Two-sided decaying exponential function $f(t) = e^{-\alpha\|t\|}$, $\quad \alpha > 0$	$\dfrac{2\alpha}{\alpha^2 + \omega^2}$	
Gate function $g_\tau(t) = \begin{cases} 1, & \|t\| < \dfrac{\tau}{2} \\ 0, & \|t\| > \dfrac{\tau}{2} \end{cases}$	$\tau\,\mathrm{Sa}(\dfrac{\omega\tau}{2})$	
Unit impulse function $\delta(t)$	1	
Impulse function derivative $\delta^{(n)}(t)$	$(j\omega)^n$	
Constant $f(t) = 1$	$2\pi\delta(\omega)$	
Sign function $\mathrm{sgn}(t) = \begin{cases} -1, & t < 0 \\ 1, & t > 0 \end{cases}$	$\dfrac{2}{j\omega}$	
Unit step function $\varepsilon(t)$	$\pi\delta(\omega) + \dfrac{1}{j\omega}$	

4.3.3 Properties of CTFT

In this section, we present the properties of the CTFT based on the transformations of the signals. Given the CTFT of a CT function $f(t)$, we are interested in calculating the CTFT of a function produced by a linear operation on $f(t)$ in the time domain. The linear operations being considered include superposition, time shifting, scaling, differentiation and integration. We also consider some basic nonlinear operations like multiplication of two CT signals, convolution in the time and frequency domains, and Parseval's relationship. A list of CTFT properties is given in Table 4.3.

There are several important explanations regarding the properties of the CTFT.

(1) Scaling: If the signal $f(t)$ compresses in the time domain ($a > 1$), the corresponding spectral function expands in the frequency domain. On the contrary, if $f(t)$ expands in the time domain ($0 < a < 1$), the spectrum compresses. To compress the duration of the signal, we have to widen the band as a cost. Therefore, in communication systems, there is a contradiction between the communication rate and the occupied bandwidth.

Tab. 4.3: Fundamental properties of CT Fourier transform

Transformation	$f(t)$	$F(j\omega)$
Linearity	$af_1(t) + bf_2(t)$	$aF_1(j\omega) + bF_2(j\omega)$
Duality	$F(jt)$	$2\pi f(-\omega)$
Scaling	$f(at), \quad a \in R, \quad a \neq 0$	$\dfrac{1}{|a|}F(j\dfrac{\omega}{a})$
Time shifting	$f(t \pm t_0)$	$e^{\pm j\omega t_0} F(j\omega)$
Frequency shifting	$f(t)e^{\pm j\omega_0 t}$	$F[j(\omega \mp \omega_0)]$
Time differentiation	$\dfrac{d^n f(t)}{dt^n}$	$(j\omega)^n F(j\omega)$
Time integration	$f^{(-1)}(t) = \int_{-\infty}^{t} f(x)\,dx$	$\dfrac{F(j\omega)}{j\omega} + \pi F(0)\delta(\omega)$ $F(0) = F(j\omega)|_{\omega=0} = \int_{-\infty}^{\infty} f(t)\,dt.$
Frequency differentiation	$(-jt)^n f(t)$	$\dfrac{d^n F(j\omega)}{d\omega^n}$
Frequency integration	$\pi f(0)\delta(t) + \dfrac{1}{-jt}f(t)$	$\int_{-\infty}^{\omega} F(jx)\,dx$
Time convolution	$f_1(t) * f_2(t)$	$F_1(j\omega) \cdot F_2(j\omega)$
Frequency convolution	$f_1(t) \cdot f_2(t)$	$\dfrac{1}{2\pi}F_1(j\omega) * F_2(j\omega)$

(2) Time shifting: The delayed function $f(t-t_0)$ is obtained by shifting $f(t)$ towards the right-hand side of the time axis. The time-shifting property states that if a signal is shifted by t_0 time units in the time domain, the CTFT of the original signal is modified by a multiplicative factor of $e^{-j\omega t_0}$. The magnitude spectrum of the CTFT of the time-shifted signal is unchanged, while the phase spectrum is modified by an additive factor of $-\omega t_0$.

(3) Frequency shifting: Frequency shifting the CTFT of a signal does not change the amplitude of the signal $f(t)$ in the time domain. The only change is in the phase of the signal $f(t)$, which is modified by an additive factor of $-\omega t_0$. Frequency shifting is widely used in communications systems, such as amplitude modulation, synchronous demodulation, mixing, etc. The principle is to multiply the signal $f(t)$ by the carrier signal $\cos(\omega_0 t)$ or $\sin(\omega_0 t)$.

Note: How can frequency shifting by $f(t) \times \cos(\omega_0 t)$ be achieved? ❗

The following examples are applications of CTFT properties.

Example 4.3.1. Using the linearity, calculate the CTFT of the waveform plotted in Figure 4.13 (a).

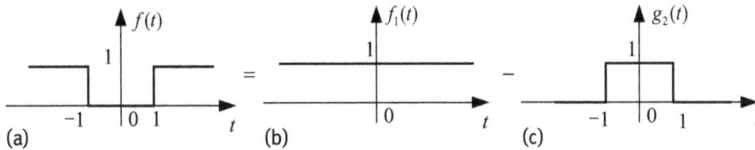

Fig. 4.13: Waveform used in Example 4.3.1

Solution: By inspection, the waveform $f(t)$ can be expressed as a linear combination of $f_1(t)$ and $g_2(t)$ from Figure 4.13 (b) and (c):

$$f(t) = f_1(t) - g_2(t)$$

Based on Table 4.2, the CTFT pairs of $f_1(t) = 1$ and $g_2(t)$ are given by:

$$f_1(t) = 1 \leftrightarrow 2\pi\delta(\omega)$$
$$g_2(t) \leftrightarrow 2\,\text{Sa}(\omega)$$

Using the linearity property, the CTFT of $f(t)$ is therefore given by:

$$F(j\omega) = 2\pi\delta(\omega) - 2\,\text{Sa}(\omega).$$

Example 4.3.2. Calculate the CTFT of the sampling function $Sa(t) = \sin t/t$.

Solution: Based on Table 4.2, the CTFT pair of $g_\tau(t)$ is given by:

$$g_\tau(t) \leftrightarrow \tau \, Sa\left(\frac{\omega\tau}{2}\right) .$$

Substituting $\tau = 2$, the above CTFT is as follows:

$$g_2(t) \leftrightarrow 2 \, Sa\,(\omega)$$

By the duality property, we obtain:

$$2 \, Sa(t) \leftrightarrow 2\pi g_2(-\omega) = 2\pi g_2(\omega) \qquad (4.36)$$

Therefore, the CTFT of $Sa(t) = \sin t/t$ is $\pi g_2(\omega)$.

Figure 4.14 shows the symmetry across the time and frequency domains in the sense that the CTFT of a gate function $g_\tau(t)$ is a sampling function $Sa(\omega)$, while the CTFT of a sampling function $Sa(t)$ is a gate function $g_\tau(\omega)$.

! **Note:** Try to prove the symmetry property.

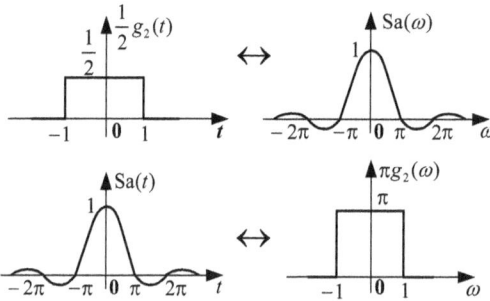

Fig. 4.14: Example of symmetry across time and frequency domains

Example 4.3.3. Calculate the CTFT of the function $f(t) = 1/jt - 1$.

Solution:

$$e^{-t}\varepsilon(t) \leftrightarrow \frac{1}{j\omega + 1}$$

Based on the duality property, we obtain:

$$\frac{1}{jt + 1} \leftrightarrow 2\pi e^{\omega}\varepsilon(-\omega)$$

Applying the time-scaling property with $a = -1$, the above transformation is given by:

$$\frac{1}{-jt + 1} \leftrightarrow 2\pi e^{-\omega}\varepsilon(\omega) .$$

So, the CTFT of $f(t)$ is as follows:

$$f(t) = \frac{1}{jt - 1} \leftrightarrow -2\pi e^{-\omega}\varepsilon(\omega) .$$

Example 4.3.4. Calculate the CTFT of the function $f(t) = 1/(1 + t^2)$.

Solution:

$$e^{-\alpha|t|} \leftrightarrow \frac{2\alpha}{\alpha^2 + \omega^2}$$

Substituting $\alpha = 1$, we obtain:

$$e^{-|t|} \leftrightarrow \frac{2}{1 + \omega^2}$$

Based on the duality property, we obtain:

$$\frac{2}{1 + t^2} \leftrightarrow 2\pi e^{-|\omega|}$$

Therefore, the CTFT of $f(t)$ is as follows:

$$f(t) = \frac{1}{1 + t^2} \leftrightarrow \pi e^{-|\omega|} .$$

Example 4.3.5. Calculate the CTFT of the waveform plotted in Figure 4.15 (a).

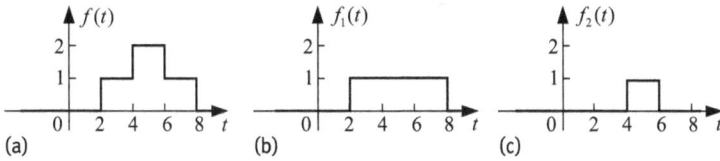

Fig. 4.15: Waveform used in Example 4.3.5

Solution: By inspection, the signal $f(t)$ can be expressed as the sum of two signals:

$$f(t) = f_1(t) + f_2(t) ,$$

where $f_1(t) = g_6(t - 5)$ and $f_2(t) = g_2(t - 5)$.
Using the time-shifting property:

$$g_6(t - 5) \leftrightarrow 6\,\text{Sa}(3\omega)e^{-j5\omega} \quad \text{and} \quad g_2(t - 5) \leftrightarrow 2\,\text{Sa}(\omega)e^{-j5\omega} .$$

Finally, by applying the linearity property, we obtain:

$$F(j\omega) = [6\,\text{Sa}(3\omega) + 2\,\text{Sa}(\omega)]e^{-j5\omega} .$$

Example 4.3.6. Calculate the CTFT of the function $f(t) = (t^2 - 2t + 3)/(t^2 - 2t + 2)$.

Solution: The function reduces to:

$$f(t) = \frac{t^2 - 2t + 3}{t^2 - 2t + 2} = 1 + \frac{1}{t^2 - 2t + 2} = 1 + \frac{1}{(t-1)^2 + 1}$$

According to the CTFT of the two-sided decaying exponential function, we have:

$$e^{-|t|} \leftrightarrow \frac{2}{\omega^2 + 1}$$

Using the duality property:

$$\frac{2}{t^2 + 1} \leftrightarrow 2\pi e^{-|-\omega|}$$

By the time-shifting property:

$$\frac{1}{(t-1)^2 + 1} \leftrightarrow \pi e^{-j\omega} e^{-|\omega|}$$

Finally, the CTFT of $f(t)$ is given by:

$$f(t) = \frac{t^2 - 2t + 3}{t^2 - 2t + 2} = 1 + \frac{1}{(t-1)^2 + 1} \leftrightarrow 2\pi\delta(\omega) + \pi e^{-j\omega} e^{-|\omega|} .$$

Example 4.3.7. Calculate the CTFT of the function $f(t) = e^{j3t}$.

Solution:

$$1 \leftrightarrow 2\pi\delta(\omega)$$

Using the frequency-shifting property:

$$f(t) = e^{j3t} \cdot 1 \leftrightarrow 2\pi\delta(\omega - 3) .$$

Example 4.3.8. Calculate the CTFT of the functions $\cos(\omega_0 t)$ and $\sin(\omega_0 t)$.

Solution:

$$1 \leftrightarrow 2\pi\delta(\omega)$$

Using the frequency-shifting property:

$$e^{j\omega_0 t} \cdot 1 \leftrightarrow 2\pi\delta(\omega - \omega_0)$$
$$e^{-j\omega_0 t} \cdot 1 \leftrightarrow 2\pi\delta(\omega + \omega_0)$$

Using Euler's formula as follows:

$$\cos(\omega_0 t) = \frac{1}{2}\left(e^{j\omega_0 t} + e^{-j\omega_0 t}\right)$$
$$\sin(\omega_0 t) = \frac{1}{2j}\left(e^{j\omega_0 t} - e^{-j\omega_0 t}\right) ,$$

and the linearity property, we obtain:

$$\cos(\omega_0 t) \leftrightarrow \pi[\delta(\omega + \omega_0) + \delta(\omega - \omega_0)]$$
$$\sin(\omega_0 t) \leftrightarrow j\pi[\delta(\omega + \omega_0) - \delta(\omega - \omega_0)]$$

(4.37)

Note: Remember the CTFT of cosine function.

Example 4.3.9. If $f(t) \leftrightarrow F(j\omega)$, calculate the CTFT of the function $f(t)\cos(\omega_0 t)$.

Solution 1. Using Euler's formula:

$$f(t)\cos(\omega_0 t) = \frac{1}{2}\left[e^{j\omega_0 t}f(t) + e^{-j\omega_0 t}f(t)\right].$$

Applying the frequency-shifting and the linearity properties:

$$f(t)\cos(\omega_0 t) \leftrightarrow \frac{1}{2}\left[F\left(j(\omega - \omega_0)\right) + F\left(j(\omega + \omega_0)\right)\right] \tag{4.38}$$

Solution 2. Using the CTFT of the periodic cosine function:

$$\cos(\omega_0 t) \leftrightarrow \pi[\delta(\omega + \omega_0) + \delta(\omega - \omega_0)].$$

Applying the frequency convolution property, we obtain:

$$f(t)\cos(\omega_0 t) \leftrightarrow \frac{1}{2\pi}F(j\omega) * \{\pi[\delta(\omega + \omega_0) + \delta(\omega - \omega_0)]\}$$
$$= \frac{1}{2}\{F[j(\omega + \omega_0)] + F[j(\omega - \omega_0)]\} \tag{4.39}$$

As we can see, if the signal $f(t)$ is multiplied by $\cos(\omega_0 t)$, the frequency spectrum $F(j\omega)$ shifts towards the left-hand and right-hand sides of the frequency axis. This is the basic theory of frequency modulating used in radio transmission systems.

Example 4.3.10. Calculate the CTFT of the function $f(t) = (\sin t / t)^2$.

Solution:
$$g_2(t) \leftrightarrow 2\,\mathrm{Sa}(\omega)$$

Using the duality property:
$$2\,\mathrm{Sa}(t) \leftrightarrow 2\pi g_2(-\omega)$$
$$\mathrm{Sa}(t) \leftrightarrow \pi g_2(\omega)$$

Using the frequency convolution property:

$$\left(\frac{\sin t}{t}\right)^2 \leftrightarrow \frac{1}{2\pi}[\pi g_2(\omega)] * [\pi g_2(\omega)] = \frac{\pi}{2}g_2(\omega) * g_2(\omega)$$

In Section 2.4.5, the convolution of two rectangular pulses is computed as a triangle pulse. Figure 4.16 gives the frequency spectrum $F(j\omega)$.

Note: The spectrum is another method of CTFT expression besides functions.

!

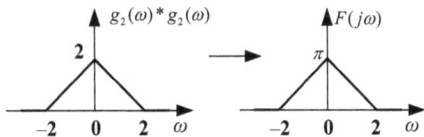

Fig. 4.16: The frequency spectrum of $f(t)$

Example 4.3.11. Calculate the CTFT of the function $f(t) = 1/t^2$.

Solution:

$$\text{sgn}(t) \leftrightarrow \frac{2}{j\omega}$$

Using the duality property:

$$\frac{2}{jt} \leftrightarrow 2\pi \, \text{sgn}(-\omega)$$

$$\frac{1}{t} \leftrightarrow -j\pi \, \text{sgn}(\omega)$$

Using the time differentiation property:

$$\frac{d}{dt}\left(\frac{1}{t}\right) \leftrightarrow -(j\omega)j\pi \, \text{sgn}(\omega)$$

$$f(t) = \frac{1}{t^2} \leftrightarrow -\pi\omega \, \text{sgn}(\omega) = -\pi|\omega|$$

Example 4.3.12. If $f'(t) \leftrightarrow F_1(j\omega)$, prove the CTFT pair of function $f(t)$ as follows:

$$f(t) \leftrightarrow \frac{1}{j\omega}F_1(j\omega) + \pi \, [f(-\infty) + f(\infty)] \, \delta(\omega)$$

Proof. By integrating the derivative function $f'(t)$ and using the time-integration property, we obtain:

$$f(t) - f(-\infty) = \int_{-\infty}^{t} \frac{df(t)}{dt} \, dt \leftrightarrow \frac{1}{j\omega}F_1(j\omega) + \pi \int_{-\infty}^{\infty} \frac{df(t)}{dt} \, dt\delta(\omega)$$

$$= \frac{1}{j\omega}F_1(j\omega) + \pi \, [f(\infty) - f(-\infty)] \, \delta(\omega)$$

Rearranging the above terms:

$$F(j\omega) - 2\pi f(-\infty)\delta(\omega) = \frac{1}{j\omega}F_1(j\omega) + \pi \, [f(\infty) - f(-\infty)] \, \delta(\omega)$$

The CTFT of $f(t)$ is given by:

$$F(j\omega) = \frac{1}{j\omega}F_1(j\omega) + \pi \, [f(\infty) + f(-\infty)] \, \delta(\omega) . \tag{4.40}$$

According to Equation (4.40), we can obtain some other conclusions to further simply the computation of CTFT.

(1) If $f(\infty) + f(-\infty) = 0$ and $f'(t) \leftrightarrow F_1(j\omega)$, then:

$$f(t) \leftrightarrow F_1(j\omega)/j\omega \tag{4.41}$$

(2) If $f(\infty) + f(-\infty) = 0$ and $f^{(n)}(t) \leftrightarrow F_n(j\omega)$, then:

$$f(t) \leftrightarrow F(j\omega) = F_n(j\omega)/(j\omega)^n \tag{4.42}$$

□

! **Note:** Given $d\varepsilon(t)/dt = \delta(t) \leftrightarrow 1$, we have $\varepsilon(t) \leftrightarrow 1/(j\omega) + \pi\delta(\omega)$.

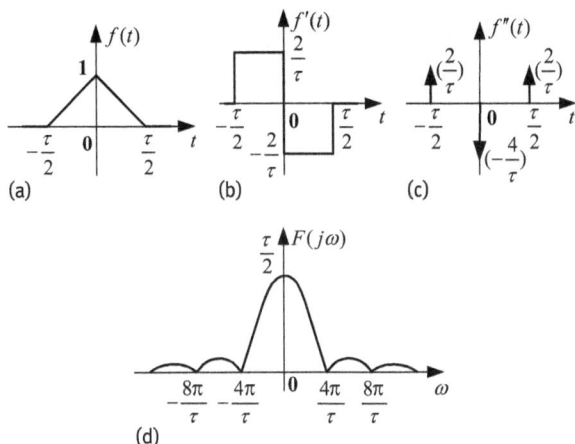

Fig. 4.17: Waveforms in Example 4.3.13

Example 4.3.13. Calculate the CTFT of function $f(t)$ plotted in Figure 4.17 (a).

Solution: The derivatives $f'(t)$ and $f''(t)$ are, respectively, shown in Figure 4.17 (b) and (c):

$$f''(t) = \frac{2}{\tau}\left[\delta\left(t+\frac{2}{\tau}\right) - 2\delta(t) + \delta\left(t-\frac{2}{\tau}\right)\right]$$

By using the time-shifting property:

$$f''(t) \leftrightarrow F_2(j\omega) = \frac{2}{\tau}\left(e^{j\omega\frac{\tau}{2}} - 2 + e^{-j\omega\frac{\tau}{2}}\right) = \frac{2}{\tau}\left[2\cos\left(\omega\frac{\tau}{2}\right) - 2\right]$$

Using Equation (4.42), the CTFT of the function is given by:

$$F(j\omega) = \frac{F_2(j\omega)}{(j\omega)^2} = \frac{\tau}{2}\,\text{Sa}^2\left(\frac{\omega\tau}{4}\right).$$

Example 4.3.14. Calculate the CTFT of function $f(t) = t\varepsilon(t)$.

Solution:

$$\varepsilon(t) \leftrightarrow \pi\delta(\omega) + \frac{1}{j\omega}$$

By using the frequency-differentiation property, we obtain:

$$-jt\varepsilon(t) \leftrightarrow \frac{d}{d\omega}\left[\pi\delta(\omega) + \frac{1}{j\omega}\right],$$

which can be expressed as:

$$f(t) = t\varepsilon(t) \leftrightarrow j\pi\delta'(\omega) - \frac{1}{\omega^2}.$$

Note: Can we use the property of $t\varepsilon(t) = \varepsilon(t) * \varepsilon(t)$ to compute the CTFT?

Example 4.3.15. Calculate the integration $\int_{-\infty}^{\infty} \sin(a\omega)/\omega \, d\omega$.

Solution:

$$g_{2a}(t) \leftrightarrow \frac{2\sin(a\omega)}{\omega}$$

By using the CTFT synthesis equation, we obtain:

$$g_{2a}(t) = \frac{1}{2\pi} \int_{-\infty}^{\infty} \frac{2\sin(a\omega)}{\omega} e^{j\omega t} \, d\omega = \frac{1}{\pi} \int_{-\infty}^{\infty} \frac{\sin(a\omega)}{\omega} e^{j\omega t} \, d\omega .$$

Given $t = 0$:

$$g_{2a}(0) = \frac{1}{\pi} \int_{-\infty}^{\infty} \frac{\sin(a\omega)}{\omega} \, d\omega$$

$$\int_{-\infty}^{\infty} \frac{\sin(a\omega)}{\omega} \, d\omega = \pi$$

The function $\sin(a\omega)/\omega$ is even, and therefore, the integration is given by:

$$\int_{0}^{\infty} \frac{\sin(a\omega)}{\omega} \, d\omega = \frac{\pi}{2} .$$

4.3.4 Fourier transforms of real-valued even and odd functions

In this section, we consider various properties of the CTFT for real-valued functions. According to the definition of CTFT, $F(j\omega)$ can be expressed as follows:

$$F(j\omega) = \int_{-\infty}^{\infty} f(t) e^{-j\omega t} \, dt = \int_{-\infty}^{\infty} f(t) \cos(\omega t) \, dt - j \int_{-\infty}^{\infty} f(t) \sin(\omega t) \, dt$$

$$= R(\omega) + jX(\omega)$$

$$= |F(j\omega)| e^{j\varphi(\omega)} \tag{4.43}$$

The real and imaginary components are obtained, respectively, as follows:

$$\left. \begin{array}{l} R(\omega) = \int_{-\infty}^{\infty} f(t) \cos(\omega t) \, dt \\[2em] X(\omega) = - \int_{-\infty}^{\infty} f(t) \sin(\omega t) \, dt \end{array} \right\} \tag{4.44}$$

The magnitude and phase of the frequency spectrum are given by:

$$\left. \begin{array}{l} |F(j\omega)| = \sqrt{R^2(\omega) + X^2(\omega)} \\[1em] \varphi(\omega) = \arctan\left(\dfrac{X(\omega)}{R(\omega)}\right) \end{array} \right\} \tag{4.45}$$

Tab. 4.4: Symmetry properties of CTFT

Type of function $f(t)$	Real part $R(\omega)$	Imaginary part $X(\omega)$	Magnitude $|F(j\omega)|$	Phase $\varphi(\omega)$
Real-valued	Even	Odd	Even	Odd
Real-valued and even	Real even	0	$|F(j\omega)| = |R(\omega)|$	0
Real-valued and odd	0	Imaginary odd	$|F(j\omega)| = |X(\omega)|$	$\pi/2$

Table 4.4 lists the symmetry properties of CTFT, from which we can conclude that the CTFT of a real-valued signal $f(t)$ satisfies the following:

$$f(-t) \leftrightarrow F(-j\omega) = F^*(j\omega) , \qquad (4.46)$$

where $F^*(j\omega)$ denotes the complex conjugate of $F(j\omega)$.

Note: Here, the signal is supposed to be a real-valued signal. !

4.3.5 Parseval's energy theorem

Parseval's theorem relates the energy of a signal in the time domain to the energy of its CTFT in the frequency domain. It shows that the CTFT is a lossless transform as there is no loss of energy if a signal is transformed by the CTFT.

For an energy signal $f(t)$, Parseval's energy theorem is given by:

$$E = \int_{-\infty}^{\infty} |f(t)|^2 \, dt = \frac{1}{2\pi} \int_{-\infty}^{\infty} |F(j\omega)|^2 \, d\omega \qquad (4.47)$$

Example 4.3.16. Calculate the energy of the function $f(t) = 2\cos(997t)\sin 5t/(\pi t)$.

Solution:

$$g_{10}(t) \leftrightarrow 10\,\text{Sa}(5\omega)$$

Using the duality property:

$$10\,\text{Sa}(5t) \leftrightarrow 2\pi g_{10}(-\omega)$$

$$\frac{\sin 5t}{\pi t} \leftrightarrow g_{10}(\omega) .$$

Based on the time convolution property, we have:

$$2\cos(997t)\frac{\sin 5t}{\pi t} \leftrightarrow \frac{1}{2\pi} 2\pi[\delta(\omega + 997) + \delta(\omega - 997)] * g_{10}(\omega)$$

$$= g_{10}(\omega - 997) + g_{10}(\omega + 997)$$

Using Parseval's energy theorem:

$$E = \frac{1}{2\pi} \int_{-\infty}^{\infty} |F(j\omega)|^2 \, d\omega = \frac{1}{2\pi}(10 + 10) = \frac{10}{\pi} .$$

Note: Parseval's energy and power theorems are easier than the calculation in the time domain. !

4.3.6 CTFT of periodic functions

In Example 4.3.8, we computed the CTFT of the period cosine and sine functions usingSection 4.1 the frequency shifting property:

$$\cos(\omega_0 t) = (e^{j\omega_0 t} + e^{-j\omega_0 t})/2 \leftrightarrow \pi[\delta(\omega + \omega_0) + \delta(\omega - \omega_0)] \qquad (4.48)$$

$$\sin(\omega_0 t) = (e^{j\omega_0 t} - e^{-j\omega_0 t})/(2j) \leftrightarrow j\pi[\delta(\omega + \omega_0) - \delta(\omega - \omega_0)] . \qquad (4.49)$$

For a general periodic signal $f_T(t)$ with a fundamental period of T, its exponential CTFS expression is given by:

$$f_T(t) = \sum_{n=-\infty}^{\infty} F_n e^{jn\Omega t} , \qquad (4.50)$$

where $\Omega = 2\pi/T$ is the fundamental frequency of the periodic signal, and F_n denotes the exponential CTFS coefficients.

Calculating the CTFT of both sides of Equation (4.50), we obtain the CTFT of a period function as follows:

$$f_T(t) = \sum_{n=-\infty}^{\infty} F_n e^{jn\Omega t} \leftrightarrow F_T(j\omega) = 2\pi \sum_{n=-\infty}^{\infty} F_n \delta(\omega - n\Omega) . \qquad (4.51)$$

Note: The CTFT of a periodic signal is composed of a train of impulses at the harmonic frequencies $n\Omega (n = 0, \pm 1, \pm 2, \dots)$.

Example 4.3.17. Calculate the exponential CTFS coefficients and the CTFT representation of the periodic impulses of $\delta_T(t) = \sum_{m=-\infty}^{\infty} \delta(t - mT)$ shown in Figure 4.18 (a).

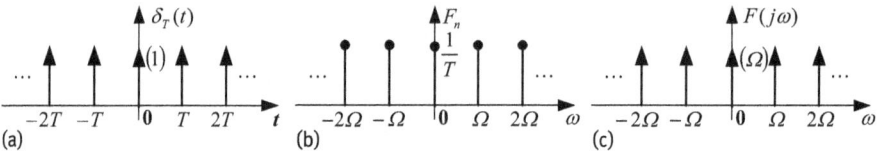

Fig. 4.18: The waveforms in Example 4.3.17

Solution:

$$F_n = \frac{1}{T} \int_{-\frac{T}{2}}^{\frac{T}{2}} f(t) e^{-jn\Omega t} \, dt = \frac{1}{T}$$

Using Equation (4.51), the CTFT expression is given by:

$$\delta_T(t) \leftrightarrow \frac{2\pi}{T} \sum_{n=-\infty}^{\infty} \delta(\omega - n\Omega) = \Omega \sum_{n=-\infty}^{\infty} \delta(\omega - n\Omega) = \Omega \delta_\Omega(\omega) \qquad (4.52)$$

The CTFS coefficients F_n and the CTFT $F(j\omega)$ of the periodic impulse function are plotted in Figure 4.18 (b) and (c).

Example 4.3.18. Calculate the CTFT representation of the periodic waveform shown in Figure 4.19 (a).

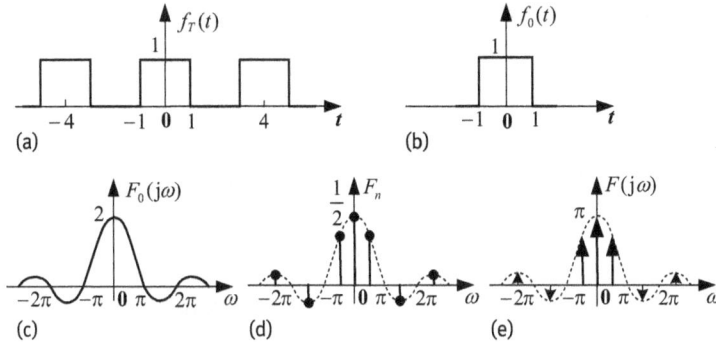

Fig. 4.19: The waveforms in Example 4.3.18

Solution: The periodic signal $f_T(t)$ is considered as a periodic extension of aperiodic signal $f_0(t)$ within the range $-T/2 \leq t \leq T/2$, as shown in Figure 4.19 (b):

$$f_T(t) = \delta_T(t) * f_0(t) \tag{4.53}$$

According to the time convolution property and Equation (4.52), the CTFT $F(j\omega)$ is given by:

$$f_T(t) \leftrightarrow F(j\omega) = \Omega\delta_\Omega(\omega)F_0(j\omega) = \Omega \sum_{n=-\infty}^{\infty} F_0(jn\Omega)\delta(\omega - n\Omega), \tag{4.54}$$

where $\Omega = 2\pi/T$ is the fundamental frequency of the periodic signal, and $F_0(j\omega)$ shown in Figure 4.19 (c) denotes the CTFT of the aperiodic signal $f_0(t)$.

In this example, the aperiodic signal $f_0(t)$ is shown in Figure 4.19(b) and the CTFT $F_0(j\omega)$ is shown in Figure 4.19 (c). The corresponding pair of CTFT is obtained by

$$f_0(t) = g_2(t) \leftrightarrow F_0(j\omega) = 2\,\text{Sa}(\omega).$$

Substituting $\Omega = 2\pi/T = \pi/2$ in Equation (4.54) results in the following expression for the CTFT:

$$F(j\omega) = \Omega \sum_{n=-\infty}^{\infty} 2\,\text{Sa}(n\Omega)\delta(\omega - n\Omega) = \pi \sum_{n=-\infty}^{\infty} \text{Sa}\left(n\frac{\pi}{2}\right)\delta\left(\omega - n\frac{\pi}{2}\right).$$

Comparing Equations (4.51) and (4.54), we can derive the exponential CTFS coefficients of a periodic signal with period T from the CTFT using the following expression:

$$F_n = \frac{\Omega}{2\pi}F_0(jn\Omega) = \frac{1}{T}F_0(j\omega)|_{\omega=n\Omega} \tag{4.55}$$

! **Note:** This is another way of computing CTFS coefficients.

The CTFS coefficients F_n and the CTFT $F(j\omega)$ of the periodic function are plotted in Figures 4.19 (d) and (e).

In conclusion, the exponential CTFS of a periodic signal is obtained using the following steps:

(1) Compute CTFT $F_0(j\omega)$ of the aperiodic signal $f_0(t)$ obtained from one period of $f_T(t)$ as:

$$f_0(t) = \begin{cases} f_T(t) & -T/2 \leq t \leq T/2 \\ 0 & \text{elsewhere .} \end{cases}$$

(2) The exponential CTFS coefficients F_n of the periodic signal are given by:

$$F_n = \frac{1}{T} F_0(j\omega)|_{\omega=n\Omega} ,$$

where $\Omega = 2\pi/T$ denotes the fundamental frequency of the periodic signal $f_T(t)$. In a word, CTFS coefficients F_n are samples of CTFT $F_0(j\omega)$.

4.4 LTIC systems analysis using CTFT and CTFS

In the frequency domain, the signal is decomposed as the sum of imaginary exponential functions with different frequencies. For a periodic signal, the CTFS expression is given by:

$$f(t) = \sum_{n=-\infty}^{\infty} F_n e^{jn\Omega t} ,$$

while the CTFT expression is given by:

$$f(t) = \frac{1}{2\pi} \int_{-\infty}^{\infty} F(j\omega) e^{j\omega t} \, d\omega . \tag{4.56}$$

The elementary signals of the above two decompositions are $e^{jn\Omega t}$ and $e^{j\omega t}$, respectively.

4.4.1 Response of the LTIC system to the complex exponential function

In Chapter 2, we showed that the zero-state response of the LTIC system is the convolution of the input signal and the impulse response $h(t)$. When the input signal is the complex exponential function $e^{j\omega t}$, its response is given by:

$$y(t) = h(t) * e^{j\omega t} . \tag{4.57}$$

According to the definition of convolution:

$$y(t) = \int_{-\infty}^{\infty} h(\tau)e^{j\omega(t-\tau)}\,d\tau = e^{j\omega t} \int_{-\infty}^{\infty} h(\tau)e^{-j\omega\tau}\,d\tau \qquad (4.58)$$

Obviously, the integration $\int_{-\infty}^{\infty} h(\tau)e^{-j\omega\tau}\,d\tau$ is exactly the CTFT of the impulse response $h(t)$ and it is referred to as $H(j\omega)$; $H(j\omega)$ is the Fourier transfer function of the LTIC system and provides meaningful insights into the behavior of the system. Equation (4.58) simplifies to:

$$y(t) = H(j\omega)e^{j\omega t} . \qquad (4.59)$$

4.4.2 Response of the LTIC system to an arbitrary signal

Using the definition of inverse CTFT, an arbitrary CT signal $f(t)$ can be represented as follows:

$$f(t) = \frac{1}{2\pi} \int_{-\infty}^{\infty} F(j\omega)e^{j\omega t}\,d\omega$$

Note: Here, $f(t)$ is considered as a linear combination of $e^{j\omega t}$. ❗

When the input signal is $e^{j\omega t}$, the zero-state response is obtained in Equation (4.59):

$$e^{j\omega t} \rightarrow H(j\omega)e^{j\omega t}$$

Using the homogeneity property, we have:

$$\frac{1}{2\pi}F(j\omega)\,d\omega e^{j\omega t} \rightarrow \frac{1}{2\pi}F(j\omega)H(j\omega)\,d\omega e^{j\omega t} .$$

According to the additive property:

$$\frac{1}{2\pi} \int_{-\infty}^{\infty} F(j\omega)e^{j\omega t}\,d\omega \rightarrow \frac{1}{2\pi} \int_{-\infty}^{\infty} H(j\omega)F(j\omega)e^{j\omega t}\,d\omega . \qquad (4.60)$$

The above expression simplifies to:

$$f(t) \rightarrow y(t) = \mathcal{F}^{-1}[F(j\omega)H(j\omega)] \qquad (4.61)$$

Mathematically, the CTFT of the output response $y(t)$ is given by:

$$Y(j\omega) = F(j\omega)H(j\omega) . \qquad (4.62)$$

4.4.3 The Fourier transfer function of an LTIC system

It was mentioned in Chapter 2 that the impulse response relates the zero-state response $y(t)$ of an LTIC system to its input $f(t)$ using:

$$y(t) = h(t) * f(t)$$

Calculating the CTFT of both sides of the equation, we obtain:

$$Y(j\omega) = H(j\omega)F(j\omega)$$

The *Fourier transfer function* $H(j\omega)$ can be defined as the ratio of the CTFT of the zero-state output response and the CTFT of the input signal. Mathematically, the transfer function $H(j\omega)$ is given by:

$$H(j\omega) = \frac{Y(j\omega)}{F(j\omega)} = |H(j\omega)|\, e^{j\theta(\omega)}. \tag{4.63}$$

The magnitude spectrum $|H(j\omega)|$ is also referred to as the *gain response* of the system, while the phase spectrum $\theta(\omega)$ is referred to as the *phase response* of the system; $|H(j\omega)|$ is an even function of ω, while $\theta(\omega)$ is an odd function of ω.

! **Note:** The transfer function relates the system output and input.

Example 4.4.1. The circuit is shown in Figure 4.20 (a) with $R = 1\,\Omega$ and $C = 1\,\text{F}$. Considering $u_C(t)$ as the output, calculate the impulse response $h(t)$.

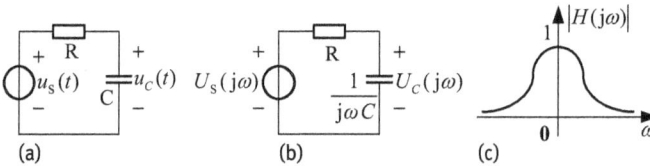

Fig. 4.20: The circuit of Example 4.4.1

Solution: Draw the circuit model in the frequency domain, which is shown in Figure 4.20 (b).

The transfer function is, therefore, given by the ratio of the impedance:

$$H(j\omega) = \frac{U_C(j\omega)}{U_S(j\omega)} = \frac{\frac{1}{j\omega C}}{R + \frac{1}{j\omega C}} = \frac{1}{j\omega + 1}$$

Taking the inverse CTFT, the impulse response is obtained:

$$h(t) = e^{-t}\varepsilon(t)$$

The magnitude spectrum $|H(j\omega)|$ is plotted in Figure 4.20 (c). As the frequency ω increases, the magnitude $|H(j\omega)|$ drops and approaches zero when $\omega \to \infty$.

! **Note:** This circuit is a first-order RC low-pass filter.

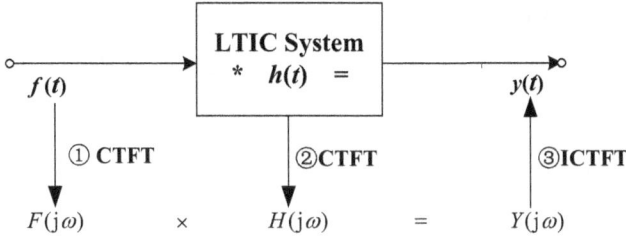

Fig. 4.21: Block diagram of CTFT analysis for the LTIC system

4.4.4 Steps of calculating the response with CTFT

As shown in Figure 4.21, the response of the LTIC system with the input $f(t)$ can be derived using the following the steps.
(1) Compute the CTFT of the input signal $f(t)$ to obtain $F(j\omega)$.
(2) Obtain the Fourier transfer function $H(j\omega)$ of the LTIC system and multiply with $F(j\omega)$ to obtain $Y(j\omega) = F(j\omega)H(j\omega)$.
(3) Calculate the inverse CTFT of $Y(j\omega)$ to obtain the response $y(t)$.

Example 4.4.2. The differential equation of an LTIC system is given by:

$$y'(t) + 2y(t) = f(t) .$$

Determine the transfer function $H(j\omega)$ and compute the zero-state response when the input is given by: $f(t) = e^{-t}\varepsilon(t)$.

Solution: Taking the CTFT of both sides of the differential equation and applying the time-differentiation property yields:

$$j\omega Y(j\omega) + 2Y(j\omega) = F(j\omega) .$$

Making $Y(j\omega)$ common on the left-hand side of the above expression, we obtain:

$$(j\omega + 2)Y(j\omega) = F(j\omega)$$

Based on Equation (4.63), the transfer function is, therefore, given by:

$$H(j\omega) = \frac{Y(j\omega)}{F(j\omega)} = \frac{1}{j\omega + 2} .$$

The CTFT of the input signal is calculated as follows:

$$f(t) = e^{-t}\varepsilon(t) \leftrightarrow F(j\omega) = \frac{1}{j\omega + 1} .$$

The CTFT of the output response is given by:

$$Y(j\omega) = H(j\omega)F(j\omega) = \frac{1}{(j\omega + 1)(j\omega + 2)} \qquad (4.64)$$

By partial fraction expansion, it reduces to:

$$Y(j\omega) = \frac{1}{j\omega + 1} - \frac{1}{j\omega + 2} \cdot$$

! **Note:** Partial fraction expansion will be detailed in Chapter 5.

Taking the inverse CTFT:

$$y(t) = (e^{-t} - e^{-2t})\varepsilon(t)$$

4.4.5 Steps of calculating the response with CTFS

For the periodic signal, the output response of an LTIC system can be derived with CTFS. The CTFS expression of a periodic signal is given by:

$$f_T(t) = \sum_{n=-\infty}^{\infty} F_n e^{jn\Omega t}$$

In Section 4.4.1, the response of the complex exponential function $e^{j\omega t}$ is $y(t) = H(j\omega)e^{j\omega t}$. Likewise, the response of $e^{jn\Omega t}$ is $y(t) = H(jn\Omega)e^{jn\Omega t}$. Therefore, the response of a periodic signal $f_T(t)$ is given by:

$$y(t) = h(t) * f_T(t) = \sum_{n=-\infty}^{\infty} F_n \left[h(t) * e^{jn\Omega t} \right] = \sum_{n=-\infty}^{\infty} F_n H(jn\Omega)e^{jn\Omega t} . \tag{4.65}$$

The exponential CTFS coefficients of $y(t)$ can be computed as:

$$Y_n = F_n H(jn\Omega) \tag{4.66}$$

where

$$H(jn\Omega) = H(j\omega)|_{\omega=n\Omega} . \tag{4.67}$$

As is shown in Figure 4.22, the response of the LTIC system with the periodic input $f_T(t)$ can be derived using the following the steps.

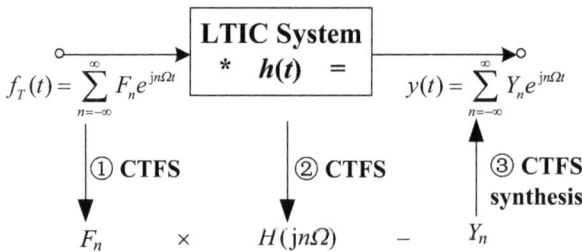

Fig. 4.22: Block diagram of CTFS analysis for the LTIC System

(1) Compute the exponential CTFS coefficients F_n of the input signal $f_T(t)$.
(2) Obtain the Fourier transfer function $H(j\omega)$ of the LTIC system and determine $H(jn\Omega) = H(j\omega)|_{\omega=n\Omega}$ at each harmonic frequency.
(3) Compute the exponential CTFS coefficients $Y_n = F_n H(jn\Omega)$ to obtain the response $y(t)$.

Especially when the input periodic signal is in the form of triangular CTFS expansion, the output $y(t)$ can be expressed as follows:

$$f_T(t) = \frac{A_0}{2} + \sum_{n=1}^{\infty} A_n \cos(n\Omega t + \varphi_n), \tag{4.68}$$

$$H(j\omega) = |H(j\omega)| \, e^{j\theta(\omega)}, \tag{4.69}$$

and

$$y(t) = \frac{A_0}{2} H(0) + \sum_{n=1}^{\infty} A_n |H(jn\Omega)| \cos\left[n\Omega t + \varphi_n + \theta(n\Omega)\right], \tag{4.70}$$

where $|H(jn\Omega)|$ and $\theta(n\Omega)$ are the magnitude and phase of $H(j\omega)$ evaluated at $\omega = n\Omega$.

Example 4.4.3. The magnitude spectrum $|H(j\omega)|$ and phrase spectrum $\theta(\omega)$ of an LTIC system are shown in Figure 4.23. Given the input $f(t) = 2 + 4\cos(5t) + 4\cos(10t)$, calculate the response of the system.

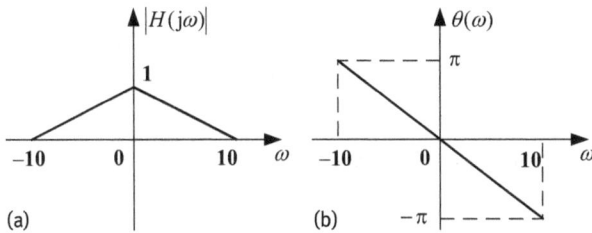

Fig. 4.23: The transfer function in Example 4.4.3

Solution 1.
(1) Calculate the CTFT of the input signal:

$$F(j\omega) = 4\pi\delta(\omega) + 4\pi\left[\delta(\omega - 5) + \delta(\omega + 5)\right] + 4\pi\left[\delta(\omega - 10) + \delta(\omega + 10)\right]$$

(2) The CTFT of the output response is obtained by multiplying $F(j\omega)$ by $H(j\omega)$:

$$Y(j\omega) = F(j\omega)H(j\omega) = 4\pi\delta(\omega)H(0) + 4\pi\left[\delta(\omega - 5)H(j5) + \delta(\omega + 5)H(-j5)\right]$$
$$+ 4\pi\left[\delta(\omega - 10)H(j10) + \delta(\omega + 10)H(-j10)\right]$$
$$= 4\pi\delta(\omega) + 4\pi\left[-j0.5\delta(\omega - 5) + j0.5\delta(\omega + 5)\right]$$

(3) Taking the inverse CTFT, the output is given by:

$$y(t) = \mathcal{F}^{-1}\left[Y(j\omega)\right] = 2 + 2\sin(5t).$$

Solution 2.
(1) The input periodic signal is in the form of a triangular CTFS expansion with the fundamental frequency $\Omega = 5$ rad/s:

$$f(t) = 2 + 4\cos(5t) + 4\cos(10t)$$

(2) The $|H(jn\Omega)|$ and $\theta(n\Omega)$ at $\omega = n\Omega$ are obtained from Figure 4.22.

$$H(0) = 1, \quad H(j\Omega) = 0.5e^{-j0.5\pi}, \quad H(j2\Omega) = 0$$

(3) According to Equation (4.70), the output response is given by:

$$y(t) = 2 \times 1 + 4 \times 0.5 \cos(5t - 0.5\pi) = 2 + 2\sin(5t).$$

❗ Note: Which solution process is easier?

4.4.6 Response computation with MATLAB

MATLAB provides the function for solving the zero-state response of the LTIC system. The function is *lsim*, and its method is as follows:

```
y=lsim(b,a,ft, t)
```

where, t indicates the sampling point vector of the system response, ft is the input signal of system, and b and a are the coefficients of $j\omega$ in the numerator and denominator of $H(j\omega)$, respectively.

Example 4.4.4. The transfer function of an LTIC system is given by:

$$H(j\omega) = \frac{1 - j\omega}{1 + j\omega}.$$

Determine the zero-state response when the input is given by $f(t) = \sin t + \sin 3t - \infty < t < \infty$ with MATLAB.

Solution:

```
t=0:pi/100:4*pi;      % define time instants
b=[-1], [1];          % the coefficients of the numerator of the
                      % transfer function,
a=[1 1];              % the coefficients of the denominator of the
                      % transfer function
ft=sin(t)+sin(3*t);   % the input signal
yt=lsim(b,a,ft,t);    % solve the response and plot it in Figure 4.24
```

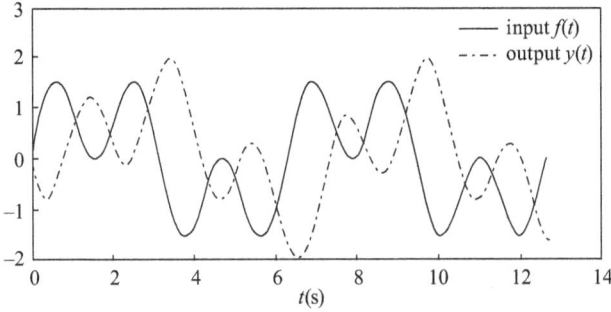

Fig. 4.24: Wave of the output response with MATLAB

4.5 Applications of transmission and filtering

4.5.1 The undistorted transmission system

Definition. The *undistorted transmission* refers to the fact that the output of the system is simply a time shift of the input with amplitude change. The output $y(t)$ can be expressed as:

$$y(t) = Kf(t - t_d) \tag{4.71}$$

Taking the CTFT of both sides of the equation and applying the time-shifting property yields:

$$Y(j\omega) = Ke^{-j\omega t_d}F(j\omega) \tag{4.72}$$

Therefore, for undistorted transmission, the requirements of $h(t)$ and $H(j\omega)$ are as follows:

(1) The impulse response is given by:

$$h(t) = K\delta(t - t_d) . \tag{4.73}$$

(2) The Fourier transfer funcion $H(j\omega)$ is given by:

$$H(j\omega) = Y(j\omega)/F(j\omega) = Ke^{-j\omega t_d} . \tag{4.74}$$

$$|H(j\omega)| = K , \quad \theta(\omega) = -\omega t_d \tag{4.75}$$

By observing the gain response $|H(j\omega)| = K$, we know that the amplitude of each of the frequency component changes uniformly. That is, the gain of the system is a constant for all frequencies. The phase response $\theta(\omega) = -\omega t_d$ is a linear phase with an integer slope $-t_d$. The system imparts a time shift of $-t_d$, or equivalently, a group delay of t_d. Figure 4.25 presents the amplitude and phase spectrum of the undistorted system.

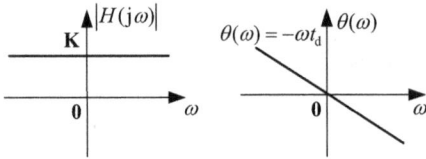

Fig. 4.25: The amplitude and phrase spectra of the undistorted transmission system

The above condition $H(j\omega) = Ke^{-j\omega t_d}$ is the *ideal condition* of undistorted transmission. In real applications, the bandwidth of the transmitted signal is limited. The transfer function should meet the condition for all the frequency components of the signal.

Note: How do we judge whether a system is undistorted?

Example 4.5.1. The amplitude spectrum $|H(j\omega)|$ and the phrase spectrum $\theta(\omega)$ of the LTIC system are plotted in Figures 4.26 (a) and (b), respectively. Judge the distortion of four given input signals to the system:

(1) $f_1(t) = \cos(t) + \cos(8t)$ (2) $f_2(t) = \sin(2t) + \sin(4t)$
(3) $f_3(t) = \sin(2t)\sin(4t)$ (4) $f_4(t) = \cos^2(4t)$

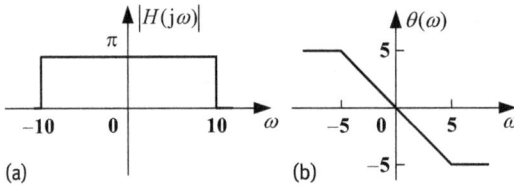

(a)   (b)

Fig. 4.26: Amplitude and phase spectra of Example 4.5.1

Solution:
(1) The input signal $f_1(t) = \cos(t) + \cos(8t)$ has two frequencies at $\omega = 0, 8$. By observing Figure 4.26 (b), we can see that it has phase distortion.
(2) The input signal $f_2(t) = \sin(2t) + \sin(4t)$ has two frequencies at $\omega = 2, 4$. By observing Figure 4.26 (a)(b), we can see that it does not have any distortion.
(3) The input signal $f_3(t) = \sin(2t)\sin(4t)$ has two frequencies at $\omega = 2, 6$. By observing Figure 4.26 (b), we can see that it has phase distortion.
(4) The input signal $f_4(t) = \cos^2(4t)$ has two frequencies at $\omega = 0, 8$. By observing Figure 4.26 (b), we can see that it has phase distortion.

Note: We should find out all the frequency components included in the signal.

4.5.2 Frequency characteristics of an ideal low-pass filter

An ideal frequency-selective filter is a system that passes a prespecified range of frequency components without any attenuation but completely rejects the remaining frequency components. The range of input frequencies that is left unaffected by the filter is referred to as the *passband*, while the range of input frequencies that are blocked from the output is referred to as the *stopband*. Depending upon the range of frequencies within the passband and stopband, the ideal frequency-selective filter is categorized into four different categories, including the low-pass, high-pass, band,-pass and bandstop filters.

The transfer function $H_{lp}(j\omega)$ of an ideal low-pass filter is defined as follows:

$$H_{lp}(j\omega) = \begin{cases} e^{-j\omega t_d}, & |\omega| < \omega_c \\ 0, & |\omega| > \omega_c \end{cases} \tag{4.76}$$

where ω_c is referred to as the cut-off frequency of the filter. The passband of the low-pass filter is given by $|\omega| \le \omega_c$, while the stopband of the low-pass filter is given by $\omega_c < |\omega| < \infty$.

The frequency characteristics of an ideal low-pass filter are plotted in Figure 4.27. We observe that the magnitude toggles between the values of 1 within the passband and zero within the stopband.

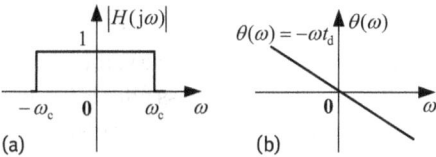

Fig. 4.27: The amplitude and phase spectra of the low-pass filter

4.5.3 Impulse and step response of an ideal low-pass filter

The transfer function $H_{lp}(j\omega)$ of an ideal low-pass filter can be expressed as follows:

$$H_{lp}(j\omega) = g_{2\omega_c}(\omega)e^{-j\omega t_d} \tag{4.77}$$

Taking the inverse CTFT of the above equation, we obtain:

$$h(t) = \mathcal{F}^{-1}\left[g_{2\omega_c}(\omega)e^{-j\omega t_d}\right] = \frac{\omega_c}{\pi} \text{Sa}\left[\omega_c(t - t_d)\right] \tag{4.78}$$

Note: Readers can prove it . . .

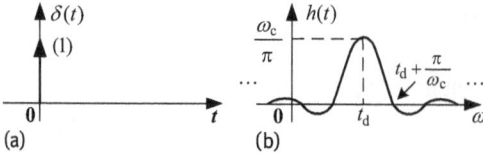

Fig. 4.28: Impulse response of an ideal low-pass filter

Figure 4.28 gives the impulse response of an ideal low-pass filter. The system is actually a noncausal system and, therefore, it is not physically realizable.

The unit step response is then computed as follows:

$$g(t) = h(t) * \varepsilon(t) = \int_{-\infty}^{t} h(\tau)\, d\tau = \int_{-\infty}^{t} \frac{\omega_c}{\pi} \frac{\sin [\omega_c(\tau - t_d)]}{\omega_c(\tau - t_d)}\, d\tau \qquad (4.79)$$

It can be derived as:

$$g(t) = \frac{1}{2} + \frac{1}{\pi} \int_{0}^{\omega_c(t-t_d)} \frac{\sin x}{x}\, dx, \qquad (4.80)$$

where $\mathrm{Si}(y) = \int_{0}^{y} \sin x/x\, dx$ is referred to as the *sinusoidal integral*:

$$g(t) = \frac{1}{2} + \frac{1}{\pi} \mathrm{Si}\,[\omega_c(t - t_d)] \qquad (4.81)$$

As shown in Figure 4.29, the step response output has significant distortion. The overshoot amplitude of the first peak at $t_d + \pi/\omega_c$ is about 9% higher than the stable value. This phenomenon caused by the frequency cutoff effect of the low-pass filter is known as the Gibbs phenomenon, which was introduced in Section 4.1.1. The overshoot is constant no matter how large N becomes. Substituting $t_d + \pi/\omega_c$ into Equation (4.81), we can compute the value of the overshoot:

$$g_{\max} = 0.5 + \mathrm{Si}(\pi)/\pi = 1.0895 \qquad (4.82)$$

! **Note:** The ideal low-pass filter is not physically realizable.

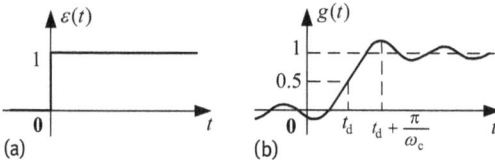

Fig. 4.29: Step response of an ideal low-pass filter

4.5.4 Conditions of physically realizable systems

With respect to the time-domain characteristic, the physically realizable system must meet the causality condition as follows:

$$h(t) = 0, \quad t < 0 \tag{4.83}$$

With respect to the frequency-domain characteristic, Paley and Wiener [11] have proven that the amplitude frequency response must satisfy:

$$\int_{-\infty}^{\infty} |H(j\omega)|^2 \, d\omega < \infty \quad \text{and} \quad \int_{-\infty}^{\infty} \frac{|\ln|H(j\omega)||}{1 + \omega^2} \, d\omega < \infty. \tag{4.84}$$

The Paley–Wiener theorem is the necessary condition for the physically realizable system. The amplitude frequency response $|H(j\omega)|$ of a physically realizable system cannot be zero, except at a finite set of points in the frequency. It cannot be constant in any finite range of frequencies, and the transition from passband to stopband cannot be infinitely sharp, i.e., $|H(j\omega)|$ cannot drop from value 1 to 0 abruptly. Actually, a small amount of ripple in the passband and a small nonzero value or a small amount of ripple in the stopband are tolerable.

Note: Refer to other documents about the Paley–Wiener theorem.

4.5.5 Nonideal low-pass filter

To obtain a physically realizable fiter, it is necessary to relax some of the requirements of ideal filters. For example, Figure 4.30 shows a first-order RC circuit of a low-pass filter. The voltage source is the input signal, and the voltage on the capacitor is the output.

The differential equation between the output $u_C(t)$ and the input $u_S(t)$ is established as follows:

$$\frac{du_C}{dt} + \frac{1}{RC} u_C = \frac{1}{RC} u_S$$

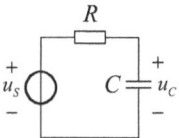

Fig. 4.30: First-order RC filter

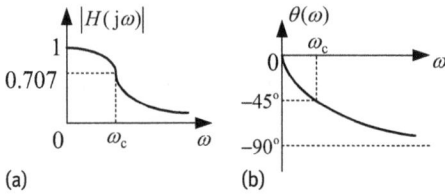

(a)

(b)

Fig. 4.31: Amplitude and phase frequency response of the firs- order RC filter

Calculating the CTFT of both sides of the equation, the transfer function is given by:

$$H(j\omega) = \frac{\frac{1}{j\omega C}}{R + \frac{1}{j\omega C}} = \frac{1}{1 + j\omega RC} = |H(j\omega)| \, e^{j\theta(\omega)}$$

$$|H(j\omega)| = \frac{1}{\sqrt{1 + (\omega CR)^2}}, \qquad \theta(\omega) = -\arctan(\omega CR)$$

The frequency response is plotted in Figure 4.31. It is observed that $|H(j\omega)| = 1$ at the frequecny of $\omega = 0$, while it decreases gently with the increase of ω. The cut-off frequency ω_c of the filter is defined as the frequency at which the gain of the filter drops to 0.707 times its maximun value.

4.5.6 Application of the amplitude modulation system

Amplitude modulation (AM) is a modulation technique used in electronic communication, most commonly for transmitting information via a radio carrier wave. Here, we use the CTFT to analyze a typical AM-based communication system.

Figure 4.32 (a) presents a schematic diagram of an AM system. The sinusoidal carrier is $s(t) = \cos\omega_0 t$ with $\omega_0 = 500\,\text{rad/s}$ and the input signal $f(t) = \sin t/\pi t$ $(-\infty < t < \infty)$. The transfer function of the low-pass filter is shown in Figure 4.32 (b), and its phase response is $\varphi(\omega) = 0$. Calculate the output signal $y(t)$.

The first modulated signal $f_a(t)$ is given by:

$$f_a(t) = f(t) \times s(t) = \frac{\sin t}{\pi t} \cdot \cos 500t \,. \tag{4.85}$$

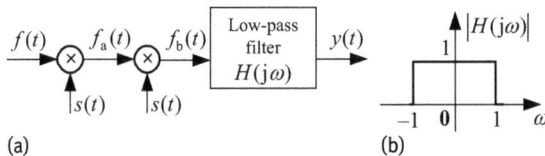

(a)

(b)

Fig. 4.32: The process of amplitude modulation

The CTFT of $f(t)$ is $F(j\omega) = g_2(\omega)$ with the maximum frequency $\omega_{max} = 1$. Calculating the CTFT of both sides of Equation (4.85), we obtain:

$$F_a(j\omega) = \frac{1}{2\pi} F(j\omega) * S(j\omega)$$

$$= \frac{1}{2\pi} g_2(\omega) * \pi [\delta(\omega + 500) + \delta(\omega - 500)]$$

$$= \frac{1}{2} [g_2(\omega + 500) + g_2(\omega - 500)]$$

The spectrum of the modulated signal $f_a(t)$ is the sum of two components: the scaled replica of $F(j\omega)$ shifted to $+\omega_0$ and the scaled replica of $F(j\omega)$ shifted to $-\omega_0$. These two replicas of $F(j\omega) = g_2(\omega)$ are referred to as the sidebands of the AM signal.

Now, we condider demodulation to reconstruct $f(t)$ from $f_a(t)$. The second modulated signal $f_b(t)$ is given by $f_b(t) = f_a(t) \times s(t)$. Using the frequency-convolution property of CTFT, we have:

$$F_b(j\omega) = \frac{1}{2\pi} F_a(j\omega) * S(j\omega)$$

$$= \frac{1}{2\pi} \left\{ \frac{1}{2} [g_2(\omega + 500) + g_2(\omega - 500)] \right\} * \pi [\delta(\omega + 500) + \delta(\omega - 500)]$$

$$= \frac{1}{4} [g_2(\omega + 1000) + 2g_2(\omega) + g_2(\omega - 1000)]$$

The spectrum $F_a(j\omega)$ of the modulated signal is further frequency shifted by ω_0. One of the sidebands is shifted to zero frequency and to $2\omega_0 = 1,000$. The second sideband is shifted to zero frequency and to $-2\omega_0 = -1,000$. In order to remove the side band shifted to frequency $\pm 2\omega_0$, a low-pass filter with $-\omega_{max} \le \omega \le \omega_{max}$ is applied on the signal $f_b(t)$.

According to Figure 4.32(b), the transfer function of the low-pass filter is expressed as $H(j\omega) = g_2(\omega)$. Therefore, the CTFT of the output signal is given by:

$$Y(j\omega) = F_b(j\omega) \cdot H(j\omega)$$

$$= \frac{1}{4} [g_2(\omega + 1000) + 2g_2(\omega) + g_2(\omega - 1000)] \cdot g_2(\omega)$$

$$= \frac{1}{2} g_2(\omega)$$

Note: Plot the spectrum of each signal in the diagram. ❗

Taking the inverse CTFT of the above equation, the output signal of the low-pass filter is a scaled version of $f(t)$:

$$y(t) = \frac{\sin t}{2\pi t} = \frac{Sa(t)}{2\pi} = \frac{1}{2} f(t)$$

4.6 Sampling theorem

Under certain conditions, a CT signal can be completely represented by and recoverable from knowledge of its values, or samples, at points equally spaced in time. The sampling is to convert a CT signal to a DT signal. A DT sequence may occur naturally. Examples are the one-dimensional (1D) hourly measurements made with an electronic thermometer, or the two-dimensional (2D) image recorded with a digital camera. Alternatively, a DT sequence may be derived from a CT signal by the process known as sampling.

4.6.1 Model of ideal impulse-train sampling

In this section, we consider the sampling of a CT signal $f(t)$, whose CTFT $F(j\omega)$ is frequency limited within ω_m:

$$F(j\omega) = 0 \quad \text{for} \quad |\omega| > \omega_m . \tag{4.86}$$

Figure 4.33 gives the model of CT signal sampling. To derive the DT version of the CT signal $f(t)$, we multiply $f(t)$ by a switching function $s(t)$ to obtain the sampling signal as follows:

$$f_s(t) = f(t)s(t) . \tag{4.87}$$

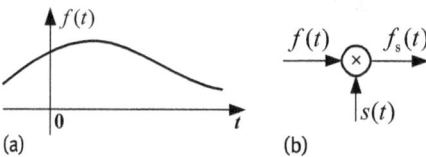

Fig. 4.33: Model of CT signal sampling; (a) a CT signal, (b) sampling model

Definition. If the switching function is an impulse-train $s(t) = \delta_{T_s}(t)$, the sampling process is defined as *ideal impulse-train sampling*:

$$s(t) = \delta_{T_s}(t) = \sum_{n=-\infty}^{\infty} \delta(t - nT_s) \tag{4.88}$$

! (**Note:** Here, we use ideal impulse-train sampling to analyze the theory of sampling.)

Where T_s denotes the separation between two consecutive impulses and is called the sampling interval. Another related parameter is the sampling rate $\omega_s = 2\pi/T_s$ with units radians/s. The time-domain representation of the process of impulse-train sampling is shown in Figure 4.34.

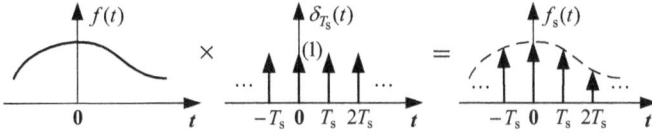

Fig. 4.34: Time-domain process of impulse-train sampling

4.6.2 CTFT of the sampled signal

In impulse-train sampling, the resulting sampled signal is given by:

$$f_s(t) = f(t) \sum_{n=-\infty}^{\infty} \delta(t - nT_s) = \sum_{n=-\infty}^{\infty} f(nT_s)\delta(t - nT_s) \tag{4.89}$$

Calculating the CTFT of Equation (4.89), the CTFT $F_s(j\omega)$ of the sampled signal $f_s(t)$ is given by:

$$F_s(j\omega) = \frac{1}{2\pi}F(j\omega) * \omega_s\delta_{\omega_s}(\omega) = \frac{1}{2\pi}F(j\omega) * \omega_s \sum_{n=-\infty}^{\infty} \delta(\omega - n\omega_s)$$

$$= \frac{1}{T_s} \sum_{n=-\infty}^{\infty} F[j(\omega - n\omega_s)] , \tag{4.90}$$

(**Note:** What is the CTFT of the impulse train $\delta_{T_s}(t)$?)

where $*$ denotes the convolution operator. In deriving Equation (4.90), we used the following CTFT pair:

$$s(t) = \delta_{T_s}(t) = \sum_{n=-\infty}^{\infty} \delta(t - nT_s) \leftrightarrow \omega_s\delta_{\omega_s}(\omega) = \omega_s \sum_{n=-\infty}^{\infty} \delta(\omega - n\omega_s) \tag{4.91}$$

Figure 4.35 illustrates the frequency-domain interpretation of Equation (4.90). The spectrum of the original signal $f(t)$ is assumed to be an arbitrary triangular waveform within the frequency range $-\omega_m \le \omega \le \omega_m$ and is shown in Figure 4.35 (a). The spectrum $F_s(j\omega)$ of the sampled signal $f_s(t)$ is plotted in Figures 4.35 (c) and (e) for the following cases:

case I $\quad \omega_s \ge 2\omega_m$;

case II $\quad \omega_s < 2\omega_m$.

When the sampling rate $\omega_s \ge 2\omega_m$, no overlap exists between consecutive replicas in $F_s(j\omega)$. However, as the sampling rate ω_s is decreased such that $\omega_s < 2\omega_m$, adjacent replicas overlap with each other. The overlapping of replicas is referred to as aliasing, which distorts the spectrum of the original signal $f(t)$ such that $f(t)$ cannot be reconstructed from its samples. To prevent aliasing, the sampling rate must satisfy the criterion $\omega_s \ge 2\omega_m$. This condition is referred to as the sampling theorem and is stated in the following.

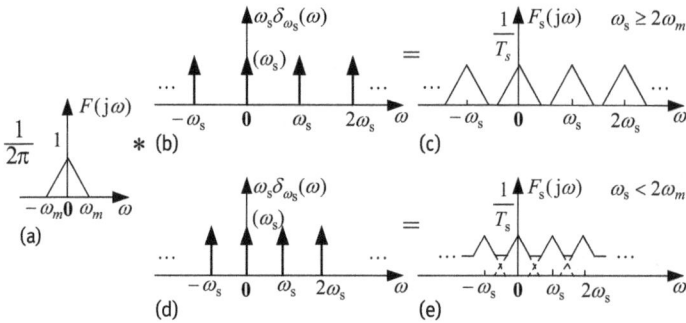

Fig. 4.35: Frequency-domain illustration of impulse-train sampling

4.6.3 Sampling theorem

A CT signal $f(t)$, band-limited to ω_m, can be reconstructed accurately from its samples $f(nT_s)$ if the sampling rate ω_s, satisfies the following condition:

$$\omega_s \geq 2\omega_m \tag{4.92}$$

The minimum sampling rate $\omega_s = 2\omega_m$ required for perfect reconstruction of the original band-limited signal is referred to as the *Nyquist rate*.

Note: Why is the CT signal band limited?

4.6.4 Reconstruction of a band-limited signal from its samples

In the condition of $\omega_s \geq 2\omega_m$, the reconstruction is accomplished by applying the sampled signal $f_s(t)$ to the input of an ideal low-pass filter (LPF) with the following transfer function:

$$H(j\omega) = \begin{cases} T_s, & |\omega| < \omega_c \\ 0, & |\omega| > \omega_c, \end{cases} \tag{4.93}$$

where $\omega_m < \omega_c < \omega_s - \omega_m$. The CTFT $Y(j\omega)$ of the output $y(t)$ of the LPF is given by $Y(j\omega) = F_s(j\omega)H(j\omega)$, and, therefore, all shifted replicas at frequencies $\omega > \omega_c$ are eliminated. All frequency components within the passband $\omega \leq \omega_c$ of the LPF are amplified by a factor of T_s to compensate for the attenuation of $1/T_s$ introduced during sampling. The process of reconstructing $f(t)$ from its samples in the frequency domain is illustrated in Figure 4.36. We now proceed to analyze the reconstruction process in the time domain.

The transfer function of the LPF in Equation (4.93) is expressed as follows:

$$H(j\omega) = T_s \cdot g_{2\omega_c}(\omega). \tag{4.94}$$

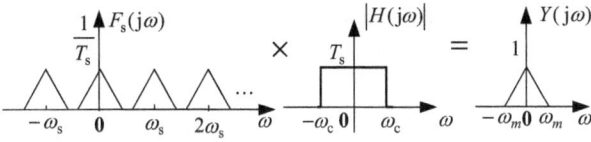

Fig. 4.36: Reconstruction of the original baseband signal in the frequency domain

Calculating the inverse CTFT of $H(j\omega)$ yields:

$$H(j\omega) = T_s \cdot g_{2\omega_c}(\omega) \leftrightarrow h(t) = T_s\frac{\omega_c}{\pi} \cdot \text{Sa}(\omega_c t) \tag{4.95}$$

Note: Readers can prove it. !

For convenience, substituting $\omega_c = 0.5\omega_s$ in the above equation yields:

$$h(t) = \text{Sa}\left(\frac{\omega_s t}{2}\right) \tag{4.96}$$

Convolving the impulse response $h(t)$ with the sampled signal $f_s(t)$ shown in Equation (4.89) yields:

$$y(t) = f_s(t) * h(t) = \sum_{n=-\infty}^{\infty} f(nT_s)\delta(t - nT_s) * \text{Sa}\left(\frac{\omega_s t}{2}\right), \tag{4.97}$$

which reduces to:

$$y(t) = \sum_{n=-\infty}^{\infty} f(nT_s)\,\text{Sa}\left[\frac{\omega_s}{2}(t - nT_s)\right]. \tag{4.98}$$

The above equation indicates that the original CT signal $f(t)$ is reconstructed by adding a series of time-shifted Sa functions, whose amplitudes are scaled according to the values of the samples at the center location of the Sa functions. The time-domain interpretation of the reconstruction of the original band-limited signal $f(t)$ is illustrated in Figure 4.37. At $t = nT_s$, only the n-th Sa function, with amplitude $f(nT_s)$, is nonzero. The remaining Sa functions are all zero. The value of the reconstructed signal at $t = nT_s$ is, therefore, given by $f(nT_s)$. In other words, the values of the reconstructed

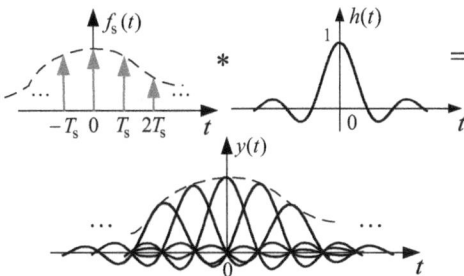

Fig. 4.37: Reconstruction of the original signal in the time domain

signal at the sampling instants are given by the respective samples. The values in between two samples are interpolated using a linear combination of the time-shifted Sa
functions.

Example 4.6.1. The maximum frequency of band-limited signals $f_1(t)$ and $f_2(t)$ are
ω_{m1} and ω_{m2}, respectively. Calculate the Nyquist frequency ω_s of each of the following
signals:

$$(1)\ f_1(\alpha t)\,,\quad \alpha \neq 0\,; \qquad (2)\ f_1(t) + f_2(t)\,;$$
$$(3)\ f_1(t) * f_2(t)\,; \qquad (4)\ f_1(t)f_2(t)\,;$$

Solution:

(1) Using the scaling property of the CTFT:

$$f_1(\alpha t)\,,\quad \alpha \neq 0 \leftrightarrow \frac{1}{|\alpha|}F_1\left(j\frac{\omega}{\alpha}\right).$$

The maximum frequency of band-limited signals $f_1(t)$ is ω_{m1}, and we obtain:

$$\left|\frac{\omega}{\alpha}\right| \leq \omega_{m1} \rightarrow \omega_m = |\alpha|\,\omega_{m1}$$

Thus, the Nyquist frequency is:

$$\omega_s = 2\omega_m = 2\,|\alpha|\,\omega_{m1}\,.$$

(2) Using the linearity of the CTFT,

$$f_1(t) + f_2(t) \leftrightarrow F_1(j\omega) + F_2(j\omega)\,.$$

Its maximum frequency is given by:

$$\omega_m = \max\{\omega_{m1}, \omega_{m2}\}\,.$$

Thus the Nyquist frequency is:

$$\omega_s = 2\omega_m = 2\max\{\omega_{m1}, \omega_{m2}\}$$

(3) Using the time convolution property of the CTFT:

$$f_1(t) * f_2(t) \leftrightarrow F_1(j\omega)F_2(j\omega)$$

Its maximum frequency is given by:

$$\omega_m = \min\{\omega_{m1}, \omega_{m2}\}\,.$$

Thus, the Nyquist frequency is:

$$\omega_s = 2\omega_m = 2\min\{\omega_{m1}, \omega_{m2}\}$$

(4) Using the frequency convolution property of the CTFT:

$$f_1(t)f_2(t) \leftrightarrow \frac{1}{2\pi} F_1(j\omega) * F_2(j\omega) . \qquad (4.99)$$

Its maximum frequency is given by:

$$\omega_m = \omega_{m1} + \omega_{m2} .$$

Thus, the Nyquist frequency is:

$$\omega_s = 2\omega_m = 2(\omega_{m1} + \omega_{m2}) . \qquad (4.100)$$

Note: To determine the Nyquist frequency, we should first compute the maximum frequency of each CT signal.

4.6.5 Sampling with MATLAB

Example 4.6.2. The CTFT of the signal $f(t) = \mathrm{Sa}(t)$ is

$$F(j\omega) = \begin{cases} \pi, & |\omega| \leq 1 \\ 0, & |\omega| > 1 \end{cases}$$

with the band limited to $B = 1$. The Nyquist sampling frequency is $\omega_s = 2B$, and the cut-off frequency of the LPF is $\omega_c = B$. Use MATLAB to simulate the sampling and reconstruction process.

Solution:

```
B=1;                    % maximum frequency of the original signal
wc=B;                   % cut-off frequency of the LPF
Ts=pi/B;                % sampling interval
ws=2*pi/Ts              % sampling rate
N=100;                  % sampling points of the filter in the time domain
n=-N:N;
nTs=n.*Ts;              % sampling instants
fs=sinc(nTs/pi);        % sampling value of the function
Dt=0.005;               % sampling interval to recover the signal
t=-15:Dt:15;            % range of the signal to be recovered
fa=fs*Ts*wc/pi*sinc((wc/pi)*(ones(length(nTs),1)*t-nTs'*ones(1,length
    (t)))); % reconstruct the signal
error=abs(fa-sinc(t/pi));  % normalized error between the recon-
                           % structed signal and the original signal
```

The sampled signal, the reconstructed signal and the corresponding error are plotted in Figure 4.38. It can be seen from Figure 4.38 (c) that the error between the recon-

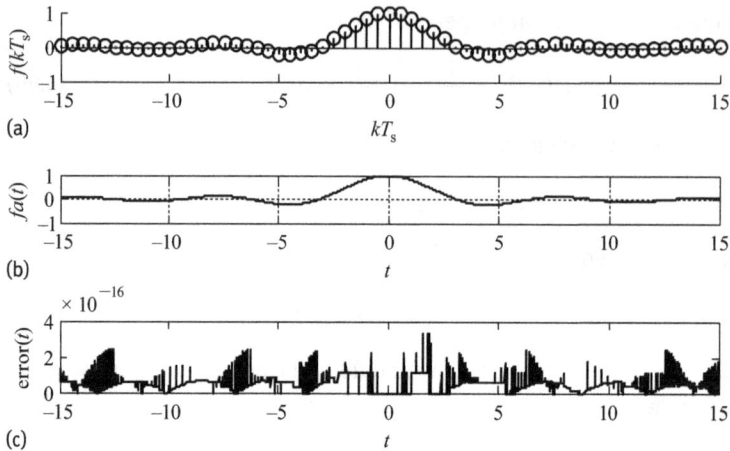

Fig. 4.38: Sampling and reconstruction of the signal; (a) the sampled DT signal, (b) the reconstructed signal, (c) the error between the original and the reconstructed signal

structed signal and the original signal is rather small. This example indicates that a CT signal can be reconstructed from the DT signal sampled with the Nyquist frequency.

! **Note:** The error comes from the numerical computation.

4.7 Summary

This chapter mainly introduced frequency-domain analysis methods for LTIC systems and some applications. The elementary signals $\cos(n\Omega t+\varphi_n)$, $e^{jn\Omega t}$ and $e^{j\omega t}$ were used to obtain CTFS and CTFT in Section 4.1 and 4.3, respectively. Section 4.4 introduced the transfer function $H(j\omega) = Y(j\omega)/F(j\omega)$. For an arbitrary input signal, the zero-state response can be calculated by CTFT using the equation $y(t) = \mathcal{F}^{-1}[F(j\omega)H(j\omega)]$. In Section 4.5, the requirement of the undistorted transmission sysytem was given as $H(j\omega) = Ke^{-j\omega t_d}$. Section 4.6 gave the sampling theorem to convert a CT signal to a DT signal and the Nyquist rate is $\omega_s = 2\omega_m$.

Chapter 4 problems

4.1 Using symmetry and harmonic characteristics, judge the frequency components contained in the CTFS of the plotted signal.

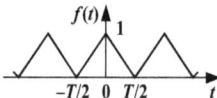

Fig. P4.1: Periodic signal in Problem 4.1

4.2 The frequency spectra of the periodic signal $f(t)$ are shown in Figure P4.2. Determine the trigonometric CTFS expression.

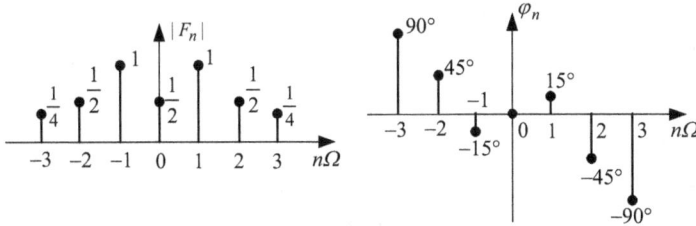

Fig. P4.2: Spectra in Problem 4.2

4.3 The first one-fourth period of the periodic function $f(t)$ is shown in Figure P4.3. Plot the entire wave form of $f(t)$ in a period $(0 < t < T)$ according to the following requirements:

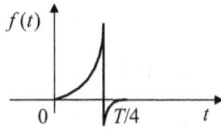

Fig. P4.3: Waveform in Problem 4.3

(1) $f(t)$ is even, contains only even harmonics;
(2) $f(t)$ is even, contains only odd harmonics;
(3) $f(t)$ is odd, contains only even harmonics;
(4) $f(t)$ is odd, contains only odd harmonics;

4.4 For each of the following CT functions, calculate the expression for the CTFT:

(1) $\dfrac{\sin t \cdot \sin 2t}{t^2}$;

(2) $g_{2\pi}(t) \cdot \cos 5t$;

(3) $e^{-(2+2t)}\delta(t)$;

(4) $\operatorname{Sgn}(t) \cdot g_2(t)$.

4.5 The waveform of $f(t)$ is plotted in Figure P4.4. Determine $F(0)$ and the integral $\int_{-\infty}^{\infty} F(j\omega)\,d\omega$.

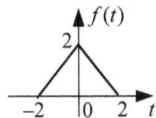

Fig. P4.4: Waveform in Problem 4.5

4.6 Given:

$$f(t) = 2 \cos 997t \cdot \frac{\sin 5t}{\pi} \quad \text{and} \quad h(t) = 2 \cos 1000t \cdot \frac{\sin 4t}{\pi t},$$

calculate $f(t) * h(t)$ using CTFT.

4.7 Determine the CTFT of the signal shown in Figure P4.5.

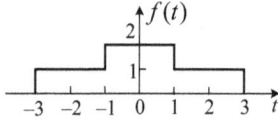

Fig. P4.5: Waveform in Problem 4.7

4.8 Determine the CTFT of each of the following signals:

(1) $f_1(t) = \dfrac{\sin 2\pi(t-2)}{\pi(t-2)}$; (2) $f_2(t) = \dfrac{2\alpha}{\alpha^2 + t^2}$; (3) $f_3(t) = \left(\dfrac{\sin 2\pi t}{2\pi t}\right)^2$.

4.9 Determine the inverse CTFT $f(t)$ of the following CTFTs:

(1) $F(j\omega) = \delta(\omega - \omega_0)$

(2) $F(j\omega) = \delta(\omega + \omega_0) - \delta(\omega - \omega_0)$

(3) $F(j\omega) = \varepsilon(\omega + \omega_0) - \varepsilon(\omega - \omega_0)$

(4) $F(j\omega) = \begin{cases} \dfrac{\omega_0}{\pi}, & \omega > \omega_0 \\ 0, & \omega > \omega_0 \end{cases}$

4.10 Given the pair $F[f(t)] = F(j\omega)$, derive the CTFT for the following set of functions:

(1) $tf(2t)$ (2) $(t-2)f(t)$ (3) $t\dfrac{df(t)}{dt}$

(4) $f(1-t)$ (5) $(1-t)f(1-t)$ (6) $f(2t-5)$

4.11 The transfer function is $H(j\omega) = (2 - j\omega)/(2 + j\omega)$. Calculate the resulting output with the input: $f(t) = \cos(2t)$.

4.12 Suppose the CT signal $f(t) = \varepsilon(t-1)$ is applied as input to a causal LTIC system modeled by the impulse response $h(t) = e^{-t}\varepsilon(t+1)$. Calculate the resulting output $y_{zs}(t)$.

4.13 The gain and phase responses for the LTIC system are given in Figure P4.6. Calculate the resulting output with the input $f(t) = 1 + 2 \cos(4t) + 5 \cos(8t)$.

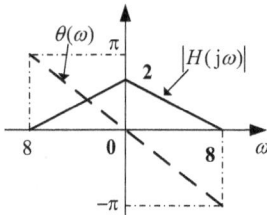

Fig. P4.6: Gain and phase responses in Problem 4.13

4.14 The relation between the input $f(t)$ and the zero-stare response $y_{zs}(t)$ is given as follows:

$$y_{zs}(t) = \frac{1}{\pi} \int_{-\infty}^{\infty} \frac{f(\tau)}{t-\tau} \, d\tau$$

(1) Determine the impulse response $h(t)$ and the transfer function $H(j\omega)$.

(2) Proof that the energy of $y_{zs}(t)$ equals that of $f(t)$.

4.15 The transfer function of the LTIC system is given by $H(j\omega) = e^{-j4\omega} g_4(\omega)$. Judge the distortion with the input $f(t) = \sin(t) + \sin(3t)$.

4.16 Calculate the ideal Nyquist sampling rate and interval of the following CT signals:

(1) $Sa(100t)$

(2) $Sa^2(100t)$

(3) $Sa(100t) + 3\,Sa^2(60t)$

(4) $Sa(50t)\,Sa(100t)$

4.17 The AM system diagram is shown in Figure P4.7. The input signal is $f(t) = \sin t/(\pi t)\cos(1000t)$ and the sinusoidal carrier is $s(t) = \cos(1000t)$. Determine the output response $y(t)$.

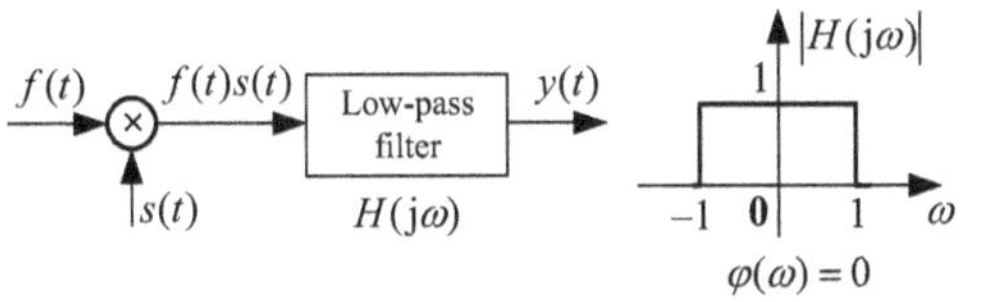

Fig. P4.7: The diagram of the AM system and the spectra of the LPF

4.18 MATLAB practice: The signal is formed according to the following program:

```
t=-2*pi:0.001:2*pi;
y=sawtooth(0.5*t,1);
plot(t,y)
```

Write the time-domain function of the above signal formed by periodical extension. Calculate the exponential CTFS coefficients (up to the 11-th harmonic), plot the accumulative waveform of the first 11 harmonics and point out where the Gibbs phenomenon appears.

5 Laplace transform and complex frequency-domain analysis

Please focus on the following key questions.
1. Why is the analysis of LTIC systems in the complex frequency domain introduced?
2. What is the elementary signal and response of complex frequency-domain analysis?
3. What is the decomposition equation of signals in complex frequency domain? What is the method in the complex frequency domain to analyze LTIC systems?
4. How can we apply the Mason formula to compute the system transfer function of the signal flow graph?

5.0 Introduction

In Chapter 4, the continuous-time Fourier series (CTFS) for periodic signals and the continuous-time Fourier transform (CTFT) were introduced to determine the output of an LTIC system. However, the CTFT is not defined for all aperiodic signals. In this chapter, an alternative analysis method based on the Laplace transform is introduced to analyze the LTIC systems. The CTFT expands an aperiodic signal as a linear combination of the complex exponential functions $e^{j\omega t}$, which are referred to as its basis functions. The Laplace transforms uses e^{st} as the basis function, where the independent Laplace variable $s = \sigma + j\omega$ is complex. The corresponding analysis method is called the complex frequency-domain analysis or S-domain analysis.

In this chapter, we first introduce the bilateral and unilateral Laplace transform definition. The properties of the Laplace transform and the inverse Laplace transform are proposed. Section 5.4 gives applications of the Laplace transform in a circuit system. Section 5.5 applies the Laplace transform to solve the differential equation. The transfer function and the stability analysis of the LTIC system in the S-domain is presented in Section 5.6. The signal flow graph and system simulation are introduced in Section 5.7. Finally, the chapter is concluded in Section 5.8.

5.1 Analytical development

5.1.1 From CTFT to the bilateral laplace transform

The CTFT is not defined for all aperiodic signals, such as e^{2t}, which does not satisfy the absolute integrable condition. For this reason, the signal is multiplied with an

https://doi.org/10.1515/9783110593907-005

attenuation factor $e^{-\sigma t}$, $\sigma \in R$. The expression for the bilateral Laplace transform is derived by considering the CTFT of the modified version $f(t)e^{-\sigma t}$:

$$F_b(\sigma + j\omega) = \mathcal{F}[f(t)e^{-\sigma t}] = \int_{-\infty}^{\infty} f(t)e^{-\sigma t}e^{-j\omega t}\,dt = \int_{-\infty}^{\infty} f(t)e^{-(\sigma+j\omega)t}\,dt \tag{5.1}$$

The corresponding inverse CTFT is as follows:

$$f(t)e^{-\sigma t} = \frac{1}{2\pi}\int_{-\infty}^{\infty} F_b(\sigma + j\omega)e^{j\omega t}\,d\omega$$

$$f(t) = \frac{1}{2\pi}\int_{-\infty}^{\infty} F_b(\sigma + j\omega)e^{(\sigma+j\omega)t}\,d\omega \tag{5.2}$$

We substitute $s = \sigma + j\omega$ and $d\omega = ds/j$ in Equations (5.1) and (5.2) to obtain the following definition:

The *Laplace analysis equation*:

$$F_b(s) = \mathcal{L}[f(t)] = \int_{-\infty}^{\infty} f(t)e^{-st}\,dt \tag{5.3}$$

The *Laplace synthesis equation*:

$$f(t) = \mathcal{L}^{-1}[F(s)] = \frac{1}{2\pi j}\int_{\sigma-j\infty}^{\sigma+j\infty} F_b(s)e^{st}\,ds \tag{5.4}$$

The above two equations form the bilateral Laplace transforms pair, which is denoted by:

$$f(t) \leftrightarrow F_b(s)\,.$$

5.1.2 Region of convergence

From the definition of the bilateral Laplace transform in Equation (5.3), the range of values of $\text{Re}(s) = \sigma$ to make the bilateral Laplace transform exist is referred to as the region of convergence(ROC). Therefore, the ROC is given by finding the appropriate area of σ in the following equation:

$$\lim_{t\to\infty} |f(t)|e^{-\sigma t} = 0 \tag{5.5}$$

To illustrate the different cases of ROC in computing the bilateral Laplace transform, we consider the following examples.

Example 5.1.1. Calculate the bilateral Laplace transform of the causal exponential function:

$$f_1(t) = e^{at}\varepsilon(t)\,.$$

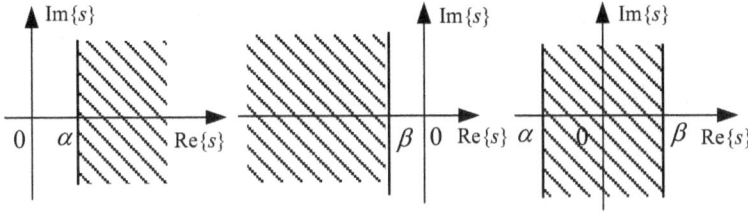

Fig. 5.1: Different cases of ROC

Solution:

$$F_{1b}(s) = \int_{-\infty}^{\infty} e^{\alpha t} \varepsilon(t) e^{-st} \, dt = \int_{0}^{\infty} e^{\alpha t} e^{-st} \, dt = \frac{e^{-(s-\alpha)t}}{-(s-\alpha)} \Big|_{0}^{\infty}$$

$$= \frac{1}{(s-\alpha)} \left[1 - \lim_{t \to \infty} e^{-(\sigma-\alpha)t} e^{-j\omega t} \right]$$

At the lower limit, $t \to 0$, $e^{-(s-\alpha)t} = 1$. At the upper limit, $t \to \infty$, $e^{-(s-\alpha)t} = 0$ if $\text{Re}(s-\alpha) > 0$ or $\text{Re}(s) > \alpha$. If $\text{Re}(s) \le \alpha$, then the value of $e^{-(s-\alpha)t}$ is infinite at the upper limit $t \to \infty$. Therefore,

$$F_{1b}(s) = \begin{cases} \frac{1}{s-\alpha}, & \text{Re}[s] = \sigma > \alpha \\ \text{undefined}, & \sigma \le \alpha \end{cases}$$

The ROC is given by $\text{Re}(s) > \alpha$ for the bilateral Laplace transform of the exponential function $f_1(t) = e^{\alpha t} \varepsilon(t)$. Figure 5.1 (a) highlights the ROC by shading the appropriate area in the complex S-plane. For $\alpha > 0$, the exponential function $f_1(t) = e^{\alpha t} \varepsilon(t)$ is not absolutely integrable, and hence its CTFT does not exist. This is an important distinction between the CTFT and the bilateral Laplace transform.

Note: What is the ROC of a causal function?

Example 5.1.2. Calculate the bilateral Laplace transform of the noncausal exponential function:

$$f_2(t) = e^{\beta t} \varepsilon(-t) .$$

Solution:

$$F_{2b}(s) = \int_{-\infty}^{0} e^{\beta t} e^{-st} \, dt = \frac{e^{-(s-\beta)t}}{-(s-\beta)} \Big|_{-\infty}^{0} = \frac{1}{-(s-\beta)} \left[1 - \lim_{t \to -\infty} e^{-(\sigma-\beta)t} e^{-j\omega t} \right]$$

At the upper limit, $t \to 0$, $e^{-(s-\beta)t} = 1$. At the lower limit, $t \to -\infty$, $e^{-(s-\beta)t} = 0$ only if $\text{Re}(s-\beta) < 0$ or $\text{Re}(s) < \beta$. Therefore, the bilateral Laplace transform is given by:

$$F_{2b}(s) = \begin{cases} \frac{1}{-(s-\beta)}, & \text{Re}[s] = \sigma < \beta \\ \text{undefined}, & \sigma \ge \alpha . \end{cases}$$

Figure 5.1 (b) illustrates the ROC, $\text{Re}(s) < \beta$, for the bilateral Laplace transform of $f_2(t) = e^{\beta t}\varepsilon(-t)$.

! **Note:** What is the ROC of a noncausal function?

Example 5.1.3. Calculate the bilateral Laplace transform of the bilateral exponential function:

$$f_3(t) = f_1(t) + f_2(t) = \begin{cases} e^{\beta t}, & t < 0 \\ e^{\alpha t}, & t > 0 \end{cases}, \quad \beta > \alpha.$$

Solution: From the conclusions of Examples 5.1.1 and 5.1.2, we know that:

$$F_{3b}(s) = F_{1b}(s) + F_{2b}(s) = \frac{1}{s - \alpha} - \frac{1}{s - \beta}, \quad \alpha < \text{Re}[s] = \sigma < \beta.$$

Figure 5.1(c) illustrates the ROC, $\alpha < \text{Re}(s) < \beta$, for the bilateral Laplace transform of $f_3(t)$.

! **Note:** What is the ROC of a bilateral function (if it exists)? What is the conclusion if $\alpha > \beta$?

Example 5.1.4. Calculate the bilateral Laplace transform of the following functions:

$$f_1(t) = e^{-3t}\varepsilon(t) + e^{-2t}\varepsilon(t)$$
$$f_2(t) = -e^{-3t}\varepsilon(-t) - e^{-2t}\varepsilon(-t)$$
$$f_3(t) = e^{-3t}\varepsilon(t) - e^{-2t}\varepsilon(-t)$$

Solution:

$$f_1(t) \leftrightarrow F_1(s) = \frac{1}{s+3} + \frac{1}{s+2}, \quad \text{Re}[s] = \sigma > -2 \tag{5.6}$$

$$f_2(t) \leftrightarrow F_2(s) = \frac{1}{s+3} + \frac{1}{s+2}, \quad \text{Re}[s] = \sigma < -3 \tag{5.7}$$

$$f_3(t) \leftrightarrow F_3(s) = \frac{1}{s+3} + \frac{1}{s+2}, \quad -3 < \sigma < -2 \tag{5.8}$$

It can be seen that the three different functions have the same bilateral Laplace transforms but with different ROCs. By specifying the ROC with the bilateral Laplace transform expression, we can completely determine the original function in the time domain.

5.1.3 Unilateral Laplace transform

In signal processing, most physical systems and signals are causal. The Laplace transform for causal signals is referred to as the unilateral Laplace transform.

Definition. The *unilateral Laplace transform* of the signal $f(t)$ is defined as follows:

$$F(s) = \int_{0-}^{\infty} f(t)e^{-st}\,dt\,,\qquad(5.9)$$

where the initial conditions of the system are incorporated by the lower limit $t = 0_-$. In this book, we will mostly use the unilateral Laplace transform. In subsequent discussion, the Laplace transform implies the unilateral Laplace transform. If we want to use the bilateral Laplace transform, the term "bilateral" will be explicitly stated.

Note: For causal signals, the unilateral and bilateral Laplace transforms are the same.　❗

5.1.4 Relationship between CTFT and the Laplace transform

The definitions of CTFT and unilateral Laplace transform are given, respectively, as follows:

$$F(s) = \int_{0_-}^{\infty} f(t)e^{-st}\,dt\,,\quad \mathrm{Re}[s] > \sigma_0 \qquad(5.10)$$

$$F(j\omega) = \int_{-\infty}^{\infty} f(t)e^{-j\omega t}\,dt \qquad(5.11)$$

The CTFT is expressed as $F(j\omega)$ to emphasize that the CTFT is computed on the imaginary $j\omega$-axis. The Laplace transform is expressed as $F(s)$ in the S-plane, where $s = \sigma + j\omega$. The CTFT is used for the frequency-domain analysis, while the Laplace transform is for the complex frequency-domain analysis. The basic function changes from $e^{j\omega t}$ of CTFT to e^{st} of the bilateral Laplace transform. The CTFT is a special case of the Laplace transforms when $s = j\omega$, i.e., $\sigma = 0$.

To discuss their relationship, the signal must be causal. According to the value range of σ_0, the relationship is detailed as follows:

(1) $$\sigma_0 < 0$$

The $j\omega$-axis is included within the ROC of the Laplace transform. Therefore, the CTFT can be obtained from the Laplace transform by substituting $s = j\omega$, i.e.,

$$F(j\omega) = F(s)|_{s=j\omega}\,. \qquad(5.12)$$

For example,
$$f(t) = e^{-2t}\varepsilon(t) \leftrightarrow F(s) = 1/(s+2)\,,\quad \sigma > -2$$

Then,
$$f(t) = e^{-2t}\varepsilon(t) \leftrightarrow F(j\omega) = 1/(j\omega + 2)$$

(2) $$\sigma_0 = 0$$

The boundary of ROC is the $j\omega$-axis. Then, the CTFT is obtained by:

$$F(j\omega) = \lim_{\sigma \to 0} F(s) . \tag{5.13}$$

For example,

$$f(t) = \varepsilon(t) \leftrightarrow F(s) = 1/s$$

Then,

$$F(j\omega) = \lim_{\sigma \to 0} \frac{1}{\sigma + j\omega} = \lim_{\sigma \to 0} \frac{\sigma}{\sigma^2 + \omega^2} + \lim_{\sigma \to 0} \frac{-j\omega}{\sigma^2 + \omega^2} = \pi\delta(\omega) + \frac{1}{j\omega}$$

(3) $\sigma_0 > 0$

The ROC does not contain the $j\omega$-axis; the substitutions $s = j\omega$ cannot be made, and the CTFT does not exist.

For example,

$$f(t) = e^{2t}\varepsilon(t) \leftrightarrow F(s) = 1/(s-2) , \quad \sigma > 2$$

Then, the CTFT does not exist.

5.2 Basic pairs and properties of the Laplace transform

5.2.1 Laplace transform pairs for several elementary CT signals

Table 5.1 lists the Laplace transforms for elementary CT signals, which are useful for complex-frequency analysis. Almost all of the listed transforms can be calculated by definition. Readers can refer to other references and try to prove it.

The Laplace transform of the periodic signal $f_T(t)$ is deduced as follows:

The illustration of a causal periodic signal $f_T(t)$ is given in Figure 5.2. Using the definition of the Laplace transform, we obtain:

$$F(s) = \int_0^\infty f_T(t)e^{-st} dt$$

$$= \int_0^T f_T(t)e^{-st} dt + \int_T^{2T} f_T(t)e^{-st} dt + \cdots$$

$$= \sum_{n=0}^\infty \int_{nT}^{(n+1)T} f_T(t)e^{-st} dt \tag{5.14}$$

Substituting $t = t + nT$:

$$F(s) = \sum_{n=0}^\infty e^{-nsT} \int_0^T f_T(t)e^{-st} dt = \frac{1}{1 - e^{-sT}} \int_0^T f_T(t)e^{-st} dt \tag{5.15}$$

Tab. 5.1: Laplace transforms for elementary signals

	$f(t)(t > 0)$	$F(s)$
1	$\delta(t)$	1
2	$\varepsilon(t)$	$\dfrac{1}{s}$
3	$t\varepsilon(t)$	$\dfrac{1}{s^2}$
4	$t^n \varepsilon(t)$	$\dfrac{n!}{s^{n+1}}$
5	$e^{s_0 t}$	$\dfrac{1}{s - s_0}$
6	$\cos(\omega_0 t)$	$\dfrac{s}{s^2 + \omega_0^2}$
7	$\sin(\omega_0 t)$	$\dfrac{\omega_0}{s^2 + \omega_0^2}$
8	Period signal $f_T(t)$, $f_0(t)$ is the aperiodic signal within the range $0 \le t \le T$	$\dfrac{F_0(s)}{1 - e^{-sT}}$, $F_0(s)$ is the Laplace transform of $f_0(t)$

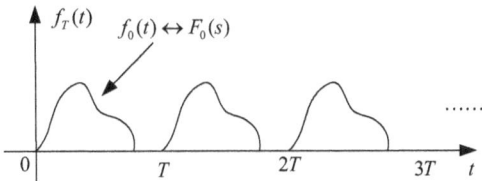

Fig. 5.2: Illustration of a causal periodic signal

The function in the integral interval $0 \sim T$ of $f_T(t)$ is referred to as $f_0(t)$. The Laplace transform of $f_0(t)$ is $F_0(s)$. Equation (5.15) reduces to:

$$f_T(t) \leftrightarrow \frac{F_0(s)}{1 - e^{-sT}} \tag{5.16}$$

Special case:

$$\delta_T(t) \leftrightarrow \frac{1}{1 - e^{-sT}} \tag{5.17}$$

5.2.2 Properties of the Laplace transform

In this section, we present the properties of the Laplace transform based on the transformations of the signals. These properties are similar to the properties of the CTFT covered in Section 4.4.3. A list of the Laplace transform properties is included in Table 5.2.

Tab. 5.2: Properties of the Laplace transform

Properties	$f(t)(t > 0)$	$F(s)$
Linearity	$af_1(t) + bf_2(t)$	$aF_1(s) + bF_2(s)$
Time scaling	$f(at), \ a \in R, \ a > 0$	$\dfrac{1}{a}F\left(\dfrac{s}{a}\right)$
Time shifting	$f(t - t_0)\varepsilon(t - t_0), \ t_0 > 0$	$e^{-st_0}F(s)$
S-domain shifting	$f(t)e^{s_a t}$	$F(s - s_a)$
Time differentiation	$f'(t)$	$sF(s) - f(0_-)$
	$f''(t)$	$s[sF(s) - f(0_-)] - f'(0_-)$
	$f^{(n)}(t)$ of casual signal $f(t)$	$s^n F(s)$
Time integration	$f^{(-1)}(t) = \int_{-\infty}^{t} f(x)\,dx$	$s^{-1}F(s) + s^{-1}f^{(-1)}(0_-)$
	$\left(\int_{0_-}^{t}\right)^n f(x)\,dx$	$\dfrac{1}{s^n}F(s)$
S-domain differentiation	$(-t)f(t)$	$\dfrac{dF(s)}{ds}$
S-domain integration	$\dfrac{f(t)}{t}$	$\int_{s}^{\infty} F(\eta)\,d\eta$
Time convolution	$f_1(t) * f_2(t)$	$F_1(s) \cdot F_2(s)$
Initial value	$f(0_+) = \lim_{t \to 0_+} f(t)$	$f(0_+) = \lim_{s \to \infty} sF(s)$
Final value	if $f(\infty) = \lim_{t \to \infty} f(t)$ exists	$f(\infty) = \lim_{s \to 0} sF(s)$

The following examples are applications of the Laplace transform properties.

Example 5.2.1. Given the pair of $f_1(t) \leftrightarrow F_1(s)$, calculate the Laplace transform of $f_2(t)$ plotted in Figure 5.3.

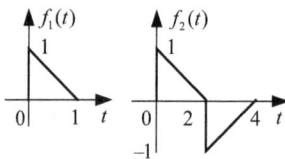

Fig. 5.3: Waveform used in Example 5.2.1

Solution: By inspection, the waveform $f_2(t)$ can be expressed as:

$$f_2(t) = f_1(0.5t) - f_1[0.5(t - 2)]$$

By the time-scaling property:

$$f_1(0.5t) \leftrightarrow 2F_1(2s)$$

By the time-shifting property:

$$f_1[0.5(t-2)] \leftrightarrow 2F_1(2s)e^{-2s}$$

Using the linearity property, the Laplace transform of $f_2(t)$ is given by:

$$f_2(t) \leftrightarrow 2F_1(2s)(1-e^{-2s}).$$

Example 5.2.2. Calculate the Laplace transform of the causal function $f(t)$ shown in Figure 5.4 (a).

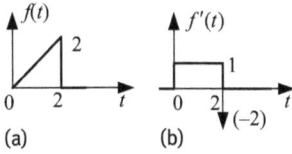

(a) (b)

Fig. 5.4: Waveform used in Example 5.2.2

Solution: Taking the time differentiation of $f(t)$, its waveform is plotted in Figure 5.4 (b). The Laplace transform of $f'(t)$ is given by:

$$f'(t) = \varepsilon(t) - \varepsilon(t-2) - 2\delta(t-2) \leftrightarrow F_1(s) = \frac{1}{s}(1-e^{-2s}) - 2e^{-2s}.$$

Calculating the integration of $f'(t)$ yields:

$$\int_{0-}^{t} f'(x)\,dx = f(t) - f(0_-)$$

As $f(t)$ is a causal signal with $f(0_-) = 0$, the above expression reduces to:

$$f(t) = \int_{0-}^{t} f'(x)\,dx.$$

Using the time-integration property, the Laplace transform of $f(t)$ is given by:

$$f(t) = \int_{0-}^{t} f'(x)\,dx \leftrightarrow F(s) = \frac{F_1(s)}{s} = \frac{1}{s^2}(1-e^{-2s}) - \frac{2}{s}e^{-2s}$$

From this example, we can draw the following conclusion:

When the Laplace transform pair of the derivative function $f^{(n)}(t)$ is known as:

$$f^{(n)}(t) \leftrightarrow F_n(s),$$

the Laplace transform of the original causal function $f(t)$ is:

$$f(t) \leftrightarrow \frac{F_n(s)}{s^n}.$$

Note: Readers can prove it.

Example 5.2.3. Using the Laplace transform pair:

$$\varepsilon(t) \leftrightarrow \frac{1}{s} \, ,$$

calculate the Laplace transform of $f(t) = 5 \cos 2t\varepsilon(t)$.

Solution: Using Euler's formula:

$$\cos(2t) = \frac{1}{2}\left(e^{j2t} + e^{-j2t}\right)$$

Using the frequency-shifting property:

$$e^{j2t}\varepsilon(t) \leftrightarrow \frac{1}{s-2}$$
$$e^{-j2t}\varepsilon(t) \leftrightarrow \frac{1}{s+2}$$

Based on the linearity property:

$$(e^{j2t} + e^{-j2t})\varepsilon(t) \leftrightarrow \frac{1}{s-2} + \frac{1}{s+2} = \frac{s}{s^2+4} \, .$$

Therefore, the Laplace transform of $f(t) = 5 \cos 2t\varepsilon(t)$ is given by:

$$f(t) = \frac{5}{2}\left(e^{j2t} + e^{-j2t}\right)\varepsilon(t) \leftrightarrow F(s) = \frac{5s}{s^2+4} \, .$$

Example 5.2.4. Calculate the initial and the final value of the function $f(t)$. The Laplace transform is specified as $F(s) = s^2/(s^2 + 2s + 2)$.

Solution: Before applying the initial-value theorem, its application conditions in the time domain and the complex frequency domain should be specified first.

In the time domain, the function $f(t)$ does not contain $\delta(t)$ or its derivatives. In the S-domain, the Laplace transform $F(s)$ is a proper fraction. Obviously, $F(s)$ in this example should be rewritten as:

$$F(s) = 1 - \frac{2s+2}{s^2+2s+2} = 1 + F_1(s) \tag{5.18}$$

The original function of 1 is $\delta(t)$, which is zero at $t = 0+$. The initial value of $f(t)$ only depends on $F_1(s)$. Applying the initial-value theorem to $F_1(s)$, we obtain:

$$f(0+) = \lim_{s\to\infty} sF_1(s) = \lim_{s\to\infty} \frac{-2s^2-2s}{s^2+2s+2} = -2$$

Applying the final-value theorem to $F(s)$, we obtain:

$$f(\infty) = \lim_{s\to0} sF(s) = \lim_{s\to0} \frac{s^3}{s^2+2s+2} = 0$$

! **Note:** The final-value theorem is valid if the $j\omega$-axis is included in the ROC.

5.3 Inverse Laplace transformation

5.3.1 Characteristic roots of the Laplace transform

The inverse Laplace transform can be calculated directly by solving the complex integral in the synthesis equation Equation (5.4). This method involves contour integration, which is beyond the scope of this book. In this section, Heaviside's partial fraction expansion is used to calculate the inverse Laplace transform.

Suppose that the function $F(s)$ is a rational proper fraction of s. If not, it should be decomposed into the sum of rational polynomials and rational proper fractions:

$$F(s) = P(s) + \frac{B(s)}{A(s)} \tag{5.19}$$

where $P(s)$ is the rational polynomial and $B(s)/A(s)$ is the rational proper fraction. The inverse Laplace transform of the polynomial $P(s)$ is composed of the impulse function and its derivatives. We now consider the rational proper fraction $B(s)/A(s)$:

$$F(s) = \frac{B(s)}{A(s)} = \frac{b_m s^m + b_{m-1} s^{m-1} + \cdots + b_1 s + b_0}{s^n + a_{n-1} s^{n-1} + \cdots + a_1 s + a_0}, \quad m < n \tag{5.20}$$

Definition. The equation $A(s) = 0$ in Equation (5.21) is referred to as the *characteristic equation*, and its roots are defined as the *characteristic roots*, or the *poles* of $F(s)$:

$$A(s) = s^n + a_{n-1} s^{n-1} + \cdots + a_1 s + a_0 = 0 \tag{5.21}$$

For an n-th-order characteristic equation, there will be n roots. Characteristic roots may be real valued or complex valued. We will further discuss the inverse Laplace transform with different cases of poles.

5.3.2 Real-valued and first-order poles

The n first-order roots of the characteristic equation are real valued. The Laplace transform $F(s)$ is represented as:

$$F(s) = \frac{B(s)}{A(s)} = \frac{B(s)}{(s - p_1)(s - p_2) \cdots (s - p_n)} \tag{5.22}$$

Using Heaviside's partial fraction expansion formula, $F(s)$ is decomposed into a summation of the first-order fractions:

$$F(s) = \frac{B(s)}{A(s)} = \frac{K_1}{s - p_1} + \frac{K_2}{s - p_2} + \cdots + \frac{K_i}{s - p_i} + \cdots + \frac{K_n}{s - p_n}, \tag{5.23}$$

where the coefficients are obtained from:

$$K_i = (s - p_i)F(s)|_{s=p_i} . \tag{5.24}$$

From Table 5.1,

$$e^{p_i t}\varepsilon(t) \leftrightarrow \frac{1}{s - p_i} .$$

Using the above transform pair, the inverse Laplace transform of $F(s)$ is given by:

$$f(t) = \sum_{i=1}^{n} K_i e^{p_i t}\varepsilon(t) . \tag{5.25}$$

Example 5.3.1. Calculate the inverse Laplace transform of

$$F(s) = \frac{s^3 + 5s^2 + 9s + 7}{(s + 1)(s + 2)} .$$

Solution: The function $F(s)$ is decomposed into the sum of rational polynomials and rational proper fraction.

$$F(s) = s + 2 + \frac{s + 3}{(s + 1)(s + 2)} = s + 2 + F_1(s)$$

Using the partial fraction expansion, the function $F_1(s)$ is expressed as

$$F_1(s) = \frac{s + 3}{(s + 1)(s + 2)} = \frac{k_1}{s + 1} + \frac{k_2}{s + 2} .$$

The coefficients are obtained from

$$k_1 = (s + 1) \cdot \frac{s + 3}{(s + 1)(s + 2)}\bigg|_{s=-1} = 2 , \quad k_2 = (s + 2) \cdot \frac{s + 3}{(s + 1)(s + 2)}\bigg|_{s=-2} = -1 .$$

The partial fraction expansion of the Laplace transform $F(s)$ is therefore given by:

$$F(s) = s + 2 + \frac{2}{s + 1} - \frac{1}{s + 2} .$$

Using the pair of elementary signals, the inverse Laplace transform is:

$$f(t) = \delta'(t) + 2\delta(t) + (2e^{-t} - e^{-2t})\varepsilon(t) .$$

5.3.3 Complex-valued and first-order poles

The characteristic equation of $F(s) = B(s)/A(s)$ is given by:

$$A(s) = (s + \alpha)^2 + \beta^2 = 0 ,$$

which has complex roots at $s = -\alpha \pm j\beta$. The partial fraction expansion roots of $F(s)$ is, therefore, given by:

$$F(s) = \frac{K_1}{s + \alpha - j\beta} + \frac{K_2}{s + \alpha + j\beta} . \tag{5.26}$$

Note that in this case, there are two complex-conjugate poles at $s = -\alpha \pm j\beta$. Using Heaviside's formula, the residues K_i are given by:

$$K_1 = [(s + \alpha - j\beta)F(s)]|_{s=-\alpha+j\beta} = |K_1|e^{j\theta}, \quad K_2 = K_1^*. \tag{5.27}$$

Here, the coefficients K_i of the complex-valued poles occur in conjugate pairs. The partial fraction expansion of the Laplace transform $F(s)$ is, therefore, given by:

$$F(s) = \frac{|K_1|e^{j\theta}}{s + \alpha - j\beta} + \frac{|K_1|e^{-j\theta}}{s + \alpha + j\beta}. \tag{5.28}$$

Using the pair of $e^{s_0 t}\varepsilon(t) \leftrightarrow 1/(s - s_0)$, the inverse Laplace transform of $F(s)$ is determined by the following deduction:

$$\begin{aligned} f(t) &= \left[|K_1|e^{j\theta}e^{(-\alpha+j\beta)t} + |K_1|e^{-j\theta}e^{(-\alpha-j\beta)t}\right]\varepsilon(t) \\ &= \left[|K_1|e^{j\theta}e^{-\alpha t}e^{j\beta t} + |K_1|e^{-j\theta}e^{-\alpha t}e^{-j\beta t}\right]\varepsilon(t) \\ &= |K_1|e^{-\alpha t}\left[e^{j(\beta t+\theta)} + e^{-j(\beta t+\theta)}\right]\varepsilon(t) \\ &= 2|K_1|e^{-\alpha t}\cos(\beta t + \theta)\varepsilon(t) \end{aligned} \tag{5.29}$$

In addition, the partial fraction expansion of $F(s)$ can be expressed as:

$$F(s) = \frac{A + jB}{s + \alpha - j\beta} + \frac{A - jB}{s + \alpha + j\beta}. \tag{5.30}$$

The inverse Laplace transform of $F(s)$ can be derived as follows:

$$f_1(t) = 2e^{-\alpha t}[A\cos(\beta t) - B\sin(\beta t)]\varepsilon(t) \tag{5.31}$$

Note: Readers can prove it.   !

Example 5.3.2. Calculate the inverse Laplace transform of

$$F(s) = \frac{s^2 + 3}{(s^2 + 2s + 5)(s + 2)}.$$

Solution: Using the partial fraction expansion, the function $F(s)$ is expressed as:

$$F(s) = \frac{s^2 + 3}{(s + 1 - j2)(s + 1 + j2)(s + 2)} = \frac{k_1}{s + 1 - j2} + \frac{k_1^*}{s + 1 + j2} + \frac{k_0}{s + 2} \tag{5.32}$$

The coefficients are given by:

$$k_1 = \left.\frac{s^2 + 3}{(s + 1 + j2)(s + 2)}\right|_{s=-1+j2} = \frac{-1 + j2}{5} \tag{5.33}$$

$$k_0 = \left.\frac{s^2 + 3}{(s + 1 + j2)(s + 1 - j2)}\right|_{s=-2} = \frac{7}{5} \tag{5.34}$$

Substituting the values of the partial fraction coefficients, we obtain:

$$F(s) = \frac{-\frac{1}{5} + j\frac{2}{5}}{s + 1 - j2} + \frac{-\frac{1}{5} - j\frac{2}{5}}{s + 1 + j2} + \frac{7}{5(s + 2)}. \tag{5.35}$$

Using Equation (5.31), the inverse Laplace transform of $F(s)$ is given by:

$$f(t) = \left\{2e^{-t}\left[-\frac{1}{5}\cos(2t) - \frac{2}{5}\sin(2t)\right] + \frac{7}{5}e^{-2t}\right\}\varepsilon(t) \tag{5.36}$$

5.3.4 Real-valued and repeated poles

The above discussion is based on the partial fraction when the poles are not repeated. However, when there are multiple poles at the same location, we cannot directly calculate the coefficients corresponding to the fractions at multiple pole locations. To illustrate the partial fraction expansion for repeated poles, consider a Laplace transform $F(s)$ with r repeated poles at $s = p_1$. The partial fraction expansion of the Laplace transform $F(s)$ can be expressed as follows:

$$F(s) = \frac{B(s)}{A(s)} = \frac{K_{11}}{(s - p_1)^r} + \frac{K_{12}}{(s - p_1)^{r-1}} + \cdots + \frac{K_{1r}}{(s - p_1)} \tag{5.37}$$

The coefficients can be calculated as follows:

$$K_{11} = \left[(s - p_1)^r F(s)\right]\big|_{s=p_1} \tag{5.38}$$

$$K_{12} = \frac{d}{ds}\left[(s - p_1)^r F(s)\right]\Big|_{s=p_1} \tag{5.39}$$

$$K_{1r} = \frac{1}{(r-1)!}\frac{d^{r-1}}{ds^{r-1}}\left[(s - p_1)^r F(s)\right]\Big|_{s=p_1} \tag{5.40}$$

From Table 5.1,

$$t^n \varepsilon(t) \leftrightarrow \frac{n!}{s^{n+1}}.$$

Using the above transform pair and S-domain shifting property, we obtain:

$$\frac{1}{n!}t^n e^{p_1 t}\varepsilon(t) \leftrightarrow \frac{1}{(s - p_1)^{n+1}}. \tag{5.41}$$

Example 5.3.3. Calculate the inverse Laplace transform of

$$F(s) = \frac{s-2}{s(s+1)^3}.$$

Solution: Using the partial fraction expansion, the function $F(s)$ is expressed as:

$$F(s) = \frac{k_{11}}{(s+1)^3} + \frac{k_{12}}{(s+1)^2} + \frac{k_{13}}{(s+1)} + \frac{k_2}{s} \tag{5.42}$$

The partial fraction coefficient k_2 is calculated using the Heaviside formula:

$$k_2 = sF(s)|_{s=0} = \frac{s-2}{(s+1)^3}\Big|_{s=0} = -2 \tag{5.43}$$

For simplicity, we introduce $F_1(s)$ as follows:

$$F_1(s) = (s+1)^3 F(s) = \frac{s-2}{s}. \tag{5.44}$$

The remaining partial fraction coefficients are calculated using Equation (5.40):

$$k_{11} = F_1(s)|_{s=p_1} = \left.\frac{s-2}{s}\right|_{s=-1} = 3$$

$$k_{12} = \left.\frac{d}{ds}F_1(s)\right|_{s=p_1} = \left.\frac{s-(s-2)\cdot 1}{s^2}\right|_{s=-1} = 2$$

$$k_{13} = \left.\frac{1}{2}\frac{d^2}{ds^2}F_1(s)\right|_{s=p_1} = \left.\frac{1}{2}\frac{-4s}{s^4}\right|_{s=-1} = 2$$

Therefore, the partial fraction expansion is given by:

$$F(s) = \frac{3}{(s+1)^3} + \frac{2}{(s+1)^2} + \frac{2}{(s+1)} - \frac{2}{s}. \tag{5.45}$$

Using Equation (5.41), the inverse Laplace transform is given by:

$$f(t) = \left(\frac{3}{2}t^2 e^{-t} + 2te^{-t} + 2e^{-t} - 2\right)\varepsilon(t). \tag{5.46}$$

5.3.5 Calculation with MATLAB

In MATLAB, the tool functions of *laplace* and *ilaplace* can be used to compute the Laplace transforms and the inverse transforms, respectively.

Example 5.3.4. Calculate the convolution of

$$f_1(t) = e^{-t}\varepsilon(t), \quad f_2(t) = te^{-\frac{1}{2}t}\varepsilon(t)$$

using the time-convolution property of the Laplace transform with MATLAB.

Solution:

```
syms t;t=sym('t','positive');   % define t as a variable
fs1=laplace(exp(-t));           % Laplace transform of f_1(t)
fs2=laplace(t*exp(-t/2));       % Laplace transform of f_2(t)
yt=simple(ilaplace(fs1* fs2))   % inverse Laplace transform
                                % of F_1(s)F_2(s)
```

The operation result is:

```
yt =2 * (t-2) * exp (-1 / 2 * t) + 4 * exp (-t)
```

5.4 Application of the Laplace transform in circuit analysis

In this section, we will apply the Laplace transform to analyze circuit systems. The RCL circuit system is composed of linear time-invariant components of resistor, capacitor, inductor, independent power supply and linear controlled source. In the time domain, the differential equation is established to compute the output voltage or current. This section focuses on the S-domain models of the circuit. Considering a circuit as an impedance network, the VCR (voltage–current relation) is expressed by their Laplace transform. We can directly establish algebraic equations describing the circuit (loops or nodes) in the S-domain.

5.4.1 S-domain models of circuit

1. Resistor
Assuming that reference directions of the voltage and current of the linear time-invariant resistor are related, the VCRs in the time domain and S-domain are shown in Equation (5.47). The circuit models are shown in Figure 5.5:

$$u(t) = Ri(t) \leftrightarrow U(s) = RI(s) \tag{5.47}$$

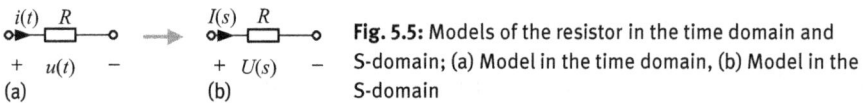

Fig. 5.5: Models of the resistor in the time domain and S-domain; (a) Model in the time domain, (b) Model in the S-domain

2. Inductor
The reference directions of voltage and current of the linear time-invariant inductor are related. Using the VCR in the time domain and the time-differentiation property of the Laplace transform, the VCR in the S-domain is shown in Equation (5.48). The corresponding circuit models are shown in Figure 5.6. The series model in Figure 5.6 (b) and the parallel model in Figure 5.6 (c) are equivalent.

$$u(t) = L\frac{di_L(t)}{dt} \leftrightarrow U(s) = sL \cdot I_L(s) - L \cdot i_L(0_-)$$
$$I_L(s) = \frac{1}{sL}U(s) + \frac{i_L(0_-)}{s} \tag{5.48}$$

! **Note:** Simplify the models with zero-state in the S-domain.

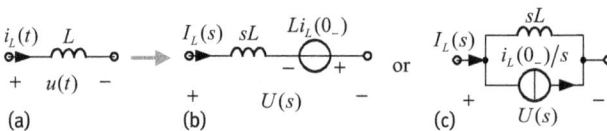

Fig. 5.6: Models of inductor in the time domain and S-domain, (a) Model in time domain (b) Series model in S-domain (c) Parallel model in S-domain

3. Capacitor

The reference directions of voltage and current of the linear time-invariant capacitor are related. Using the VCR in the time domain and the time-differentiation property of Laplace transform, the VCR in the S-domain is shown in Equation (5.49). The corresponding circuit models are shown in Figure 5.7. The series model in Figure 5.7 (b) and the parallel model in Figure 5.7 (c) are equivalent:

$$i(t) = C\frac{du_C(t)}{dt} \leftrightarrow I(s) = sCU_C(s) - Cu_C(0_-)$$

$$U_C(s) = \frac{1}{sC}I(s) + \frac{u_C(0_-)}{s}$$

(5.49)

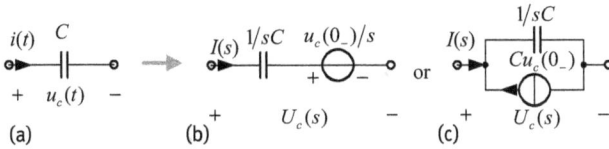

Fig. 5.7: Models of the capacitor in the time domain and S-domain; (a) Model in time domain, (b) Series model in S-domain, (c) Parallel model in S-domain

Note: Simplify the models with zero-state in the S-domain.

When the system is under the condition of zero state, the VCR of the capacitor and the inductor can be reduced to as follows:

$$U_L(s) = sL \cdot I_L(s)$$

$$U_C(s) = \frac{1}{sC}I_C(s)$$

(5.50)

Therefore, the S-domain model of the inductor and the capacitor can be expressed as the impedance of Ls and $1/Cs$, respectively.

4. Independent source

For an independent voltage source or current source, the time-domain model is directly converted to the S-domain model, i.e., $u_S(t) \leftrightarrow U_S(s)$, $i_S(t) \leftrightarrow I_S(s)$.

5. Kirchhoff's theorem

Using the linearity of the Laplace transform, the KCL and KVL in the S-domain can be directly expressed as follows:

$$\text{KCL}: \quad \sum_k i_k(t) = 0 \leftrightarrow \sum_k I_k(s) = 0$$

$$\text{KVL}: \quad \sum_k u_k(t) = 0 \leftrightarrow \sum_k U_k(s) = 0$$

5.4.2 Analysis in the S-domain of the circuit system

After establishing the S-domain model for the circuit system, the analysis is made in the S-domain to establish the algebraic equations. The time-domain differential equation is replaced and, therefore, the process of output calculation is obviously simplified.

Example 5.4.1. An RCL circuit is shown in Figure 5.8. The input signal is the voltage source $u(t) = 10\varepsilon(t)$, and the current through the capacitor is the output signal $i_1(t)$. The values of all components are $C = 1\,\text{F}$, $R_{12} = 0.2\,\Omega$, $R_2 = 1\,\Omega$, $L = 0.5\,\text{H}$. Calculate the zero-state response of $i_1(t)$.

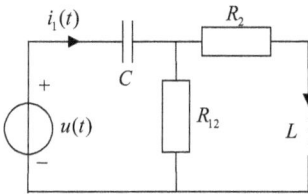

Fig. 5.8: Circuit in Example 5.4.1

Solution: The S-domain circuit model in condition of zero-state is drawn according to Equation (5.50) in Figure 5.9. The mesh analysis is used with two loop currents $I_1(s)$, $I_2(s)$ as the unknown variables:

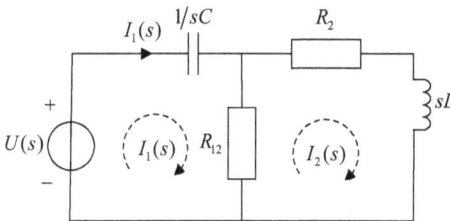

Fig. 5.9: Circuit model in the S-domain of Example 5.4.1

The mesh-current equations are established as follows:

$$\begin{cases} Z_{11}(s)I_1(s) - Z_{12}(s)I_2(s) = U(s) \\ -Z_{12}(s)I_1(s) + Z_{22}(s)I_2(s) = 0 \end{cases} \tag{5.51}$$

Using the determinant method, the output $I_1(s)$ is given by:

$$I_1(s) = \cfrac{Z_{22}}{\begin{vmatrix} Z_{11} & -Z_{12} \\ -Z_{12} & Z_{22} \end{vmatrix}} \cdot U(s) = \frac{Z_{22} \cdot U(s)}{Z_{11} \cdot Z_{22} - Z_{12}^2} \cdot \tag{5.52}$$

! **Note:** This solution process is only one illustration, other methods can also find the results.

Substituting all the following parameters into Equation (5.52) yields:

$$Z_{11}(s) = R_{12} + 1/Cs = 0.2 + s^{-1}(\Omega)$$

$$Z_{12}(s) = 0.2(\Omega)$$

$$Z_{22}(s) = R_{12} + R_2 + Ls = 0.2 + 1 + 0.5s = 1.2 + 0.5s(\Omega) ,$$

and we have:

$$I_1(s) = \frac{(1.2 + 0.5s)10s^{-1}}{(0.2 + s^{-1})(1.2 + 0.5s) - (0.2)^2} = \frac{50s + 120}{s^2 + 7s + 12} . \tag{5.53}$$

Using the partial fraction expansion, the above equation is expressed as follows:

$$I_1(s) = \frac{-30}{s + 4} + \frac{40}{s + 3} \tag{5.54}$$

Therefore, the zero-state response $i_1(t)$ is given by:

$$i_1(t) = (-30e^{-3t} + 80e^{-4t})\varepsilon(t) . \tag{5.55}$$

5.5 Application of solutions of differential equations

In this section, the Laplace transform is applied to solve linear, constant-coefficient differential equations. The basic steps involved in the Laplace-transform-based approach are as follows:

Step 1: Using its time-differentiation property, the Laplace transform of both sides of differential equation is made.

Step 2: Substituting the initial conditions into the algebraic equations, the zero-input and zero-state solutions in the S-domain are obtained.

Step 3: Using the partial fraction expansion, the inverse Laplace transform is obtained to compute the response in the time domain.

Now we illustrate the steps through examples of solving the second-order differential equations.

5.5.1 Analysis of computing zero-input and zero-state response

Example 5.5.1. The LTIC system is modeled by the linear, constant-coefficient differential equation:

$$y''(t) + 5y'(t) + 6y(t) = 2f(t) , \tag{5.56}$$

for the initial condition $y(0_-) = 1$, $y'(0_-) = -1$ and $f(t) = 5 \cos t\varepsilon(t)$ applied at the input. Determine the zero-input and zero-state responses of the system, respectively.

Solution:

(1) Zero-input response:

To calculate the zero-input response, the input signal is assumed to be zero. The differential equation reduces to:

$$y''_{zi}(t) + 5y'_{zi}(t) + 6y_{zi}(t) = 0$$

Taking the Laplace transform of the above differential equation yields:

$$s[sY_{zi}(s) - y(0_-)] - y'(0_-) + 5[sY_{zi}(s) - y(0_-)] + 6Y_{zi}(s) = 0. \qquad (5.57)$$

or:

$$Y_{zi}(s) = \frac{sy(0_-) + y'(0_-) + 5y(0_-)}{s^2 + 5s + 6} \qquad (5.58)$$

Note: Here, we have $y'_{zi}(0_-) = y'(0_-)$, $y_{zi}(0_-) = y(0_-)$.

Substituting the initial conditions $y(0_-) = 1$, $y'(0_-) = -1$ and using the partial fraction expansion, the above equation is expressed as follows:

$$Y_{zi}(s) = \frac{2}{s+2} + \frac{-1}{s+3}. \qquad (5.59)$$

Taking the inverse Laplace transform, the zero-input response is given by:

$$y_{zi}(t) = (2e^{-2t} - e^{-3t})\varepsilon(t).$$

(2) Zero-state response:

To obtain the zero-state response, the initial conditions are assumed to be zero, i.e., $y(0_-) = y'(0_-) = 0$. Taking the Laplace transform of the differential equation Equation (5.56) yields:

$$s^2 Y_{zs}(s) + 5sY_{zs}(s) + 6Y_{zs}(s) = 2F(s). \qquad (5.60)$$

Note: Think about the difference between Equations (5.60) and (5.57).

or

$$Y_{zs}(s) = \frac{2}{(s+2)(s+3)} \frac{5s}{s^2+1} \qquad (5.61)$$

Using the partial fraction expansion, the above equation is expressed as follows:

$$Y_{zs}(s) = \frac{-4}{s+2} + \frac{3}{s+3} + \frac{\frac{1}{\sqrt{2}}e^{-j\frac{\pi}{4}}}{s-j} + \frac{\frac{1}{\sqrt{2}}e^{j\frac{\pi}{4}}}{s+j}. \qquad (5.62)$$

Taking the inverse Laplace transform, the zero-state response is given by:

$$y_{zs}(t) = \left[-4e^{-2t} + 3e^{-3t} + \sqrt{2}\cos\left(t - \frac{\pi}{4}\right)\right]\varepsilon(t).$$

5.5.2 Analysis of computing the overall response

The overall response is the sum of the zero-input response and the zero-state response, which can be computed, respectively, using the method in Section 5.5.1. Here, we give a more efficient method to compute the overall response directly and acquire the components of zero-input and zero-state response.

Example 5.5.2. The LTIC system is modeled by the linear, constant-coefficient differential equation:

$$y''(t) + 5y'(t) + 6y(t) = 2f(t),$$

for the initial state $y(0_-) = 1$, $y'(0_-) = -1$ and $f(t) = 5\cos t\varepsilon(t)$ applied at the input. Determine the overall response and extract components of the zero-input and zero-state response.

Solution: Taking the Laplace transform directly on both sides of the differential equation yields:

$$s[sY(s) - y(0_-)] - y'(0_-) + 5[sY(s) - y(0_-)] + 6Y(s) = 2F(s). \quad (5.63)$$

Rearranging and collecting the terms corresponding to $Y(s)$ on the left-hand side of the equation results in the following:

$$(s^2 + 5s + 6)Y(s) = sy(0_-) + y'(0_-) + 5y(0_-) + 2F(s) \quad (5.64)$$

or:

$$Y(s) = \frac{sy(0_-) + y'(0_-) + 5y(0_-)}{s^2 + 5s + 6} + \frac{2}{s^2 + 5s + 6}F(s) \quad (5.65)$$

Obviously, in Equation (5.65), the former term is the Laplace transform of the zero-input response caused by $y(0_-)$ and $y'(0_-)$. The latter term is the Laplace transform of the zero-state response caused by $F(s)$. Substituting the initial conditions $y(0_-) = 1$, $y'(0_-) = -1$ and the Laplace transform $F(s) = 5s/(s^2 + 1)$ of the input signal yields:

$$Y(s) = Y_{zi}(s) + Y_{zs}(s) = \frac{s+4}{(s+2)(s+3)} + \frac{2}{(s+2)(s+3)}\frac{5s}{s^2+1}$$

Taking the partial fraction expansion, we obtain:

$$Y(s) = \underbrace{\frac{2}{s+2} + \frac{-1}{s+3}}_{Y_{zi}(s)} + \underbrace{\frac{-4}{s+2} + \frac{3}{s+3} + \frac{\frac{1}{\sqrt{2}}e^{-j\frac{\pi}{4}}}{s-j} + \frac{\frac{1}{\sqrt{2}}e^{j\frac{\pi}{4}}}{s+j}}_{Y_{zs}(s)}. \quad (5.66)$$

Calculating the inverse Laplace transform, we obtain the overall response as follows:

$$y(t) = y_{zi}(t) + y_{zs}(t) = (2e^{-2t} - e^{-3t})\varepsilon(t) + \left[-4e^{-2t} + 3e^{-3t} + \sqrt{2}\cos\left(t - \frac{\pi}{4}\right)\right]\varepsilon(t), \quad (5.67)$$

or:

$$y(t) = \left[-2e^{-2t} + 2e^{-3t} + \sqrt{2}\cos\left(t - \frac{\pi}{4}\right)\right]\varepsilon(t).$$

From Equation (5.67), the zero-input response is extracted as:

$$y_{zi}(t) = (2e^{-2t} - e^{-3t})\varepsilon(t),$$

and the zero-state response is extracted as:

$$y_{zs}(t) = \left[-4e^{-2t} + 3e^{-3t} + \sqrt{2}\cos\left(t - \frac{\pi}{4}\right)\right]\varepsilon(t).$$

! **Note:** Think about the differences of the CTFT and the Laplace transform in solving the response of the LTIC system.

5.6 Laplace transfer function

5.6.1 Definition of the Laplace transfer function

As was mentioned in Chapter 2, the impulse response relates the zero-state response $y(t)$ of an LTIC system to its input $f(t)$ using:

$$y_{zs}(t) = h(t) * f(t)$$

Calculating the Laplace transform of both sides of the equation, we obtain:

$$Y_{zs}(s) = H(s)F(s) \tag{5.68}$$

The *Laplace transfer function* $H(s)$ can be defined as the ratio of the Laplace transform of the zero-state output response and the Laplace transform of the input signal. Mathematically, the Laplace transfer function $H(s)$ is given by:

$$H(s) = \frac{Y_{zs}(s)}{F(s)}. \tag{5.69}$$

The Laplace transfer function is only related to the system structure and component parameters, but not to the input signal and initial conditions. Given the algebraic expression and the ROC, the inverse of the Laplace transfer function $H(s)$ leads to the impulse response $h(t)$ of the LTIC system.

Example 5.6.1. The impulse response of the LTIC system is:

$$h(t) = 2(e^{-2t} - e^{-3t})\varepsilon(t).$$

Determine the zero-state response for $f(t) = 5\cos t\varepsilon(t)$ applied at the input.

Solution: Taking the Laplace transform of the input signal and the impulse response yields:

$$f(t) = 5\cos t\varepsilon(t) \leftrightarrow F(s) = \frac{5s}{s^2 + 1}, \tag{5.70}$$

$$h(t) = 2(e^{-2t} - e^{-3t})\varepsilon(t) \leftrightarrow H(s) = 2\left(\frac{1}{s+2} - \frac{1}{s+3}\right) = \frac{2}{(s+2)(s+3)}. \tag{5.71}$$

Using Equation (5.68), the Laplace transform of the zero-state response is given by:

$$Y_{zs}(s) = F(s) \times H(s) = \frac{5s}{s^2 + 1} \times \frac{2}{(s+2)(s+3)} .$$

Using the partial fraction expansion, we obtain:

$$Y_{zs}(s) = \frac{-4}{s+2} + \frac{3}{s+3} + \frac{0.5 + 0.5j}{s+j} + \frac{0.5 - 0.5j}{s-j} . \tag{5.72}$$

Taking the inverse Laplace transform, the zero-state response is as follows:

$$y_{zs}(t) = [-4e^{-2t} + 3e^{-3t} + \cos t - \sin t]\varepsilon(t) . \tag{5.73}$$

Example 5.6.2. The zero-state response of the LTIC system produced by the input $f(t) = e^{-t}\varepsilon(t)$ is:

$$y_{zs}(t) = (3e^{-t} - 4e^{-2t} + e^{-3t})\varepsilon(t) .$$

Determine the impulse response and the differential equation of the system.

Solution: Taking the Laplace transform of the input signal and the zero-state response yields:

$$f(t) = e^{-t}\varepsilon(t) \leftrightarrow F(s) = \frac{1}{s+1} \tag{5.74}$$

$$y_{zs}(t) = (3e^{-t} - 4e^{-2t} + e^{-3t})\varepsilon(t) \leftrightarrow Y_{zs}(s) = \frac{3}{s+1} + \frac{-4}{s+2} + \frac{1}{s+3} \tag{5.75}$$

Using Equation (5.69), the Laplace transfer function is given by:

$$H(s) = \frac{Y_{zs}(s)}{F(s)} = \frac{2s+8}{s^2 + 5s + 6} = \frac{2(s+4)}{(s+2)(s+3)} = \frac{4}{s+2} + \frac{-2}{s+3} \tag{5.76}$$

Taking the inverse Laplace transform, the impulse response is as follows:

$$h(t) = (4e^{-2t} - 2e^{-3t})\varepsilon(t) \tag{5.77}$$

Using Equation (5.76), the relationship between $Y_{zs}(s)$ and $F(s)$ is expressed as follows:

$$(s^2 + 5s + 6)Y_{zs}(s) = (2s+8)F(s) \tag{5.78}$$

Based on the time-differentiation property of the Laplace transform, the differential equation of the system is given by:

$$y''(t) + 5y'(t) + 6y(t) = 2f'(t) + 8f(t) \tag{5.79}$$

Note: For causal signals, we have $f'(t) \leftrightarrow sF(s)$, $f''(t) \leftrightarrow s^2F(s)$.

5.6.2 Characteristic equation, zeros and poles

In Section 5.3.1, we introduced the characteristic equation and roots of the Laplace transform to compute the inverse Laplace transform. In this section, we will further discuss the poles of the Laplace transfer function. It is useful for analyzing the stability of the LTIC systems.

The LTIC system is assumed with a rational transfer function $H(s)$ of the following form:

$$H(s) = \frac{N(s)}{D(s)} = \frac{b_m s^m + b_{m-1} s^{m-1} + \cdots + b_1 s + b_0}{s^n + a_{n-1} s^{n-1} + \cdots + a_1 s + a_0}, \qquad m < n \qquad (5.80)$$

Characteristic equation: The *characteristic equation* for the transfer function in Equation (5.80) is defined as follows:

$$D(s) = s^n + a_{n-1} s^{n-1} + \cdots + a_1 s + a_0 = 0. \qquad (5.81)$$

Zeros: The *zeros* of the transfer function $H(s)$ of an LTIC system are the finite locations in the complex S-plane where $|H(s)| = 0$. For the transfer function in Equation (5.80), the location of the zeros can be obtained by solving the following equation:

$$N(s) = b_m s^m + b_{m-1} s^{m-1} + \cdots + b_1 s + b_0 = 0. \qquad (5.82)$$

Poles: The *poles* of the transfer function $H(s)$ of an LTIC system are the locations in the complex S-plane where $H(s)$ has an infinite value. For the transfer function in Equation (5.80), the location of the poles can be obtained by solving the characteristic equation, Equation (5.81).

In order to calculate the zeros and poles, the transfer function is factorized and is typically represented as follows:

$$H(s) = \frac{N(s)}{D(s)} = \frac{K \prod_{j=1}^{m} (s - z_j)}{\prod_{i=1}^{n} (s - p_i)}. \qquad (5.83)$$

Note that the transfer function has m zeros and n poles. The transfer function must be finite within its ROC. On the other hand, the magnitude of the transfer function is infinite at the location of a pole.

! **Note:** Can the ROC of a system include poles?

5.6.3 Nature of the shape of impulse response for different poles

In this section, we carry out an analysis on the relation between the shape of the impulse response $h(t)$ and the typical locations of poles. Table 5.3 shows the shape of the impulse response corresponding to the first-order poles in the left-half S-plane, $j\omega$- axis and right-half S-plane, respectively.

Tab. 5.3: Distribution of first-order poles and the shape of $h(t)$

Poles on S-plane	$H(s)$	$h(t)$, $t \geq 0$	Shape in time domain
pole at $-a$	$\dfrac{1}{s+a}$	e^{-at}	decaying exponential
poles at $-a \pm j\beta$	$\dfrac{1}{(s+a)^2+\beta^2}$	$e^{-at}\cos(\beta t + \theta)$	decaying sinusoid
pole at origin	$\dfrac{1}{s}$	$\varepsilon(t)$	step
poles at $\pm j\beta$	$\dfrac{1}{s^2+\beta^2}$	$\cos(\beta t + \theta)$	sinusoid
pole at a	$\dfrac{1}{s-a}$	e^{at}	growing exponential
poles at $a \pm j\beta$	$\dfrac{1}{(s-a)^2+\beta^2}$	$e^{at}\cos(\beta t + \theta)$	growing sinusoid

Table 5.4 shows the shape of the impulse response corresponding to the second-order poles in the left-half S-plane, the $j\omega$-axis and the right-half S-plane, respectively.

According to Tables 5.3 and 5.4, the relationship between the poles' distribution and the shape of $h(t)$ is summarized as follows:

(i) If the poles lie in the left- half S-plane, the waveform is decaying.

(ii) The shape of $h(t)$ is a step function or sinusoidal function corresponding to the first-order poles on the $j\omega$-axis. For repeated poles on the $j\omega$-axis, the shape of the waveform is increasing.

(iii) If the poles lie in the right-half S-plane, the shape of the waveform is increasing.

Tab. 5.4: Distribution of the second-order poles and the shape of $h(t)$

Poles on S-plane	$H(s)$	$h(t), t \geq 0$	Shape in time domain
	$\dfrac{1}{(s+a)^2}$	te^{-at}	
	$\dfrac{1}{s^2}$	$t\varepsilon(t)$	
	$\dfrac{1}{(s^2+\beta^2)^2}$	$t\cos(\beta t + \theta)$	

5.6.4 Stability conditions in the S-plane

In Chapter 1, the BIBO (bounded-input, bounded-output) stable LTIC system was introduced. The time-domain stability condition is that the unit impulse response is absolutely integrable:

$$\int_{-\infty}^{\infty} |h(t)|\, dt < \infty . \tag{5.84}$$

According to the discussion on the shape of the impulse response of different poles' location, the S-domain *stability condition* is stated as follows:

(i) An LTIC system will be absolutely BIBO stable if the ROC includes the $j\omega$-axis.

(ii) A causal LTIC system will be absolutely BIBO stable and causal if the ROC occupies the entire right half of the S-plane, including the $j\omega$-axis. In other words, a causal LTIC system will be absolutely BIBO stable if and only if all the poles lie in the left half of the S-plane (i.e., to the left of the $j\omega$-axis).

Example 5.6.3. Determine if the following causal LTIC systems are BIBO stable:

(1) $H_1(s) = \dfrac{(s+3)(s+6)}{s^2(s+1)(s-1)}$;

(2) $H_2(s) = \dfrac{s+11}{s^3 + 5s^2 + 17s + 13}$.

Solution:
(1) For $H_1(s)$, the poles locate at $s = -1, 0, 0, 1$. Since not all the poles lie in the left half of the S-plane, the transfer function does not represent an absolutely BIBO stable and causal system. It can be easily verified that the impulse response has the component of the rising exponential function $e^t\varepsilon(t)$.
(2) The LTIC system with the transfer function $H_2(s)$ has three poles located at $s = -1, -2 \pm j3$. Since all the poles lie in the left-half S-plane, the transfer function represents an absolutely BIBO stable and causal system.

5.6.5 Laplace transfer function with the frequency response function

In Chapter 4, the frequency response function $H(j\omega)$ was introduced. In Section 5.1.4, the relationship between CTFT and Laplace transform was discussed. When the ROC of the Laplace transform includes the $j\omega$-axis, the CTFT can be obtained from the Laplace transform by substituting $s = j\omega$, i.e.,

$$H(j\omega) = H(s)|_{s=j\omega} . \qquad (5.85)$$

The condition for the existence of frequency response function is the same with the stability condition of the LTIC system. The following conclusion is drawn:
(i) If the ROC includes the $j\omega$-axis, the frequency response function exists.
(ii) If all the poles of a causal LTIC system lie in the left half of the S-plane, the frequency response function exists as $H(j\omega) = H(s)|_{s=j\omega}$.

Example 5.6.4. Determine the frequency response functions of the following systems:
(1) The ideal delayer with T seconds delay.
(2) The ideal first-order differentiator.
(3) The ideal first-order integrator.

Solution:
(1) The Laplace transfer function of the system delayed by T seconds is given by:

$$H(s) = e^{-sT} .$$

Substituting $s = j\omega$, the frequency response function is given by:

$$H(j\omega) = e^{-j\omega T} .$$

The amplitude response is $|H(j\omega)| = 1$. The phase response is $\varphi(\omega) = -\omega T$.
It can be seen that if signal $f(t) = \cos \omega t$ goes through an ideal T seconds delayed system, the output is $y(t) = \cos \omega(t - T)$. Obviously, the output has a phase shift $-\omega T$ relative to the input, and their amplitudes are the same.

(2) The Laplace transfer function of the ideal first-order differentiator is given by:

$$H(s) = s .$$

Substituting $s = j\omega$, the frequency response function is given by:

$$H(j\omega) = j\omega .$$

The amplitude response is $|H(j\omega)| = \omega$. The phase response is $\varphi(\omega) = \pi/2$. If the signal $f(t) = \cos \omega t$ goes through a first-order differentiator, the output is $y(t) = -\omega \sin \omega t = \omega \cos(\omega t + \pi/2)$. Obviously, the output has a phase shift $\pi/2$ relative to the input. The output amplitude is ω times the input amplitude, which will amplify the high-frequency noise. So, the ideal differentiator is avoided in practical applications.

(3) The Laplace transfer function of the ideal first-order integrator is given by:

$$H(s) = \frac{1}{s} .$$

Substituting $s = j\omega$, the frequency response function is given by:

$$H(j\omega) = \frac{1}{j\omega} = \frac{1}{\omega}e^{-j\frac{\pi}{2}} .$$

The amplitude response is $|H(j\omega)| = 1/\omega$. The phase response is $\varphi(\omega) = -\pi/2$. If the signal $f(t) = \cos \omega t$ goes through a first-order integrator, the output is $y(t) = (1/\omega)\sin \omega t = (1/\omega)\cos(\omega t - \pi/2)$. The amplitude response is $|H(j\omega)| = 1/\omega$, which is inversely proportional to ω. It will suppress and smooth the high-frequency component of the noise signal. So, the integrator has better noise immunity than the differentiator.

Note: This is why we use the integrator as a basic unit to model the CT system in Section 2.1.2.

5.6.6 Calculation with MATLAB

MATLAB provides the function to compute and plot the zero-pole distributive diagram of the LTIC system [25]. The function is *pzmap*, and its method is as follows:

```
pzmap (sys)
```

where, *sys* indicates the transfer function of the system.

Example 5.6.5. Draw the zero-pole distributive diagram of the given system:

$$H(s) = \frac{s^2 + 4s + 3}{s^4 + 3s^3 + 4s^2 + 6s + 4} ,$$

and determine the stability with MATLAB.

Solution:

```
b = [1 4 3];                  % coefficients of Numerator
a = [1 3 4 6 4];              % coefficients of Denominator
sys = tf (b, a);              % transfer function
pzmap (sys); sgrid;           % plot the zero-pole distributive diagram
azp = roots (a)               % find the poles.
% Judge the stability by the value of "wd": 1 means stable; 0 means
     unstable
wd = 1;
for k = 1: length (azp)
     if real (azp (k)){\textgreater} - 0.000001  % real part of poles
                                                 % is positive
     wd = 0;
     end
if wd == 0
     title ('unstable system');
elseif wd == 1
     title ('stability system'); % real part of poles is negative
end
```

The operation result is the following:

```
Transfer function:
  S ^ 2 + 4 s + 3
-----------------------------
S ^ 4 + 3 s ^ 3 + 4 s ^ 2 + 6 s + 4
Locations of poles:
azp =
    0.0000 + 1.4142i
    0.0000 - 1.4142i
   -2.0000
   -1.0000
```

As is shown in Figure 5.10, the system has four poles and two zeros. There is a pair of conjugate poles on the imaginary axis. The system is not absolutely BIBO stable.

MATLAB provides the function to compute the frequency response of the LTIC system. The function is "freqs", and its method is as follows.

```
freqs (num, den)
```

where *num* and *den* indicate the coefficient vector of the numerator and the denominator of the transfer function.

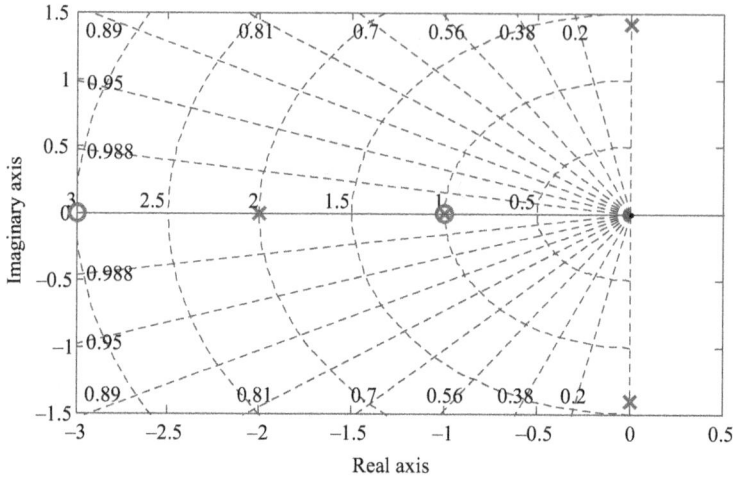

Fig. 5.10: Location of poles and zeros

Example 5.6.6. Draw the zero-pole distributive diagram of the given system:

$$H(s) = \frac{1}{s^3 + 2s^2 + 3s + 1}.$$

Calculate the unit impulse response and the frequency response of the system. Determine whether the system is stable.

Solution:

```
num = [1];                   % coefficients of Numerator
den = [1 2 3 1];             % coefficients of Denominator
sys = tf (num, den);         % transfer function
poles=roots(den);
figure(1);
pzmap(sys);                  % plot the zero-pole distributive diagram
t=0:0.02:10;
h=impulse(num,den,t);        % impulse response
figure(2);plot(t,h)
title('Impulse Response')
[H,w]=freqs(num,den);        % frequency response function
figure(3);plot(w,abs(H))     % magnitude frequency response
```

The operation result is:

```
poles= -0.4302 -0.7849+1.3071i -0.7849-1.3071i
```

Pole–Zero Map

(a)

Impulse response

(b) $t(s)$

Magnitude response

(c) ω (rad/s)

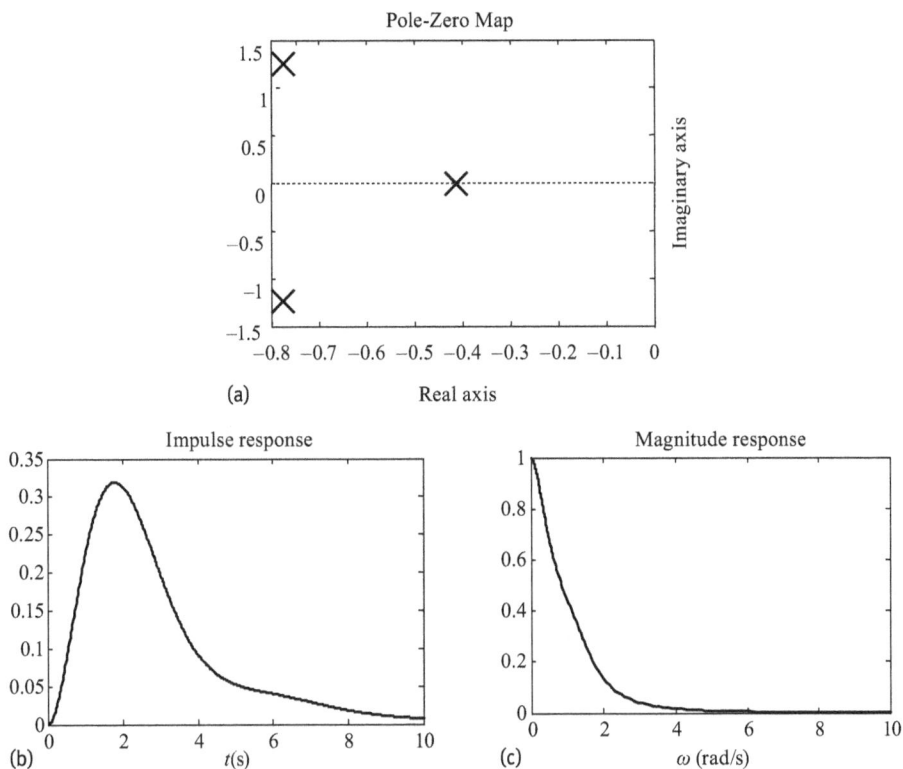

Fig. 5.11: Results of Example 5.6.6; (a) Location of poles and zeros, (b) Impulse response (c) Magnitude frequency response

Figure 5.11 shows the results of the program. The system is absolutely BIBO stable because all the poles lie in the left half S-plane in Figure 5.11 (a). The impulse response shown in Figure 5.11 (b) is decaying and absolutely integrable.

5.7 Signal-flow graph and LTIC system simulation

5.7.1 Block diagram representation

1. Cascaded configuration

A cascaded configuration between n systems is illustrated in Figure 5.12 (a). The impulse response is the convolution integral of that of the cascaded subsystem:

$$h(t) = h_1(t) * h_2(t) * \cdots * h_n(t) \tag{5.86}$$

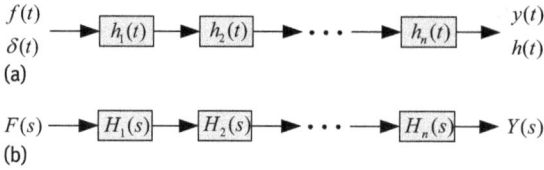

Fig. 5.12: Block diagrams for cascaded systems; (a) Cascaded configuration for connecting LTIC systems in time-domain, (b) Cascaded configuration for connecting LTIC systems in S-domain

The corresponding Laplace transform is as follows:

$$H(s) = H_1(s) \times H_2(s) \times \cdots \times H_n(s) \tag{5.87}$$

The block diagram in the S-domain is presented in Figure 5.12 (b).

2. Parallel configuration

The parallel configuration between n systems is illustrated in Figure 5.13 (a). The impulse response is the sum of that of the cascaded subsystems:

$$h(t) = h_1(t) + h_2(t) + \cdots + h_n(t) \tag{5.88}$$

The corresponding Laplace transform is as follows:

$$H(s) = H_1(s) + H_2(s) + \cdots + H_n(s) \tag{5.89}$$

The block diagram in S-domain is presented in Figure 5.13 (b).

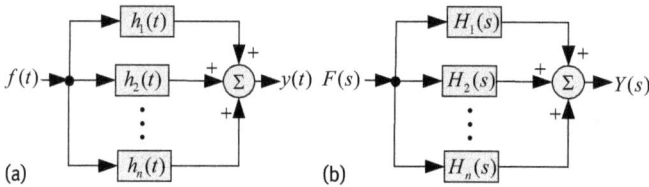

Fig. 5.13: Block diagrams for parallel systems; (a) Parallel connection in time-domain (b) Parallel connection in S-domain

5.7.2 Model of basic components of LTIC systems

In Section 2.1.2, we introduced three basic component units, including the integrator, adder and multiplier. They are connected with each other to represent the linear, constant-coefficient differential equation. As shown in Table 5.5, the models of the basic components in time-domain and S-domain are given.

Note: We use the integrator in the condition of zero state to represent the system function s^{-1}.

Tab. 5.5: Basic components of the LTIC system

Components	Time domain	S-domain
Multiplier	$f(t) \rightarrow \boxed{a} \rightarrow af(t)$ or $f(t) \xrightarrow{\quad a \quad} af(t)$	$F(s) \rightarrow \boxed{a} \rightarrow aF(s)$ or $F(s) \xrightarrow{\quad a \quad} aF(s)$
Adder	$f_1(t) \searrow^{+} \quad f_1(t) \pm f_2(t)$ $\qquad \sum$ $f_2(t) \nearrow_{\pm}$	$F_1(s) \searrow^{+} \quad F_1(s) \pm F_2(s)$ $\qquad \sum$ $F_2(s) \nearrow_{\pm}$
Integrator	$f(t) \rightarrow \boxed{\int} \rightarrow \int_{-\infty}^{t} f(x)\,dx$	$\qquad\qquad \downarrow s^{-1} f^{(-1)}(0_-)$ $F(s) \rightarrow \boxed{s^{-1}} \rightarrow \sum \xrightarrow{+}$ $\qquad\qquad\qquad\quad s^{-1}F(s) + s^{-1}f^{(-1)}(0_-)$
Integrator (zero-state)	$f(t) \rightarrow \boxed{\int} \rightarrow \int_{0}^{t} f(x)dx$	$F(s) \rightarrow \boxed{s^{-1}} \rightarrow s^{-1}F(s)$

5.7.3 The signal-flow graph of LTIC systems

A signal-flow graph (SFG) is a simplified representation for systems in which nodes represent system variables, and branches represent functional connections between pairs of nodes.

The block diagram and signal-flow graph of the basic integrator $Y(s) = s^{-1}F(s)$ are presented in Figures 5.14 (a) and (b), respectively.

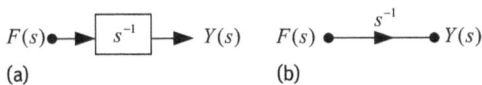

(a) $\qquad\qquad\qquad$ (b)

Fig. 5.14: Block diagram and signal-flow graph for the integrator; (a) Block diagram, (b) Signal flow graph

Example 5.7.1. The block diagram of an LTIC system is presented in Figure 5.15. Please plot the corresponding signal flow graph.

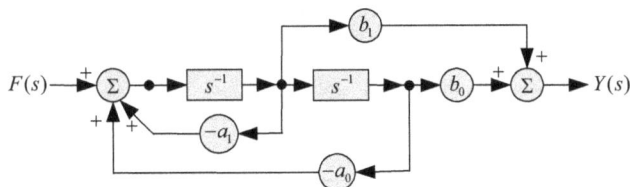

Fig. 5.15: Block diagram of an LTIC system

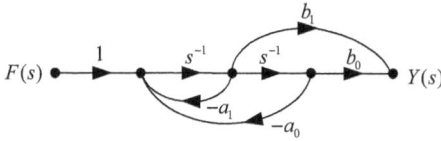

Fig. 5.16: Signal flow graph of Example 5.7.1

Solution: The signal flow graph can be plotted directly as shown in Figure 5.16.

5.7.4 Mason's rule

Mason's rule is a method for finding the transfer function from the signal flow graph. Terms are usually used in linear SFG theory as follows:
(i)   Path: A path is a continuous set of branches traversed in the direction indicated by the branch arrows. If no node is re-visited, the path is an open path.
(ii) Forward path: A path from an input node (source) to an output node (sink) that does not re-visit any node.
(iii) Path gain: the product of the gains of all the branches in the path.
(iv) Loop: A closed path. It originates and ends on the same node, and no node is touched more than once.
(v)   Loop gain: the product of the gains of all the branches in the loop.

Mason's rule is as follows:

$$H(s) = \frac{Y_{zs}(s)}{F(s)} = \frac{\sum_{i=1}^{m} p_i \Delta_i}{\Delta}$$

$$\Delta = 1 - \sum_j L_j + \sum_{m,n} L_m L_n - \sum_{p,q,r} L_p L_q L_r + \cdots$$

(5.90)

Table 5.6 gives the definition of each symbol in Mason's rule.

Tab. 5.6: Definition of symbols

Δ	The determinant of the graph
L_j	The loop gain
$L_m L_n$	The product of the loop gains of two non-touching loops
$L_p L_q L_r$	The product of the loop gains of three pairwise non-touching loops
p_i	The gain of forward path between $F(s)$ and $Y_{zs}(s)$
Δ_i	The cofactor value of i-th forward path: the determinant of the residual flow graph after removing the i-th forward path
m	Total number of forward paths between $F(s)$ and $Y_{zs}(s)$

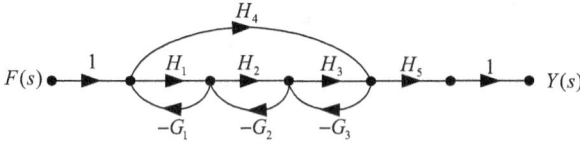

Fig. 5.17: Signal flow graph of an LTIC system in Example 5.7.2

Example 5.7.2. Compute the transfer function of the system in Figure 5.17 with Mason's rule.

Solution: All the loop gains are obtained as follows:

$$L_1 = -H_1 G_1, \quad L_2 = -H_2 G_2, \quad L_3 = -H_3 G_3, \quad L_4 = -H_4 G_1 G_2 G_3$$

The determinant of the signal flow graph is calculated as follows:

$$\Delta = 1 - \sum_j L_j + \sum_{m,n} L_m L_n$$

$$= 1 - (-H_1 G_1 - H_2 G_2 - H_3 G_3 - H_4 G_1 G_2 G_3) + H_1 G_1 H_3 G_3 \qquad (5.91)$$

The gain of the forward path and its corresponding cofactor value are given by:

$$p_1 = H_4 H_5, \qquad \Delta_1 = 1 - (-H_2 G_2)$$
$$p_2 = H_1 H_2 H_3 H_5, \qquad \Delta_2 = 1$$

Based on Mason's rule, the transfer function is computed as follows:

$$H(s) = \frac{p_1 \Delta_1 + p_2 \Delta_2}{\Delta} = \frac{H_4 H_5 (1 + H_2 G_2) + H_1 H_2 H_3 H_5}{1 + H_1 G_1 + H_2 G_2 + H_3 G_3 + H_4 G_1 G_2 G_3 + H_1 G_1 H_3 G_3} \qquad (5.92)$$

Example 5.7.3. Compute the Laplace transfer function of the system in Figure 5.18 with Mason's rule.

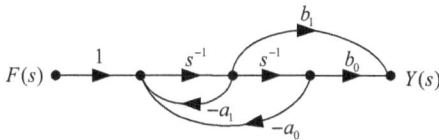

Fig. 5.18: Signal flow graph of the LTIC system in Example 5.7.3

Solution: The determinant of the signal flow graph is calculated as follows:

$$\Delta = 1 - \left(-a_1 \frac{1}{s} - a_0 \frac{1}{s^2} \right)$$

The gain of the forward path and its corresponding cofactor value are given by:

$$p_1 = \frac{b_1}{s}, \qquad \Delta_1 = 1$$

$$p_2 = \frac{1}{s^2}b_0, \qquad \Delta_2 = 1$$

Based on Mason's rule, the Laplace transfer function is computed as follows:

$$H(s) = \frac{\sum_{i=1}^{2} p_i \Delta_i}{\Delta} = \frac{\frac{b_1}{s} + \frac{b_0}{s^2}}{1 + \frac{a_1}{s} + \frac{a_0}{s^2}} = \frac{b_0 + b_1 s}{s^2 + a_1 s + a_0} \tag{5.93}$$

5.7.5 Simulation of the LTIC system

System simulation is to set up a system with basic components including integrator, adder and multiplier. When the Laplace transfer function $H(s)$ is given, the signal flow graph can also be plotted by Mason's rule to model the system. There are three simulation methods: the direct form, the cascaded form and the parallel form.

1. Direct form structure

Example 5.7.4. The Laplace transfer function is:

$$H(s) = \frac{b_1 s + b_0}{s + a_0}.$$

Plot the signal flow graph based on Mason's rule.

Solution: To match Mason's formula, the transfer function is reorganized as follows:

$$H(s) = \frac{b_1 s + b_0}{s + a_0} = \frac{b_1 + \frac{b_0}{s}}{1 - \left(-\frac{a_0}{s}\right)} \tag{5.94}$$

By observation, the denominator of Equation (5.94) is the determinant of signal flow graph. From the numerator, we can conclude that the signal flow graph has two separated open paths. Then the signal flow graph can be plotted as shown in Figure 5.19, in which the loop gain is $-a_0/s$, and the path gain of two open paths is b_1 and b_0/s, respectively.

! **Note:** The direct form II is the transposition of direct form I. The input and output are reversed, and all the branch directions are also reversed.

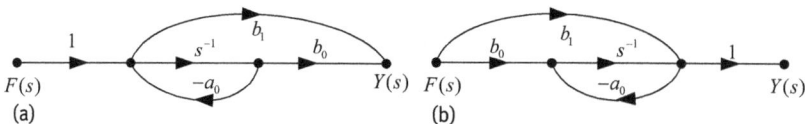

Fig. 5.19: Signal flow graph of Example 5.7.4; (a) Direct Form I, (b) Direct Form II

Example 5.7.5. The Laplace transfer function is:

$$H(s) = \frac{b_2 s^2 + b_1 s + b_0}{s^2 + a_1 s + a_0}.$$

Plot the signal flow graph based on Mason's rule.

Solution: To match Mason's formula, the transfer function is reorganized as follows:

$$H(s) = \frac{b_2 s^2 + b_1 s + b_0}{s^2 + a_1 s + a_0} = \frac{b_2 + \frac{b_1}{s} + \frac{b_0}{s^2}}{1 - \left(-\frac{a_1}{s} - \frac{a_0}{s^2}\right)} \qquad (5.95)$$

Compared with Mason's formula, the signal flow graph has three open paths and two touching loops, which are plotted in Figure 5.20.

Note: For simplicity, we usually adopt direct form I.

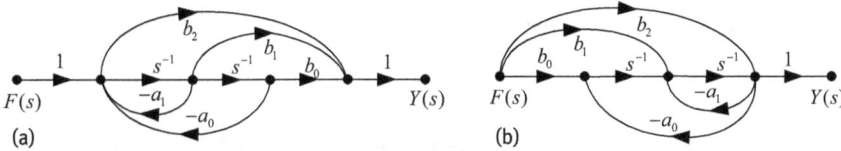

Fig. 5.20: Signal flow graph of Example 5.7.5; (a) Direct form I, (b) Direct form II

2. Cascaded form structure

In order to simulate a cascaded form structure, the system transfer function must be rewritten in the form of $H(s) = H_1(s) \cdot H_2(s) \cdot \ldots \cdot H_n(s)$. Each subsystem simulated in direct form I is then cascaded to realize a cascaded structure. Therefore, a complicated system is divided into several first-order or second-order simple subsystems.

Example 5.7.6. The Laplace transfer function is:

$$H(s) = \frac{s+1}{s^3 + 9s^2 + 26s + 24}.$$

Plot the signal flow graph in cascaded form.

Solution: The system transfer function is rewritten as:

$$H(s) = \frac{(s+1)}{(s^2 + 5s + 6)(s+4)} = \frac{(s+1)}{(s^2 + 5s + 6)} \cdot \frac{1}{(s+4)} = H_1(s) \cdot H_2(s) \qquad (5.96)$$

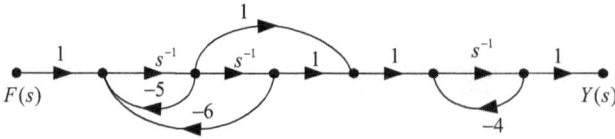

Fig. 5.21: Cascaded signal flow graph in Example 5.7.6

In the above equation, $H_1(s)$ and $H_2(s)$ represent a second-order and a first-order subsystem, respectively:

$$H_1(s) = \frac{s+1}{s^2 + 5s + 6} = \frac{\frac{1}{s} + \frac{1}{s^2}}{1 - \left(-\frac{5}{s} - \frac{6}{s^2}\right)} \tag{5.97}$$

$$H_2(s) = \frac{1}{s+4} = \frac{\frac{1}{s}}{1 - \left(-\frac{4}{s}\right)} \tag{5.98}$$

As is shown in Figure 5.21, each subsystem is simulated with direct form I, and then two signal flow graphs are cascaded.

3. Parallel form structure

In order to simulate a parallel form system, the system transfer function must be rewritten in the form of $H(s) = H_1(s) + H_2(s) + \cdots + H_n(s)$. Each subsystem simulated in direct form I is then connected in parallel to realize a parallel structure.

Example 5.7.7. The Laplace transfer function is:

$$H(s) = \frac{2s+8}{s^3 + 6s^2 + 11s + 6}.$$

Plot the signal flow graph in parallel form.

Solution:

$$H(s) = \frac{s+5}{(s+1)(s+2)(s+3)} = \frac{2}{s+1} + \frac{-3}{s+2} + \frac{1}{s+3} = H_1(s) + H_2(s) + H_3(s) \tag{5.99}$$

In the above equation, $H_1(s)$, $H_2(s)$ and $H_3(s)$ represent three subsystems:

$$H_1(s) = \frac{2}{s+1} = \frac{\frac{2}{s}}{1 - \left(-\frac{1}{s}\right)} \tag{5.100}$$

$$H_2(s) = \frac{-3}{s+2} = \frac{-\frac{3}{s}}{1 - \left(-\frac{2}{s}\right)} \tag{5.101}$$

$$H_3(s) = \frac{1}{s+3} = \frac{\frac{1}{s}}{1 - \left(-\frac{3}{s}\right)} \tag{5.102}$$

As shown in Figure 5.22, each subsystem is simulated with direct form I, and then three signal flow graphs are connected in parallel.

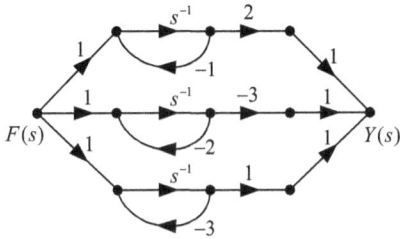

Fig. 5.22: Parallel signal flow graph in Example 5.7.7

5.8 Summary

This chapter mainly introduced S-domain analysis methods for LTIC systems. The elementary signal e^{st} was used to obtain Laplace transform and the inverse transform was defined in Section 5.3. Using the Laplace transform, the differential equation was solved in Section 5.5 to obtain the overall response consisting of zero-input and zero-state responses. Section 5.6 introduced the transfer function $H(s) = Y(s)/F(s)$ and the zero-state response was computed as $y(t) = \mathcal{L}^{-1}[F(s)H(s)]$. In Section 5.7, the Mason's rule was given to compute the transfer function and simulate the system.

Chapter 5 problems

5.1 Calculate the bilateral Laplace transform for the following signals, and determine the ROC:

(1) $(1 - e^{-2t})\varepsilon(-t)$

(2) $e^{-t}\varepsilon(t) + e^{2t}\varepsilon(-t)$

(3) $\varepsilon(t + 1) - \varepsilon(t - 1)$

(4) $e^{-|t|}$

5.2 Calculate the inverse Laplace transform of the function $H(s) = 1/(s + a)$.
5.3 Calculate the unilateral Laplace transform for the following signals:

(1) $e^{-2t}\varepsilon(t)$

(2) $\delta(t)$

(3) $e^{-j2t}\varepsilon(t)$

(4) $\sin(\omega_0 t)\varepsilon(t)$

(5) $t\varepsilon(t)$

(6) $f(t) = \begin{cases} 2t, & 0 \le t \le 1 \\ 2, & 1 \le t \le 2 \\ 0, & \text{otherwise} \end{cases}$

5.4 Calculate the unilateral Laplace transform for the following signals:

(1) te^{-2t}

(2) $2\delta(t) - e^{-t}$

(3) $3\sin t + 2\cos t$

(4) $\cos(2t + 45°)$

5.5 Given the unilateral Laplace transform pair of a causal signal $f(t) \leftrightarrow F(s) = 1/(s^2 - s + 1)$, calculate the unilateral Laplace transform for the following functions:

(1) $e^{-t}f(0.5t)$

(2) $e^{-3t}f(2t - 1)$

(3) $te^{-2t}f(3t)$

(4) $tf(2t - 1)$

5.6 Compute the initial value and the final value of $f(t)$ from its unilateral Laplace transform $F(s)$:

(1) $F(s) = \dfrac{2s + 3}{(s + 1)^2}$

(2) $F(s) = \dfrac{3s + 1}{s(s + 1)}$

5.7 Calculate the inverse Laplace transform of the right-sided signals $f(t)$ with the following transfer functions:

(1) $\dfrac{1}{(s + 2)(s + 4)}$

(2) $\dfrac{s}{(s + 2)(s + 4)}$

(3) $\dfrac{s^2 + 4s + 5}{s^2 + 3s + 2}$

(4) $\dfrac{2s + 4}{s(s^2 + 4)}$

(5) $\dfrac{1}{s(s - 1)^2}$

(6) $\dfrac{s + 5}{s(s^2 + 2s + 5)}$

5.8 Calculate the unilateral Laplace transform for the following signals:

(1) $e^{-(t-2)}[\varepsilon(t) - \varepsilon(t-2)]$

(2) $\sin(2t - 1)\varepsilon(2t - 1)$

(3) $e^{-t}\cos(t-2)\varepsilon(t-2)$

(4) $(\sin \pi t + 1)[\varepsilon(t) - \varepsilon(t-2)]$

(5) $\dfrac{d}{dt}[\sin 2t\varepsilon(t)]$

(6) $\dfrac{d^2}{dt^2}[\cos \omega_0 t]\varepsilon(t)$

(7) $\delta\left(\dfrac{1}{2}t - 1\right)$

(8) $\displaystyle\int_0^t \sin 2t\,dt$

(9) $(t-1)^2 e^{-2t}\varepsilon(t)$

(10) $te^{-3t}\cos \omega_0 t\varepsilon(t)$

5.9 Calculate the unilateral Laplace transform for the following periodic signals:

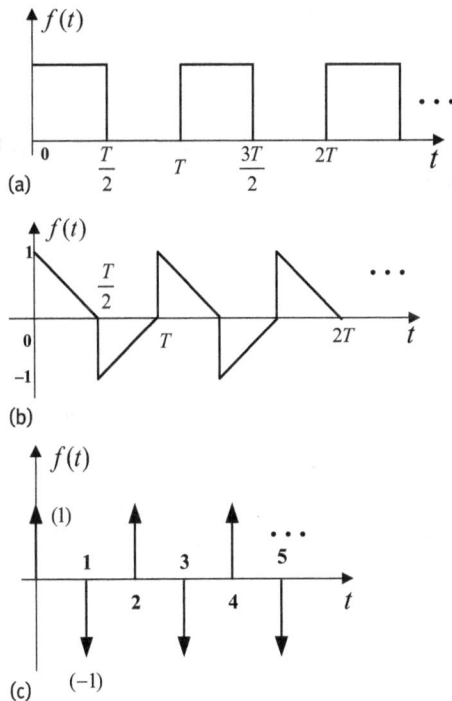

Fig. P5.1: Periodic signals in Problem 5.9

5.10 The differential equation of an LTIC system is given. Determine the zero-input, zero-state and overall response of the system produced by the input $f(t) = 2e^{-t}\varepsilon(t)$ given the initial conditions $y(0_-) = 5$ and $y'(0_-) = 0$:

$$y''(t) + 7y'(t) + 12y(t) = 12f(t) ;$$

5.11 Based on the given differential equation $y'(t) + 2y(t) = f(t)$ and the following input with the initial conditions, calculate the overall response:

(1) $f(t) = \varepsilon(t)$, $y(0_-) = 1$

(2) $f(t) = \sin(2t)\varepsilon(t)$, $y(0_-) = 0$

5.12 The differential equation of an LTIC system is given:

$$y''(t) + 5y'(t) + 6y(t) = 3f(t)$$

Determine the zero-input, zero-state and overall response of the system produced by the following inputs and the initial conditions:

(1) $f(t) = \varepsilon(t)$, $y(0_-) = 1$, $y'(0_-) = 2$

(2) $f(t) = e^{-t}\varepsilon(t)$, $y(0_-) = 0$, $y'(0_-) = 1$

5.13 The differential equation of an LTIC system is given:

$$y''(t) + 3y'(t) + 2y(t) = f'(t) + 4f(t)$$

Calculate the zero-state responses for the following inputs:

(1) $f(t) = \varepsilon(t)$

(2) $f(t) = e^{-2t}\varepsilon(t)$

5.14 The RLC circuit is shown in Figure P5.2, and the output is the capacitor voltage $u(t)$. Determine the impulse response and step response of voltage $u(t)$.

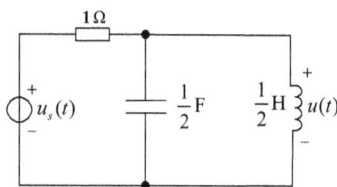

Fig. P5.2: The RLC circuit used in Problem 5.14

5.15 The RLC circuit is shown in Figure P5.3 and the output is the capacitor voltage $u(t)$. Calculate the zero-state response with the input $i_s(t) = \varepsilon(t)$:

(1) $L = 0.1\,\mathrm{H}$, $C = 0.1\,\mathrm{F}$, $G = 2.5\,\mathrm{s}$

(2) $L = 0.1\,\mathrm{H}$, $C = 0.1\,\mathrm{F}$, $G = 2\,\mathrm{s}$

(3) $L = 0.1\,\mathrm{H}$, $C = 0.1\,\mathrm{F}$, $G = 1.2\,\mathrm{s}$

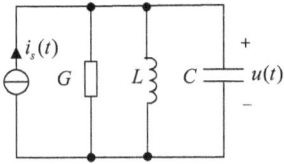

Fig. P5.3: RLC circuit used in Problem 5.15

5.16 For the LTIC system diagram in Figure P5.4, the transfer function of each subsystem is given by $H_1(s) = -e^{-2s}$ and $H_2(s) = 1/s$.
  (1) Calculate the impulse response.
  (2) Calculate the zero-state response with the input $f(t) = \varepsilon(t)$

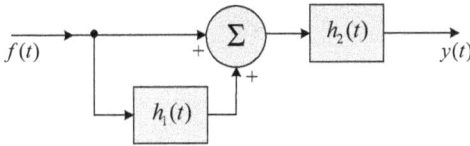

Fig. P5.4: System diagram used in Problem 5.16

5.17 Consider the following transfer functions and draw the signal-flow graph with direct form:

(1) $H(s) = \dfrac{s + 2}{(s + 1)(s + 3)}$

(2) $H(s) = \dfrac{s^2 + 2s + 1}{(s + 2)(s^2 + 5s + 6)}$

(3) $H(s) = \dfrac{s^2}{(s + 1)(s + 2)(s + 4)}$

5.18 The distribution of the zeros and poles in S-plane is given.
  (1) Compute the Laplace transfer function $H(s)$ of Figure P5.5 (a) with $H(\infty) = 1$.
  (2) Compute the Laplace transfer function $H(s)$ of Figure P5.5 (b) with $H(0) = -1/2$.
  (3) Compute the frequency response function $H(j\omega)$ and draw the corresponding amplitude and phase response curves.

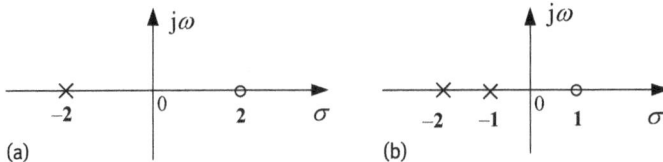

(a)

(b)

Fig. P5.5: Distribution of zeros and poles in Problem 5.18

5.19 Consider the following functions $f(t)$. Determine the CT Fourier transform $F(j\omega)$ by its Laplace transform $F(s)$.

(1) $f(t) = \varepsilon(t) - \varepsilon(t - 2)$

(2) $f(t) = t[\varepsilon(t) - \varepsilon(t - 1)]$

(3) $f(t) = \cos(\beta t) \cdot \varepsilon(t)$

(4) $f(t) = \begin{cases} 0, & t < 0 \\ t, & 0 < t < 1 \\ 1, & t > 1 \end{cases}$

5.20 Consider the following transfer functions of LTIC systems. Assuming that the systems are causal, determine if the specified systems are BIBO stable:

(1) $H(s) = \dfrac{s - 1}{s^2 + 3s + 2}$

(2) $H(s) = \dfrac{s^2 + 1}{s^2(s + 2)(s - 2)}$

(3) $H(s) = \dfrac{s + 4}{s^3 + 5s^2 + 17s + 13}$

(4) $H(s) = \dfrac{1}{e^s + 10}$

5.21 The signal flow graph of an LTIC system is given below. Calculate the transfer function $H(s)$ using Mason's rule.

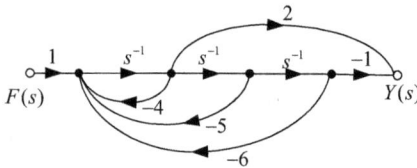

Fig. P5.6: The signal flow graph in Problem 5.21

5.22 Based on the frequency response function $H(j\omega) = (1 - j\omega)/(1 + j\omega)$ of an LTIC system, calculate the zero-state response with the following inputs:

(1) $f(t) = \varepsilon(t)$

(2) $f(t) = \sin t \cdot \varepsilon(t)$

5.23 Consider the following LTIC system. Calculate the conditions of parameter K if the system is BIBO stable.

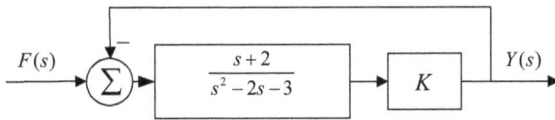

Fig. P5.7: The diagram of the LTIC system in Problem 5.23

5.24 Consider the following system diagram. Compute the Laplace transfer function $H(s)$.

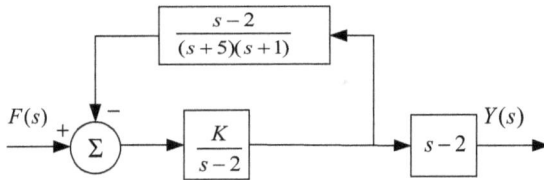

Fig. P5.8: The diagram of the LTIC system in Problem 5.24

5.25 Sketch the location of the zeros and poles for the given system:

$$H(s) = \frac{s^2 - 5s - 84}{s^4 + 7s^3 - 33s^2 - 355s - 700}$$

and determine with MATLAB whether the corresponding causal system is absolutely stable.

5.26 Sketch the location of the zeros and poles for the given system:

$$H(s) = \frac{s^2 + 3s + 2}{s^3 + 3s^2 + 2s}.$$

Calculate the unit impulse response $h(t)$ and the frequency response $H(j\omega)$ with MATLAB.

6 The Z-transform and Z-domain analysis

Please focus on the following key questions.

1. What is the mapping relation of the time-domain and Z-domain representations of discrete signals?
2. What is the method of applying the Z-transform for analyzing LTID systems?

6.0 Introduction

As illustrated in Chapter 4, the continuous-time signal can be converted to discrete-time signals by sampling. In Chapter 5, the Laplace transform is used to analyze the LTIC systems by transforming the differential equations into algebraic equations. For LTID systems, the Z-transform is applied to solve the difference equations to determine the output response. The Laplace transform uses e^{st} as the basis function. The Z-transform expresses $f(k)$ in terms of z^k, where the independent variable z is given by $z = e^{sT}$.

In this chapter, we first introduce the bilateral Z-transform definition. For causal signals, the bilateral Z-transform reduces to the unilateral Z-transform. The properties of the Z-transform are given in Section 6.2, and the inverse Z-transform is proposed in Section 6.3. The relationship between the Laplace transform and the Z-transform is discussed in Section 6.4. Section 6.5 applies the Z-transform to solve the difference equation. The Z-transfer function and the stability analysis of the LTID system in the Z-domain is presented in Section 6.6. The signal flow graph and system simulation are introduced in Section 6.7. Section 6.8 defines the frequency response, including the magnitude and phase spectra for LTID systems. Finally, the chapter is concluded in Section 6.9.

6.1 Analytical development

6.1.1 From the Laplace transform to the Z-transform

The continuous-time signal $f(t)$ is sampled by the ideal impulse-train $\delta_T(t)$ to obtain the discrete-time signal $f_s(t)$.

$$f_s(t) = f(t)\delta_T(t) = \sum_{k=-\infty}^{\infty} f(kT)\delta(t - kT) \tag{6.1}$$

Taking the bilateral Laplace transform on both sides yields:

$$F_{Sb}(s) = \sum_{k=-\infty}^{\infty} f(kT)e^{-kTs} \tag{6.2}$$

https://doi.org/10.1515/9783110593907-006

where $f(kT)$ is simplified as $f(k)$. Substituting $z = e^{sT}$ in Equation (6.2), the definition is obtained as follows:

The *bilateral Z-transform* of sequence $f(k)$:

$$F(z) = \sum_{k=-\infty}^{\infty} f(k)z^{-k} \tag{6.3}$$

! **Note:** What is the basis function of the Z-transform?

6.1.2 Region of convergence

It must be noted that the summation in the Z-transform is absolutely summable only for selected values of z. If the Z-transform has an infinite value, the Z-transform is not defined. In such cases, the region in the complex Z-plane, in which the Z-transform is defined, is referred to as the region of convergence (ROC) of the Z-transform $F(z)$. Therefore, the ROC is given by finding the appropriate area of z in the following equation:

$$\sum_{k=-\infty}^{\infty} |f(k)z^{-k}| < \infty \tag{6.4}$$

The above formula is a sufficient condition for the existence of the Z-transformation of sequence $f(k)$.

The inverse Z-transform is given by:

$$f(k) = \frac{1}{2\pi j} \oint_{c} F(z)z^{k-1}\, dz , \quad -\infty < k < \infty . \tag{6.5}$$

where c is a closed contour traversed in the counterclockwise direction within the ROC. Solving Equation (6.5) involves the technique of contour integration, which is seldom used directly. In Section 6.3, we will consider alternative approaches based on the power series expansion and partial fraction expansion to evaluate the inverse Z-transform.

To illustrate the different cases of ROC in computing the Z-transform, we consider the following examples.

Example 6.1.1. Calculate the bilateral Z-transform of the unit impulse sequence $\delta(k)$.

Solution:

$$F(z) = \sum_{k=-\infty}^{\infty} \delta(k)z^{-k} = z^0 = 1 \tag{6.6}$$

The Z-transform pair for an impulse sequence is given by:

$$\delta(k) \leftrightarrow F(z) = 1 , \quad \text{ROC: entire Z-plane} . \tag{6.7}$$

Example 6.1.2. Calculate the bilateral Z-transform of the sequence $f(k) = \varepsilon(k + 1) - \varepsilon(k - 2)$.

Solution: By definition:

$$F(z) = \sum_{k=-\infty}^{\infty} f(k)z^{-k} = \sum_{k=-1}^{1} z^{-k} = z + 1 + z^{-1} \tag{6.8}$$

It can be seen that the ROC is $0 < |z| < \infty$. For finite-duration sequences, the ROC is always the entire Z-plane, except for the possible exclusion of $z = 0$ and $z = \infty$.

Example 6.1.3. Calculate the bilateral Z-transform of the causal exponential sequence $f(k) = a^k \varepsilon(k)$.

Solution: By definition:

$$F(z) = \sum_{k=-\infty}^{\infty} a^k \varepsilon(k)z^{-k} = \sum_{k=0}^{\infty} a^k z^{-k} = \sum_{k=0}^{\infty} (az^{-1})^k$$

$$= \begin{cases} \frac{1}{1-az^{-1}} , & |az^{-1}| < 1 \\ \text{undefined} , & \text{elsewhere} \end{cases} \tag{6.9}$$

In the above expression, if $|az^{-1}| \geq 1$, the bilateral Z-transform has an infinite value. In such cases, the Z-transform is not defined. In this example, the Z-transform pair is given by:

$$f(k) = a^k \varepsilon(k) \leftrightarrow F(z) = \frac{z}{z - a} , \quad |z| > |a| \tag{6.10}$$

Figure 6.1 (a) highlights the ROC by shading the appropriate region in the complex Z-plane.

Note: What is the ROC of a causal sequence?

Example 6.1.4. Calculate the bilateral Z-transform of the noncausal exponential sequence:

$$f(k) = \begin{cases} b^k , & k < 0 \\ 0 , & k \geq 0 \end{cases} = b^k \varepsilon(-k - 1) .$$

Solution: By definition:

$$F(z) = \sum_{k=-\infty}^{\infty} b^k \varepsilon(-k - 1)z^{-k} = \sum_{k=-\infty}^{-1} (bz^{-1})^k$$

To make the limits of summation positive, we substitute $m = -k$ in the above equation to obtain:

$$F(z) = \sum_{m=1}^{\infty} (b^{-1}z)^m = \begin{cases} \frac{b^{-1}z}{1-b^{-1}z} , & |b^{-1}z| < 1 \\ \text{undefined} , & \text{elsewhere} . \end{cases}$$

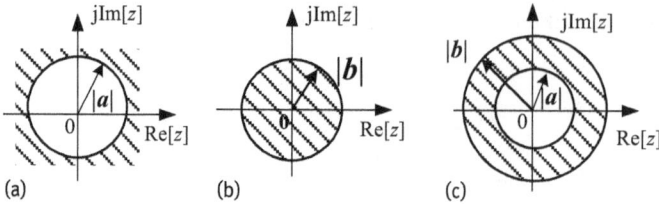

Fig. 6.1: Different cases of ROC; (a) Example 6.1.3, (b) Example 6.1.4, (c) Example 6.1.5

In this example, the Z-transform pair is given by:

$$f(k) = b^k \varepsilon(-k - 1) \leftrightarrow F(z) = -\frac{z}{z - b}, \quad |z| < |b| \tag{6.11}$$

Figure 6.1 (b) highlights the ROC by shading the appropriate region in the complex Z-plane.

! **Note:** What is the ROC of a noncausal sequence?

Example 6.1.5. Calculate the bilateral Z-transform of the bilateral exponential sequence:

$$f(k) = \begin{cases} b^k, & k < 0 \\ a^k, & k \geq 0 \end{cases}, \quad |a| < |b|$$

Solution: From the above two examples, the Z-transform can be obtained as follows:

$$F(z) = \frac{-z}{z - b} + \frac{z}{z - a}.$$

The ROC is $|a| < |z| < |b|$, which is shown in Figure 6.1 (c) as a ring-shaped region.

! **Note:** If $|a| \geq |b|$, there is no common area for ROC. The Z-transform does not exist.

From the conclusions of the above examples, the characteristics of the ROC of different sequences are summarized in Table 6.1.

Tab. 6.1: Different cases of ROC of the bilateral Z-transform

Different sequences	ROC
Finite-duration in the time domain	The entire Z-plane except $z = 0$, ∞
Causal (right-hand-sided)	Area outside the circle
Noncausal (left-hand-sided)	Area inside a circle
Bilateral (double-sided)	(If present) A ring lies between two circles

Example 6.1.6. Calculate the bilateral Laplace transform of the following functions:

$$f_1(k) = 2^k \varepsilon(k)$$
$$f_2(k) = -2^k \varepsilon(-k - 1)$$

Solution: In Examples 6.1.3 and 6.1.4, we have proved the following Z-transform pairs:

$$f_1(k) = 2^k \varepsilon(k) \leftrightarrow F_1(z) = \frac{z}{z - 2}, \quad |z| > 2 \tag{6.12}$$

$$f_2(k) = -2^k \varepsilon(-k - 1) \leftrightarrow F_2(z) = \frac{z}{z - 2}, \quad |z| < 2 \tag{6.13}$$

Although the algebraic expressions for the bilateral-transforms are the same for the two functions, the ROCs are different. This implies that a bilateral Z-transform with ROC can specify its original DT sequence.

6.1.3 Unilateral Z-transform

The bilateral Z-transform can be used to analyze both causal and noncausal LTID systems. In signal processing, most physical systems and signals are causal. The simplified bilateral Z-transform for causal signals and systems is referred to as the unilateral Z-transform, and it is obtained by assuming: $f(k) = 0, k < 0$.

Definition. The *unilateral Z-transform* of the signal $f(t)$ is defined as follows:

$$F(z) = \sum_{k=0}^{\infty} f(k) z^{-k} \tag{6.14}$$

In our subsequent discussions, we will mostly use the unilateral Z-transform. For simplicity, we will assume the "unilateral" Z-transform when referring to the Z-transform. If the bilateral Z-transform is being discussed, the term "bilateral" will be explicitly stated. The Z-transform pair is denoted by:

$$f(k) \leftrightarrow F(z) .$$

Note: For causal signals, the unilateral and bilateral Z-transforms are the same.

6.2 Basic pairs and properties of the Z-transform

6.2.1 Z-transform pairs for several elementary DT signals

Table 6.2 lists the Z-transforms for several commonly used DT sequences. Almost all of the listed transforms can be calculated by definition. Readers can refer to other references and try to prove this.

Tab. 6.2: Unilateral Z-transform pairs for several elementary sequences

f(k)	F(z)	ROC
Unit impulse $\delta(k)$	1	The entire Z-plane
Delayed unit impulse $\delta(k-m)$	z^{-m}	$|z| > 0$
Unit step $\varepsilon(k)$	$\dfrac{z}{z-1}$	$|z| > 1$
$\varepsilon(-k-1)$	$\dfrac{z}{z-1}$	$|z| < 1$
Exponential $a^k\varepsilon(k)$	$\dfrac{z}{z-a}$	$|z| > |a|$
Noncausal exponential $-a^k\varepsilon(-k-1)$	$\dfrac{z}{z-a}$	$|z| < |a|$

6.2.2 Properties of the Z-transform

In this section, we present the properties of the Z-transform. In practical applications, it is convenient to compute the Z-transform with these properties. The basic properties are given in Table 6.3. In most cases, the properties are the same for the unilateral and bilateral Z-transforms. Otherwise, it will be specified.

The following examples are the applications of the Z-transform properties.

Example 6.2.1. Calculate the bilateral Z-transform of the sequence $f(k) = 3^k[\varepsilon(k+1) - \varepsilon(k-2)]$.

Solution 1. The sequence can be expressed as:

$$f(k) = 3^k[\varepsilon(k+1) - \varepsilon(k-2)] = 3^k\varepsilon(k+1) - 3^k\varepsilon(k-2)$$

By the time-shifting property:

$$f_1(k) = 3^k\varepsilon(k+1) = 3^{-1} \cdot 3^{k+1}\varepsilon(k+1) \leftrightarrow 3^{-1} \cdot z \cdot \frac{z}{z-3}, \quad 3 < |z| < \infty$$

$$f_2(k) = 3^k\varepsilon(k-2) = 3^2 \cdot 3^{k-2}\varepsilon(k-2) \leftrightarrow 3^2 \cdot z^{-2} \cdot \frac{z}{z-3}, \quad |z| > 3$$

Using the linearity property, the Z-transform is given by:

$$f(k) = f_1(k) - f_2(k) \leftrightarrow 3^{-1} \cdot z \cdot \frac{z}{z-3} - 3^2 \cdot z^{-2} \cdot \frac{z}{z-3} = \frac{z^3 - 27}{3z(z-3)}, \quad 3 < |z| < \infty$$

Solution 2. The sequence is simplified as:

$$f(k) = 3^k[\varepsilon(k+1) - \varepsilon(k-2)] = 3^k f_1(k)$$

By the time-shifting property:

$$f_1(k) = \varepsilon(k+1) - \varepsilon(k-2) \leftrightarrow F_1(z) = z \cdot \frac{z}{z-1} - z^{-2} \cdot \frac{z}{z-1}, \quad 1 < |z| < \infty$$

Tab. 6.3: Properties of the Z-transform

Properties	$f(k)$		$F(z)$
Linearity	$af_1(k) + bf_2(k)$		$aF_1(z) + bF_2(z)$
	Bilateral	$f(k \pm m)$	$z^{\pm m} F(z)$
Time shifting	Unilateral	$f(k - m)$	$z^{-m}[F(z) + \sum_{n=-1}^{-m} f(n)z^{-n}]$
		$f(k)$ is causal: $f(k - m)$	$z^{-m} F(z)$
		$f(k + m)$	$z^m [F(z) - \sum_{n=0}^{m-1} f(n)z^{-n}]$
Time differencing	$f(k)$ is causal	$f(k) - f(k - 1)$	$(1 - z^{-1})F(z)$
Z-domain scaling	$a^k f(k)$		$F(\frac{z}{a})$
k-domain reversal (bilateral only)	$f(-k)$		$F(z^{-1})$
Time convolution	$f_1(k) * f_2(k)$		$F_1(z) \cdot F_2(z)$
Time accumulation	$\sum_{i=-\infty}^{k} f(i) = f(k) * \varepsilon(k)$		$\frac{z}{z-1} \cdot F(z)$
Z-domain differentiation	$kf(k)$		$(-z)\frac{d}{dz}F(z)$
Initial-value theorem	Right-hand-sided sequence $f(0)$		$f(0) = \lim_{z \to \infty} F(z)$
Final-value theorem	$f(\infty) = \lim_{k \to \infty} f(k)$		$f(\infty) = \lim_{z \to 1}(z - 1)F(z)$ ROC contains the unit circle

By Z-domain scaling:

$$f(k) = 3^k f_1(k) \leftrightarrow F_1\left(\frac{z}{3}\right) = \frac{\left(\frac{z}{3}\right)^2 - \left(\frac{z}{3}\right)^{-1}}{\frac{z}{3} - 1} = \frac{z^3 - 27}{3z(z - 3)}, \quad 3 < |z| < \infty$$

Example 6.2.2. Calculate the bilateral Z-transform of the sequence $f(k) = 2^{-|k|}$.

Solution: The sequence can be expressed as:

$$f(k) = 2^{-|k|} = 2^k \varepsilon(-k - 1) + 2^{-k} \varepsilon(k)$$

The Z-transforms of the exponential sequence in the right-hand side of the above expression are given by:

$$2^k \varepsilon(-k - 1) \leftrightarrow -\frac{z}{z - 2}, \quad |z| < 2$$

$$2^{-k} \varepsilon(k) \leftrightarrow \frac{z}{z - \frac{1}{2}} = \frac{2z}{2z - 1}, \quad |z| > \frac{1}{2}$$

By the linearity property, the Z-transform is obtained as follows:

$$F(z) = \frac{2z}{2z - 1} - \frac{z}{z - 2} = \frac{-3z}{(2z - 1)(z - 2)}, \quad \frac{1}{2} < |z| < 2$$

Example 6.2.3. Calculate the unilateral Z-transform of the sequence $f(k) = k\varepsilon(k)$.

Solution 1. By the time-shifting property, the sequence $f(k + 1)$ is given by:

$$f(k + 1) = (k + 1)\varepsilon(k + 1) = (k + 1)\varepsilon(k) = f(k) + \varepsilon(k)$$

! **Note:** Observe the equation $(k + 1)\varepsilon(k + 1) = (k + 1)\varepsilon(k)$ and try to understand it.

Taking the Z-transform of both sides yields:

$$zF(z) - zf(0) = F(z) + \frac{z}{z - 1}$$

$$F(z) = \frac{z}{(z - 1)^2}$$

Solution 2. The convolution of two unit steps is given by:

$$\varepsilon(k) * \varepsilon(k) = (k + 1)\varepsilon(k)$$

! **Note:** See Section 3.4.2.

The sequence $f(k) = k\varepsilon(k)$ can be expresses as follows:

$$f(k) = \varepsilon(k) * \varepsilon(k) - \varepsilon(k)$$

Using convolution properties, the Z-transform is obtained:

$$F(z) = \frac{z}{z - 1} \cdot \frac{z}{z - 1} - \frac{z}{z - 1} = \frac{z}{(z - 1)^2}$$

Solution 3. The Z-transform of the unit step is given by:

$$f_1(k) = \varepsilon(k) \leftrightarrow F_1(z) = \frac{z}{z - 1}$$

Using the Z-domain differentiation, the Z-transform is obtained:

$$f(k) = k \ f_1(k) \leftrightarrow (-z)\frac{d F_1(z)}{dz} = (-z)\frac{(z - 1) - z}{(z - 1)^2} = \frac{z}{(z - 1)^2}$$

! **Note:** $k\varepsilon(k) \leftrightarrow z/(z - 1)^2$ is a commonly used Z-transform pair.

Example 6.2.4. Calculate the Z-transform of the sequence $a^{-k}\varepsilon(-k-1)$.

Solution 1. The Z-transform of the exponential sequence is given by:

$$a^k\varepsilon(k) \leftrightarrow \frac{z}{z-a}, \quad |z| > a$$

Applying the time-shifting property yields:

$$a^{k-1}\varepsilon(k-1) \leftrightarrow \frac{z^{-1}z}{z-a} = \frac{1}{z-a}, \quad |z| > a$$

Applying the time-reversal property yields:

$$a^{-k-1}\varepsilon(-k-1) \leftrightarrow \frac{1}{z^{-1}-a}, \quad |z| < \frac{1}{a}$$

Using the linearity property, the Z-transform is obtained as follows:

$$a^{-k}\varepsilon(-k-1) \leftrightarrow \frac{a}{z^{-1}-a}, \quad |z| < \frac{1}{a}$$

Solution 2. The Z-transform of the exponential sequence is given by:

$$a^k\varepsilon(k) \leftrightarrow \frac{z}{z-a}, \quad |z| > a$$

Applying the time-reversal property yields:

$$a^{-k}\varepsilon(-k) \leftrightarrow \frac{z^{-1}}{z^{-1}-a}, \quad |z| < \frac{1}{a}$$

Note: Pay attention to the change of the convergent domain. ❗

Applying the time-shifting property yields:

$$a^{-k-1}\varepsilon(-k-1) \leftrightarrow z \cdot \frac{z^{-1}}{z^{-1}-a} = \frac{1}{z^{-1}-a}, \quad |z| < \frac{1}{a}$$

Using the linearity property, the Z-transform is obtained as follows:

$$a^{-k}\varepsilon(-k-1) \leftrightarrow \frac{a}{z^{-1}-a}, \quad |z| < \frac{1}{a}$$

Example 6.2.5. A causal periodic sequence $f_N(k)$ is given in Figure 6.2. The function in the interval $0 \sim N-1$ is referred to as $f_0(k)$. If $f_0(k) \leftrightarrow F_0(z)$, calculate the Z-transform $F(z)$ of $f_N(k)$.

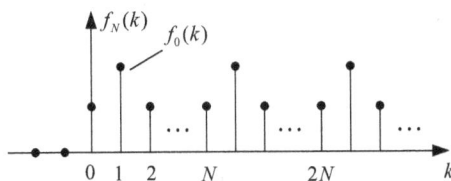

Fig. 6.2: Illustration of Example 6.2.5

Solution: The periodic sequence $f_N(k)$ is expressed as follows:

$$f_N(k) = f_0(k) + f_0(k - N) + f_0(k - 2N) + \cdots = \sum_{m=0}^{\infty} f_0(k - mN)$$

Using the time-shifting and linearity property, the Z-transform is obtained as follows:

$$\begin{aligned} F(z) &= F_0(z) + z^{-N}F_0(z) + z^{-2N}F_0(z) + \cdots \\ &= F_0(z)(1 + z^{-N} + z^{-2N} + \cdots) \\ &= \frac{F_0(z)}{1 - z^{-N}} \qquad |z| > 1 \end{aligned}$$

Note: This is similar to calculating the Laplace transform of the periodic CT signal $f_T(t) \leftrightarrow F_0(s)/(1 - e^{-sT})$.

6.3 Inverse Z-transform

The definition of the inverse Z-transform is given in Section 6.1.2. It involves the contour integration techniques and is seldom used directly. The commonly used methods to compute the inverse Z-transform are the power series method and partial fraction expansion methods.

6.3.1 Power series method

The power series method is easily applied to determine a few values of $f(k)$. If $f(k)$ is a bilateral sequence, then $F(z)$ is the power series of z and z^{-1}:

$$\begin{aligned} F(z) &= \sum_{k=-\infty}^{-1} f(k)z^{-k} + \sum_{k=0}^{\infty} f(k)z^{-k} \\ &= F_1(z) + F_2(z) \\ &= \cdots + f(-2)z^2 + f(-1)z + f(0) + f(1)z^{-1} + f(2)z^{-2} + \cdots , \quad \alpha < |z| < \beta \end{aligned} \tag{6.15}$$

This implies that the coefficients of the series determines the values of $f(k)$.

If $f(k)$ is a right-sided sequence, $F(z)$ is a power series of z^{-1}:

$$F(z) = \sum_{k=0}^{\infty} f(k)z^{-k} = f(0) + f(1)z^{-1} + f(2)z^{-2} + f(3)z^{-3} + \cdots , \quad |z| > \alpha \tag{6.16}$$

If $f(k)$ is a left-sided sequence, $F(z)$ is a power series of z:

$$F(z) = \sum_{k=-\infty}^{-1} f(k)z^{-k} = \cdots + f(-3)z^3 + f(-2)z^2 + f(-1)z , \quad |z| < \beta \tag{6.17}$$

Example 6.3.1. Given the Z-transform:

$$F(z) = \frac{z^2}{(z+1)(z-2)} = \frac{z^2}{z^2 - z - 2},$$

calculate the sequences with different ROC:

(i) $|z| > 2$

(ii) $|z| < 1$

(iii) $1 < |z| < 2$

Solution:

(i) Since ROC $|z| > 2$ is the region outside a circle, $f(k)$ is a causal sequence. Using long division (the numerator and denominator in Z-descending arrangement), the Z-transform is expanded to the power series of z^{-1}:

$$
\begin{array}{r}
1 \ + z^{-1} + 3z^{-2} + 5z^{-3} + \cdots \\
z^2 - z - 2 \overline{)\ z^2 } \\
\underline{z^2 - z \ \ -2 } \\
z \ +2 \\
\underline{z \ -1 \ \ -2z^{-1} } \\
3 \ \ +2z^{-1} \\
\cdots
\end{array}
$$

In other words:

$$F(z) = \frac{z^2}{z^2 - z - 2} = 1z^0 + z^{-1} + 3z^{-2} + 5z^{-3} + \cdots$$

Comparing this with Equation (6.16), the right-hand-sided sequence $f(k)$ is given by:

$$f(k) = \{1, 1, 3, 5, \ldots\}.$$
$$\uparrow k = 0$$

(ii) Since ROC $|z| < 1$ is the region inside a circle, $f(k)$ is a noncausal sequence. Using long division (the numerator and denominator in Z-ascending arrangement), the Z-transform is expanded to the power series of z:

$$F(z) = \frac{z^2}{z^2 - z - 2} = -\frac{1}{2}z^2 + \frac{1}{4}z^3 - \frac{3}{8}z^4 + \frac{5}{16}z^5 + \cdots$$

Comparing this with Equation (6.17), the left-hand-sided sequence $f(k)$ is given by:

$$f(k) = \left\{ \ldots, \frac{5}{16}, -\frac{3}{8}, \frac{1}{4}, -\frac{1}{2}, 0 \right\}.$$
$$\uparrow k = -1$$

(iii) Since ROC $1 < |z| < 2$ is a ring between two circles, $f(k)$ is a double-sided sequence. The Z-transform is divided two parts:

$$F(z) = \frac{z^2}{(z+1)(z-2)} = \frac{\frac{1}{3}z}{z+1} + \frac{\frac{2}{3}z}{z-2}, \quad 1 < |z| < 2.$$

According to ROC $1 < |z| < 2$, the first part is the Z-transform of right-hand side of the sequence. Likewise, the second part is the Z-transform of left-hand side of the sequence:

$$F_1(z) = \frac{\frac{1}{3}z}{z+1}, \quad |z| > 1, \quad F_2(z) = \frac{\frac{2}{3}z}{z-2}, \quad |z| < 2$$

Using long division, they are expanded to a power series of z^{-1} and z, respectively, as follows:

$$F_1(z) = \frac{1}{3} - \frac{1}{3}z^{-1} + \frac{1}{3}z^{-2} - \frac{1}{3}z^{-3} + \cdots$$

$$F_2(z) = \cdots - \frac{1}{12}z^3 - \frac{1}{6}z^2 - \frac{1}{3}z$$

Then, the double-sided sequence is given by:

$$f(k) = \left\{ \cdots, -\frac{1}{12}, -\frac{1}{6}, -\frac{1}{3}, \frac{1}{3}, -\frac{1}{3}, \frac{1}{3}, -\frac{1}{3}, \cdots \right\}$$

$$\uparrow k = 0$$

! **Note:** This method is usually used to compute the value of a given instant. We cannot obtain the mathematical function of the sequence with it.

6.3.2 Characteristic roots of the Z-transform

In LTID signals and systems analysis, the Z-transform of a sequence $f(k)$ generally takes the following rational form:

$$F(z) = \frac{N(z)}{D(z)} = \frac{b_m z^m + b_{m-1}z^{m-1} + \cdots + b_1 z + b_0}{z^n + a_{n-1}z^{n-1} + \cdots + a_1 z + a_0}, \quad m < n \quad (6.18)$$

The equation $D(z) = 0$ in Equation (6.19) is the *characteristic equation*, and its roots are defined as the *characteristic roots* or the *poles* of $F(z)$:

$$D(z) = z^n + a_{n-1}z^{n-1} + \cdots + a_1 z + a_0 = 0 \quad (6.19)$$

For an n-th-order characteristic equation, there will be n roots. Depending on the values of the coefficients, roots of the characteristic equation may be real valued or complex valued. We will further discuss the inverse Z-transform with different cases of poles.

Similar to the inverse Laplace transform, the inverse Z-transform can also be computed by the partial fraction method. The Z-transform pair of the elementary sequence in LTID system is $a^k \varepsilon(k) \leftrightarrow z/z - a$. The Z-transform is represented as $F(z)/z$ to find the factorized form by partial fraction methods:

$$\frac{F(z)}{z} = \frac{N(z)}{z(z-p_1)(z-p_2)\cdots(z-p_n)} \quad (6.20)$$

! **Note:** In Equation (6.20), we represent $F(z)/z$ not $F(z)$ in terms of its poles.

6.3.3 Real-valued and first-order poles

If none of the first-order roots are repeated, $F(z)/z$ is decomposed as:

$$\frac{F(z)}{z} = \frac{K_0}{z} + \frac{K_1}{z - p_1} + \cdots + \frac{K_n}{z - p_n} . \tag{6.21}$$

(**Note:** Why do we have $n + 1$ partial fraction coefficients?) ❗

Where the coefficients are obtained from the following expression:

$$K_i = (z - p_i)\frac{F(z)}{z}\bigg|_{z=p_i} . \tag{6.22}$$

Multiplying both sides of Equation (6.22) by z, we obtain:

$$F(z) = K_0 + \sum_{i=1}^{n} \frac{K_i z}{z - p_i} \tag{6.23}$$

The inverse transform can now be calculated by using the following transform pairs:

$$\delta(k) \leftrightarrow 1 \tag{6.24}$$

$$a^k \varepsilon(k) \leftrightarrow \frac{z}{z - a}, \quad |z| > |a| \tag{6.25}$$

$$-a^k \varepsilon(-k - 1) \leftrightarrow \frac{z}{z - a}, \quad |z| < |a| \tag{6.26}$$

The reason for performing a partial fraction expansion of $F(z)/z$ and not of $F(z)$ should now be clear. It was done so that the transform pair in Equation (6.23) can readily be applied to calculate the inverse transform. Otherwise, we would be missing the factor of z in the numerator, and the application would have been more complicated.

Example 6.3.2. Given the Z-transform:

$$F(z) = \frac{z^2}{(z + 1)(z - 2)},$$

calculate the inverse Z-transform in each case of different ROCs:
(i) $|z| > 2$
(ii) $|z| < 1$
(iii) $1 < |z| < 2$.

Solution: The characteristic roots are $z = -1$ and 2. The Z-transform can, therefore, be expressed using the fractional expansion as follows:

$$\frac{F(z)}{z} = \frac{z}{(z + 1)(z - 2)} = \frac{\frac{1}{3}}{z + 1} + \frac{\frac{2}{3}}{z - 2}.$$

$$F(z) = \frac{1}{3}\frac{z}{z + 1} + \frac{2}{3}\frac{z}{z - 2}$$

(i) Since the ROC $|z| > 2$ is the region outside a circle, $f(k)$ is a causal sequence:

$$f(k) = \left[\frac{1}{3}(-1)^k + \frac{2}{3}(2)^k \right] \varepsilon(k)$$

(ii) Since ROC $|z| < 1$ is the region inside a circle, $f(k)$ is a noncausal sequence:

$$f(k) = \left[-\frac{1}{3}(-1)^k - \frac{2}{3}(2)^k \right] \varepsilon(-k-1)$$

(iii) Since ROC $1 < |z| < 2$ is a ring between two circles, $f(k)$ is a double-sided sequence:

$$f(k) = \frac{1}{3}(-1)^k \varepsilon(k) - \frac{2}{3}(2)^k \varepsilon(-k-1)$$

Example 6.3.3. Given the Z-transform:

$$F(z) = \frac{z\left(z^3 - 4z^2 + \frac{9}{2}z + \frac{1}{2}\right)}{(z-0.5)(z-1)(z-2)(z-3)}, \quad \text{ROC}: 1 < |z| < 2 \,,$$

calculate the inverse Z-transform $f(k)$.

Solution: The Z-transform is expressed as follows:

$$\frac{F(z)}{z} = \frac{K_1}{z-0.5} + \frac{K_2}{z-1} + \frac{K_3}{z-2} + \frac{K_4}{z-3}$$

Using Equation (6.22) to calculate the coefficients, the partial fractions are given by:

$$F(z) = \underset{|z|>0.5}{\frac{-z}{z-0.5}} + \underset{|z|>1}{\frac{2z}{z-1}} + \underset{|z|<2}{\frac{-z}{z-2}} + \underset{|z|<3}{\frac{z}{z-3}}$$

where the ROC is given by noting that the first two terms are right-hand-sided sequences, and the latter two terms are left-hand-sided sequences. The inverse Z-transform is obtained as follows:

$$f(k) = -\left(\frac{1}{2}\right)^k \varepsilon(k) + 2\varepsilon(k) + (2)^k \varepsilon(-k-1) - (3)^k \varepsilon(-k-1)$$

! **Note:** Pay attention to the ROC of each partial fraction expression.

6.3.4 Complex-valued and first-order poles

Supposing there is a pair of complex poles $z_{1,2} = c \pm jd = \alpha e^{\pm j\beta}$, the partial fraction expansion is given by:

$$\frac{F(z)}{z} = \frac{K_1}{z-c-jd} + \frac{K_1^*}{z-c+jd} \qquad (6.27)$$

Using the Heaviside formula, the coefficient K_1 is given by:

$$K_1 = \left[(z - c - jd)\frac{F(z)}{z}\right]\bigg|_{z=c+jd} = |K_1|e^{j\theta}. \tag{6.28}$$

The partial fraction expansion of the Z-transform $F(z)$ is, therefore, given by:

$$F(z) = \frac{|K_1|e^{j\theta} \cdot z}{z - \alpha e^{j\beta}} + \frac{|K_1|e^{-j\theta} \cdot z}{z - \alpha e^{-j\beta}}. \tag{6.29}$$

(1) The ROC is $|z| > |\alpha|$:

Using the pair of $a^k \varepsilon(k) \leftrightarrow z/(z - a)$, $|z| > |\alpha|$, the inverse Z-transform of $F(z)$ is determined by the following deduction:

$$\begin{aligned}
f(k) &= \left[|K_1|e^{j\theta}(\alpha e^{j\beta})^k + |K_1|e^{-j\theta}(\alpha e^{-j\beta})^k\right]\varepsilon(k) \\
&= \left[|K_1|e^{j\theta}\alpha^k e^{j\beta k} + |K_1|e^{-j\theta}\alpha^k e^{-j\beta k}\right]\varepsilon(k) \\
&= |K_1|\alpha^k \left[e^{j(\beta k+\theta)} + e^{-j(\beta k+\theta)}\right]\varepsilon(k) \\
&= 2|K_1|\alpha^k \cos(\beta k + \theta)\varepsilon(k) \tag{6.30}
\end{aligned}$$

Note: Pay attention to $f(t) = 2|K_1|e^{-\alpha t}\cos(\beta t + \theta)\varepsilon(t)$ of the inverse Laplace transform. **!**

(2) The ROC is $|z| < |\alpha|$:

Using the pair of $-a^k \varepsilon(-k - 1) \leftrightarrow z/(z - a)$, $|z| < |\alpha|$, the inverse Z-transform of $F(z)$ is given by:

$$f(k) = -2|k_1|\alpha^k \cos(\beta k + \theta)\varepsilon(-k - 1) \tag{6.31}$$

Example 6.3.4. Given the Z-transform:

$$F(z) = \frac{z}{z^2 - 4z + 8}, \quad \text{ROC}: |z| > 2\sqrt{2},$$

calculate the inverse Z-transform $f(k)$.

Solution: The Z-transform is expressed as follows:

$$\frac{F(z)}{z} = \frac{K_1}{z - (2 + j2)} + \frac{K_1^*}{z - (2 - j2)}$$

Using the Heaviside formula, the coefficient K_1 is given by:

$$K_1 = \left[(z - 2 - j2))\frac{F(z)}{z}\right]\bigg|_{z=2+j2} = \frac{1}{4}e^{-j\frac{\pi}{2}}$$

The partial fraction expansion of the Z-transform $F(z)$ is, therefore, given by:

$$F(z) = \frac{1}{4}e^{-j\frac{\pi}{2}} \cdot \frac{z}{\left(z - 2\sqrt{2}e^{j\frac{\pi}{4}}\right)} + \frac{1}{4}e^{j\frac{\pi}{2}}\frac{z}{\left(z - 2\sqrt{2}e^{-j\frac{\pi}{4}}\right)}$$

The ROC is $|z| > 2\sqrt{2}$, the inverse Z-transform $f(k)$ is a causal sequence, which is obtained as follows:

$$f(k) = \frac{1}{2}(2\sqrt{2})^k \cos\left(\frac{\pi}{4}k - \frac{\pi}{2}\right)\varepsilon(k)$$

! **Note:** We can also obtain $f(k) = 1/2(2\sqrt{2})^k \cos(-\pi/4k + \pi/2)\varepsilon(k)$.

6.3.5 Real-valued and repeated poles

The above discussion is based on the partial fraction when the poles are not repeated. However, when there are multiple poles at the same location, we cannot directly calculate the coefficients corresponding to the fractions at multiple pole locations. To illustrate the partial fraction expansion for repeated poles, consider a Z-transform $F(z)$ with r repeated poles at $z = a$. The partial fraction expansion of the Z-transform $F(z)$ can be expressed as follows:

$$\frac{F(z)}{z} = \frac{K_{11}}{(z-a)^r} + \frac{K_{12}}{(z-a)^{r-1}} + \cdots + \frac{K_{1r}}{(z-a)} \tag{6.32}$$

The coefficients can be calculated as follows:

$$K_{11} = \left[(z-a)^r \frac{F(z)}{z}\right]\bigg|_{z=a} \tag{6.33}$$

$$K_{12} = \frac{d}{dz}\left[(z-a)^r \frac{F(z)}{z}\right]\bigg|_{z=a} \tag{6.34}$$

$$K_{1i} = \frac{1}{(i-1)!}\frac{d^{i-1}}{dz^{i-1}}\left[(z-a)^r \frac{F(z)}{z}\right]\bigg|_{z=a} \tag{6.35}$$

For simplicity, suppose there is a repeated pole at $z = a$ with $r = 3$. The following deduction makes it easy to remember the expressions of the inverse Z-transform. The elementary Z-transform pair is given by:

$$a^k \varepsilon(k) \longleftrightarrow \frac{z}{z-a} \tag{6.36}$$

Taking the derivation of a on both sides of Equation (6.36) yields:

$$ka^{k-1}\varepsilon(k) \longleftrightarrow \frac{z}{(z-a)^2} \tag{6.37}$$

Taking the derivation of a on both sides of Equation (6.37) yields:

$$k(k-1)a^{k-2}\varepsilon(k) \longleftrightarrow \frac{2z}{(z-a)^3} \tag{6.38}$$

And:

$$\frac{1}{2}k(k-1)a^{k-2}\varepsilon(k) \longleftrightarrow \frac{z}{(z-a)^3} \tag{6.39}$$

Example 6.3.5. The Z-transform of a right-sided sequence is given below. Calculate the inverse Z-transform $f(k)$:

$$F(z) = \frac{z^3 + z^2}{(z-1)^3}, \quad \text{ROC: } |z| > 1$$

Solution: The Z-transform is expressed as follows:

$$\frac{F(z)}{z} = \frac{z^2 + z}{(z-1)^3} = \frac{K_{11}}{(z-1)^3} + \frac{K_{12}}{(z-1)^2} + \frac{K_{13}}{z-1}$$

The coefficients are given by:

$$K_{11} = (z-1)^3 \frac{F(z)}{z}\bigg|_{z=1} = 2;$$

$$K_{12} = \frac{d}{dz}\left[(z-1)^3 \frac{F(z)}{z}\right]\bigg|_{z=1} = 3;$$

$$K_{13} = \frac{1}{2}\frac{d^2}{dz^2}\left[(z-1)^3 \frac{F(z)}{z}\right]\bigg|_{z=1} = 1.$$

The partial fraction expansion of the Z-transform $F(z)$ is obtained as follows:

$$F(z) = \frac{2z}{(z-1)^3} + \frac{3z}{(z-1)^2} + \frac{z}{z-1}$$

The inverse Z-transform is, therefore, given by:

$$f(k) = \left[\frac{2}{2!}k(k-1) + 3k + 1\right]\varepsilon(k) = (k+1)^2\varepsilon(k)$$

Note: The property of Z-transform is used to simplify the computation of the inverse Z-transform.

Example 6.3.6. The Z-transform of a right-sided sequence is given below. Calculate the inverse Z-transform $f(k)$:

$$F(z) = \frac{1}{(z-2)(z-3)}, \quad \text{ROC: } |z| > 3$$

Solution: Using the Heaviside formula, the Z-transform is directly expanded as follows:

$$F(z) = \frac{-1}{z-2} + \frac{1}{z-3}$$

The expansion is expressed as follows to match the elementary pair of $a^k\varepsilon(k) \leftrightarrow z/(z-a)$, $|z| > |a|$:

$$F(z) = z^{-1}\left(\frac{-z}{z-2} + \frac{z}{z-3}\right)$$

With the time-shifting property, the inverse Z-transform is obtained as follows:

$$f(k) = -2^{k-1}\varepsilon(k-1) + 3^{k-1}\varepsilon(k-1) = (3^{k-1} - 2^{k-1})\varepsilon(k-1)$$

6.3.6 Calculation with MATLAB

In MATLAB, the tool functions of *ztrans* and *iztrans* can be used to compute the Z-transforms and the inverse transforms, respectively.

Example 6.3.7. Calculate the Z-transform of $f(k) = \cos(ak)\varepsilon(k)$ and the inverse Z-transform of $F(z) = 1/(z+1)^2$ with MATLAB.

Solution:

(1) Compute the Z-transform:

```
f=sym('cos(a*k)');      % define f as a function
F=ztrans(f)             % compute the Z-transform
```

The operation result is:

```
F =(z-cos(a))*z/(z^2-2*z*cos(a)+1)
```

or:

$$f(k) = \cos(ak)\varepsilon(k) \leftrightarrow F(z) = \frac{z[z - \cos(a)]}{z^2 - 2z\cos(a) + 1}$$

(2) Compute the inverse Z-transform:

```
F=sym('1/(1+z)^2');     % define F
f=iztrans(F)            % compute the inverse Z-transform
```

The operation result is:

```
f =Delta(n)+(-1)^n*n-(-1)^n
```

or:

$$Z^{-1}\left\{ \frac{1}{(z+1)^2} \right\} = \delta(k) - (-1)^k\varepsilon(k) + k(-1)^k\varepsilon(k)$$

6.4 Relationship between the Laplace and Z-transforms

6.4.1 Mapping relation between S-plane and Z-plane

In Section 6.1.1, the Z-transform is defined by substituting $z = e^{sT}$. The relationship between s and z is given by:

$$z = e^{sT} \quad \text{or} \quad s = \frac{1}{T}\ln z, \tag{6.40}$$

Where T is the sampling interval, and the sampling frequency is $w_s = 2\pi/T$.

In order to illustrate the relationship between s and z, the variables are expressed as follows:

$$s = \sigma + j\omega, \quad z = re^{j\theta} \tag{6.41}$$

Substituting Equation (6.41) into Equation (6.40) yields:

$$z = re^{j\theta} = e^{sT} = e^{(\sigma+j\omega)T} = e^{\sigma T}e^{j\omega T} \tag{6.42}$$

It is clear that:

$$r = e^{\sigma T} = e^{\frac{2\pi\sigma}{\omega_S}}, \quad \theta = \omega T = 2\pi\frac{\omega}{\omega_S} \tag{6.43}$$

The mapping relationship between S-plane and Z-plane is as follows:

(1) The imaginary axis ($\sigma = 0$, $s = j\omega$) on the S-plane is mapped to the Z-plane as a unit circle $r = 1$. The right half-S-plane ($\sigma > 0$) is mapped to the area outside the unit circle ($r > 1$) of the Z-plane. The left half-S-plane ($\sigma < 0$) is mapped to the area inside the unit circle ($r < 1$) of the Z-plane.

(2) The real axis ($\omega = 0$, $s = \sigma$) of the S-plane is mapped to the positive real axis of the Z-plane. The origin ($s = 0$) is mapped to a point ($r = 1$, $\theta = 0$) at the positive real axis of the Z-plane.

(3) Since the period of $e^{j\theta}$ is ω_S, the movement along the imaginary axis in the S-plane corresponds to periodical rotation along the unit circle in the Z-plane. For each translation of ω_S along the imaginary axis in the S-plane, the mapped value z will circle around the unit circle in the Z-plane.

Note: The mapping between S- and Z-planes is not single valued.

6.4.2 Conversion from Z-transform to Laplace transform

As can be seen from Section 6.1.1, the definition of the Z-transform is derived from the Laplace transform of the ideal sampled signal. The sampled signal $f_s(t)$ is expressed as follows:

$$f_s(t) = f(t)\delta_T(t) = \sum_{k=-\infty}^{\infty} f(kT)\delta(t - kT) \tag{6.44}$$

Calculating the Laplace transform of $f_s(t)$, we obtain:

$$F(s) = \sum_{k=-\infty}^{\infty} f(kT)e^{-kTs} \tag{6.45}$$

Comparing the expression of $F(s)$ with the Z-transform:

$$F(z) = \sum_{k=-\infty}^{\infty} f(k)z^{-k}, \tag{6.46}$$

we obtain:

$$F(s) = F(z)|_{z=e^{sT}} \tag{6.47}$$

since $f(k) = f(kT)$. Equation (6.47) depicts the relationship between the Laplace transform $F(s)$ of a sampled function $f_s(t)$ and the Z-transform $F(z)$ of the DT sequence $f(k)$ obtained from the samples.

! **Note:** If we substitute $s = j\omega$ into Equation (6.47), we obtain the Fourier transform $F(j\omega) = F(z)|_{z=e^{j\omega T}}$. of the sampled function.

6.4.3 Conversion from the Laplace transform to the Z-transform

The Z-transform of the DT sequence can be obtained directly from the Laplace transform of the CT signal. If the CT signal $f(t)$ consists of exponential signals, it is expressed as follows:

$$f(t) = f_1(t) + f_2(t) + \cdots + f_N(t) = \sum_{i=1}^{N} f_i(t) = \sum_{i=1}^{N} A_i e^{p_i t} \varepsilon(t) \tag{6.48}$$

Taking the Laplace transform yields:

$$F(s) = \sum_{i=1}^{N} \frac{A_i}{s - p_i} \tag{6.49}$$

The expression of the corresponding DT sequence $f(k)$ is as follows:

$$f(k) = f_1(k) + f_2(k) + \cdots + f_N(k) = \sum_{i=1}^{N} f_i(k) = \sum_{i=1}^{N} A_i e^{p_i k T} \varepsilon(k) \tag{6.50}$$

Taking the Z-transform yields:

$$F(z) = \sum_{i=1}^{N} \frac{A_i \cdot z}{z - e^{p_i T}} \tag{6.51}$$

If $F(s)$ has n single poles p_i, the corresponding Z-transform can be directly obtained by Equation (6.51).

! **Note:** We compute single poles p_i and substitute them into Equation (6.51) to obtain the Z-transform.

Example 6.4.1. The Laplace transform of a causal signal $e^{-at}\varepsilon(t)$ is given below. Calculate the Z-transform of the corresponding sampled sequence $e^{-ak}\varepsilon(k)$:

$$F(s) = \frac{1}{s + a}$$

Solution: Given the Laplace transform:

$$F(s) = \frac{1}{s + a},$$

the pole of the Laplace transform is $p = -a$.

Using Equation (6.51), the Z-transform of $e^{-ak}\varepsilon(k)$ is directly obtained as follows:

$$F(z) = \frac{z}{z - e^{-aT}}$$

Example 6.4.2. The Laplace transform of $\sin(w_0 t)\varepsilon(t)$ is given below. Calculate the Z-transform of the corresponding sampled sequence $\sin(w_0 k)\varepsilon(k)$:

$$F(s) = \frac{w_0}{s^2 + w_0^2}$$

Solution: Given the Laplace transform:

$$F(s) = \frac{w_0}{s^2 + w_0^2} = \frac{0.5\mathrm{j}}{s + \mathrm{j}w_0} + \frac{-0.5\mathrm{j}}{s - \mathrm{j}w_0},$$

the poles in the S-plane are $p_1 = \mathrm{j}w_0$ and $p_2 = -\mathrm{j}w_0$.

Using Equation (6.51), the Z-transform of $\sin(w_0 k)\varepsilon(k)$ is directly obtained as follows:

$$F(z) = \frac{0.5\mathrm{j} \cdot z}{z - e^{-\mathrm{j}w_0 T}} + \frac{-0.5\mathrm{j} \cdot z}{z - e^{\mathrm{j}w_0 T}} = \frac{\sin(w_0 T) \cdot z}{z^2 - 2z\cos(w_0 T) + 1}$$

Table 6.4 shows the Laplace and Z-transforms of some commonly used CT and corresponding sampled DT signals.

Note: The symbol T is the sampling interval.   !

Tab. 6.4: Laplace transform and Z-transform of common signals

F(s)	f(t)	f(k) = f(kT)	F(z)
1	$\delta(t)$	$\delta(k)$	1
$\dfrac{1}{s}$	$\varepsilon(t)$	$\varepsilon(k)$	$\dfrac{z}{z-1}$
$\dfrac{1}{s^2}$	t	k	$\dfrac{z}{(z-1)^2}$
$\dfrac{1}{s+a}$	e^{-at}	e^{-aT}	$\dfrac{z}{z-e^{-aT}}$
$\dfrac{w_0}{s^2 + w_0^2}$	$\sin(w_0 t)$	$\sin(w_0 k)$	$\dfrac{\sin(w_0 T) \cdot z}{z^2 - 2z\cos(w_0 T) + 1}$
$\dfrac{s}{s^2 + w_0^2}$	$\cos(w_0 t)$	$\cos(w_0 k)$	$\dfrac{z[z - \cos(w_0 T)]}{z^2 - 2z\cos(w_0 T) + 1}$

6.5 Solution of difference equations with the Z-transform

In this section, the Z-transform is applied to solve linear, constant-coefficient difference equations. In Section 5.5, we used the Laplace transform to compute the zero-
input and zero-state responses of the differential equations. Now, we discuss a similar
solution approach based on the Z-transform to solve difference equations. The basic
steps are as follows:

Step 1: Using the time-shifting property of Z-transform, the Z-transform of both sides
of difference equation is made.

Step 2: Substituting the initial conditions in the algebraic equations, the zero-input
and zero-state solution in Z-domain are obtained.

Step 3: Using the partial fraction expansion, the inverse Z-transform is obtained to
compute the response in the time domain.

Now, we illustrate the steps through examples of solving the second-order difference
equations.

6.5.1 Analysis of computing zero-input and zero-state response

Example 6.5.1. The LTID causal system is represented by the following difference
equation:

$$y(k) + 3y(k-1) + 2y(k-2) = f(k-2) .$$

The initial condition is $y(-1) = 1$, $y(-2) = 0$ and $f(k) = \varepsilon(k)$ is applied at the input.
Determine the zero-input response $y_{zi}(k)$ and zero-state response $y_{zs}(k)$, respectively.

Solution:
(1) Zero-input response:
To calculate the zero-input response, the input signal is assumed to be zero. The difference equation reduces to:

$$y_{zi}(k) + 3y_{zi}(k-1) + 2y_{zi}(k-2) = 0 \qquad (6.52)$$

Taking the Z-transform of the difference equation with the time-shifting property
yields:

$$Y_{zi}(z) + 3z^{-1}\left[Y_{zi}(z) + y_{zi}(-1)z^{1}\right] + 2z^{-2}\left[Y_{zi}(z) + y_{zi}(-1)z^{1} + y_{zi}(-2)z^{2}\right] = 0 \quad (6.53)$$

or:

$$Y_{zi}(z) = \frac{(-3 - 2z^{-1})y_{zi}(-1) - 2y_{zi}(-2)}{1 + 3z^{-1} + 2z^{-2}} \qquad (6.54)$$

Substituting the initial conditions $y_{zi}(-1) = y(-1) = 1$, $y_{zi}(-2) = y(-2) = 0$ and using
the partial fraction expansion, the above equation is expressed as follows:

! (**Note:** Think about why $y_{zi}(-1) = y(-1)$, $y_{zi}(-2) = y(-2)$ holds.)

$$Y_{zi}(z) = \frac{-3z^2 - 2z}{(z+1)(z+2)} = \frac{z}{z+1} - \frac{4z}{z+2}, \quad |z| > 2$$

Taking the inverse Z-transform, the zero-input response is given by:

$$y_{zi}(k) = [(-1)^k - 4(-2)^k]\varepsilon(k) .$$

(2) Zero-state response:

The difference equation of the zero-state response is expressed as follows:

$$y_{zs}(k) + 3y_{zs}(k-1) + 2y_{zs}(k-2) = f(k-2)$$
$$f(k) = \varepsilon(k) , \quad y_{zs}(-1) = y_{zs}(-2) = 0$$

According to the time-shifting property, the Z-transforms on both sides of the difference equation are given by:

$$Y_{zs}(z) + 3z^{-1}y_{zs}(z) + 2z^{-2}Y_{zs}(z) = z^{-2}F(z) \tag{6.55}$$

or

$$Y_{zs}(z) = \frac{z^{-2}}{(1 + 3z^{-1} + 2z^{-2})} \cdot \frac{z}{z-1} = \frac{z}{(z+1)(z-1)(z+2)}, \quad |z| > 2 \tag{6.56}$$

Note: Think about the difference between Equations (6.55) and (6.53). !

Using the partial fraction expansion, the above equation is expressed as follows:

$$Y_{zs}(z) = \frac{1}{6}\frac{z}{(z-1)} - \frac{1}{2}\frac{z}{(z+1)} + \frac{1}{3}\frac{z}{(z+2)} .$$

Taking the inverse Z-transform, the zero-state response is given by:

$$y_{zs}(k) = \left[\frac{1}{6} \times 1^k - \frac{1}{2}(-1)^k + \frac{1}{3}(-2)^k\right]\varepsilon(k) .$$

6.5.2 Analysis of computing overall response

The overall response is the sum of the zero-input response and the zero-state response, which can be computed using the method in Section 6.5.1, respectively. Here, we give a more efficient method to compute the overall response directly and extract the component of zero-input and zero-state response.

Example 6.5.2. The LTID system is modeled by the linear, constant-coefficient difference equation:

$$y(k) + 3y(k-1) + 2y(k-2) = f(k-2) ,$$

for the initial state $y(-1) = 1$, $y(-2) = 0$ and $f(k) = \varepsilon(k)$ applied at the input. Determine the overall response.

Solution: Taking the Z-transform directly on both sides of the difference equation yields:

$$Y(z) + 3z^{-1}\left[Y(z) + y(-1)z^1\right] + 2z^{-2}\left[Y(z) + y(-1)z^1 + y(-2)z^2\right] = z^{-2}F(z) \quad (6.57)$$

Rearranging and collecting the terms corresponding to $Y(z)$ on the left-hand side of the equation results in the following:

$$(1 + 3z^{-1} + 2z^{-2})Y(z) = (-3 - 2z^{-1})y(-1) - 2y(-2) + z^{-2}F(z)$$

or:

$$Y(z) = \underbrace{\frac{(-3 - 2z^{-1})y(-1) - 2y(-2)}{1 + 3z^{-1} + 2z^{-2}}}_{Y_{zi}(z)} + \underbrace{\frac{z^{-2}}{1 + 3z^{-1} + 2z^{-2}}F(z)}_{Y_{zs}(Z)} \quad (6.58)$$

! **Note:** Refer to Section 5.5.2 and try to exact components of the zero-input and zero-state response.

Obviously, in Equation (6.58), the former term is the Z-transform of the zero-input response caused by $y(-1)$ and $y(-2)$. The latter term is the Z-transform of the zero-state response caused by $F(z)$. We can obtain the zero-input and zero-state responses, respectively, by the inverse Z-transform.

Substituting the initial conditions $y(-1) = 1$, $y(-2) = 0$ and the Z-transform $F(z) = z/(z-1)$ of the input signal yields:

$$Y(z) = \frac{-3z^3 + z^2 + 3z}{(z+1)(z-1)(z+2)}$$

Taking the partial fraction expansion, we obtain:

$$Y(z) = \frac{1}{6}\frac{z}{(z-1)} + \frac{1}{2}\frac{z}{(z+1)} - \frac{11}{3}\frac{z}{(z+2)}, \quad |z| > 2$$

The overall response is obtained by calculating the inverse Z-transformation:

$$y(k) = \left[\frac{1}{6} \times 1^k + \frac{1}{2} \times (-1)^k - \frac{11}{3}(-2)^k\right]\varepsilon(k)$$

6.6 Z-transfer function of LTID systems

6.6.1 Definition of the Z-transfer function

As was mentioned in Chapter 3, the zero-state response of the LTID system is given by the convolution sum of the impulse response $h(k)$ and the input $f(t)$:

$$y_{zs}(t) = h(t) * f(t)$$

Calculating the Z-transform of both sides of the equation, we obtain:

$$Y_{zs}(z) = H(z)F(z) \quad (6.59)$$

The *Z-transfer function* $H(z)$ can be defined as the ratio of the Z-transform of the zero-state output response and the Z-transform of the input signal. Mathematically, the Z- transfer function $H(z)$ is given by:

$$H(z) \stackrel{\text{def}}{=} \frac{Y_{zs}(z)}{F(z)} .$$ (6.60)

The Z-transfer function is only related to the system structure and component parameters, but not to the input signal and initial conditions. Given the algebraic expression and the ROC, we can compute the impulse response $h(k)$ of the LTIC system by the inverse Z-transform.

Example 6.6.1. The zero-state response of the LTID system produced by the input $f(k) = (-0.5)^k \varepsilon(k)$ is:

$$y_{zs}(k) = \left[\frac{3}{2} \left(\frac{1}{2} \right)^k + 4 \left(-\frac{1}{3} \right)^k - \frac{9}{2} \left(-\frac{1}{2} \right)^k \right] \varepsilon(k) .$$

Determine the impulse response $h(k)$ and the difference equation.

Solution: Taking the Z-transform of the input signal and the zero-state response yields:

$$f(k) = (-0.5)^k \varepsilon(k) \leftrightarrow F(z) = \frac{z}{z + 0.5}$$

$$y_{zs}(k) = \left[\frac{3}{2} \left(\frac{1}{2} \right)^k + 4 \left(-\frac{1}{3} \right)^k - \frac{9}{2} \left(-\frac{1}{2} \right)^k \right] \varepsilon(k) \leftrightarrow Y_{zs}(z) = \frac{\frac{3}{2}z}{z - \frac{1}{2}} + \frac{4z}{z + \frac{1}{3}} + \frac{-\frac{9}{2}z}{z + \frac{1}{2}}$$

Using Equation (6.60), the Z-transfer function is given by:

$$H(z) = \frac{Y_{zs}(z)}{F(z)} = \frac{z^2 + 2z}{z^2 - \frac{1}{6}z - \frac{1}{6}} = \frac{3z}{z - \frac{1}{2}} + \frac{-2z}{z + \frac{1}{3}}$$ (6.61)

Taking the inverse Z-transform, the impulse response is as follows:

$$h(k) = \left[3 \left(\frac{1}{2} \right)^k - 2 \left(-\frac{1}{3} \right)^k \right] \varepsilon(k)$$

Using Equation (6.61), the relationship between $Y_{zs}(z)$ and $F(z)$ is expressed as follows:

$$H(z) = \frac{Y_{zs}(z)}{F(z)} = \frac{z^2 + 2z}{z^2 - \frac{1}{6}z - \frac{1}{6}} = \frac{1 + 2z^{-1}}{1 - \frac{1}{6}z^{-1} - \frac{1}{6}z^{-2}}$$

$$\left(1 - \frac{1}{6}z^{-1} - \frac{1}{6}z^{-2} \right) Y_{zs}(z) = (1 + 2z^{-1})F(z)$$

Note: Pay attention to the form of z in negative exponential power. **!**

With the time-shifting property of the Z-transform, the difference equation of the system is given by:

$$y(k) - \frac{1}{6}y(k-1) - \frac{1}{6}y(k-2) = f(k) + 2f(k-1) .$$

Note: For causal signals, we have $f(k-1) \leftrightarrow z^{-1}F(z)$, $f(k-2) \leftrightarrow z^{-2}F(z)$. **!**

6.6.2 Characteristic equation, zeros and poles

In Section 5.6.2, we introduced the zeros and poles of the Laplace transfer function. In this section, we will further discuss the poles of the Z-transfer function related to the stability of LTID systems. The LTID system is assumed with a rational transfer function $H(z)$ of the following form:

$$H(z) = \frac{b_m z^m + b_{m-1} z^{m-1} + \cdots + b_1 z + b_0}{z^n + a_{n-1} z^{n-1} + \cdots + a_1 z + a_0}, \quad m < n \tag{6.62}$$

Characteristic equation: The *characteristic equation* for the transfer function in Equation (6.62) is defined as follows:

$$D(z) = z^n + a_{n-1} z^{n-1} + \cdots + a_1 z + a_0 = 0. \tag{6.63}$$

Zeros: The *zeros* of the Z-transfer function $H(z)$ are the finite locations in the complex Z-plane where $|H(z)| = 0$. For Equation (6.62), the zeros can be obtained by solving the following equation:

$$N(z) = b_m z^m + b_{m-1} z^{m-1} + \cdots + b_1 z + b_0 = 0. \tag{6.64}$$

Poles: The *poles* of the transfer function $H(z)$ are the locations in the complex Z-plane where $H(z)$ has an infinite value. For Equation (6.62), the poles can be obtained by solving the characteristic equation $D(z) = 0$.

In order to calculate the zeros and poles, the transfer function is factorized and typically represented as follows:

$$H(z) = \frac{K \prod_{j=1}^{m}(z - z_j)}{\prod_{i=1}^{n}(z - p_i)} \tag{6.65}$$

Because $D(z)$ is an n-th-order polynomial and $N(z)$ is an m-th-order polynomial, the transfer function will have a total of n poles and m zeros. However, in some cases, the location of a pole may coincide with the location of a zero. In that case, the pole and zero will cancel each other, and the actual number of poles and zeros will be reduced. The transfer function must be finite within its ROC. On the other hand, the magnitude of the transfer function is infinite at the location of a pole.

❗ Note: Can the ROC of a system include poles?

6.6.3 Nature of the shape of the impulse response for different poles

In this section, we carry out an analysis of the relation between the shape of the impulse response $h(k)$ and the typical locations of poles. Table 6.5 shows the shape of the impulse response corresponding to the first-order poles in the area of the unit circle, on the unit circle and outside the unit circle, respectively.

❗ Note: Readers can find the relation of the distribution of the second-order poles and the shape of impulse response.

Tab. 6.5: Distribution of first-order poles and the corresponding shape of $h(k)$

Poles on the Z-plane	$H(z)$	$h(k)$, $k \geq 0$	Shape of $h(k)$ in the time domain
	$\dfrac{z}{z-a}$, $\|a\| < 1$	a^k	
	$\dfrac{e^{j\theta} \cdot z}{z - \alpha e^{j\beta}} + \dfrac{e^{-j\theta} \cdot z}{z - \alpha e^{-j\beta}}$, $\|\alpha\| < 1$	$\alpha^k \cos(\beta k + \theta)$	
	$\dfrac{z}{z-1}$	$\varepsilon(k)$	
	$\dfrac{e^{j\theta} \cdot z}{z - e^{j\beta}} + \dfrac{e^{-j\theta} \cdot z}{z - e^{-j\beta}}$,	$\cos(\beta k + \theta)$	
	$\dfrac{z}{z-a}$, $\|a\| > 1$	a^k	
	$\dfrac{e^{j\theta} \cdot z}{z - \alpha e^{j\beta}} + \dfrac{e^{-j\theta} \cdot z}{z - \alpha e^{-j\beta}}$, $\|\alpha\| > 1$	$\alpha^k \cos(\beta k + \theta)$	

In Section 5.6.3, the relation between the shape of the impulse response $h(t)$ and the typical locations of poles in the S-plane was analyzed. Similarly, the relationship between the pole distribution in the Z-plane and the shape of $h(k)$ is summarized as follows:

(i) If the poles lie within the unit circle, the waveform is decaying.
(ii) The shape of $h(k)$ is a step sequence or sinusoidal sequence corresponding to the first-order poles on the unit circle. For repeated poles on the unit circle, the shape of the waveform is increasing.
(iii) If the poles lie in the region outside the unit circle, the shape of the waveform is increasing.

! **Note:** Compare this with the conclusion of the LTIC system in Section 5.6.3.

6.6.4 Stability analysis in the Z-domain

In Chapter 1, the BIBO (bounded-input, bounded-output) stable LTID system was introduced. The time-domain stability condition is as follows:

$$\sum_{n=-\infty}^{\infty} |h(k)| < \infty . \qquad (6.66)$$

According to the discussion on the shape of the impulse response with different pole locations, the Z-domain *stability condition* is stated as follows:

(i) An LTID system will be absolutely BIBO stable if the ROC includes the unit circle. The above condition does not assume the system to be causal.
(ii) A causal LTID system will be absolutely BIBO stable and causal if, and only if, all the poles lie within the unit circle. In other words, the ROC for a stable and causal system occupies the region outside and inclusive of the unit circle, which is given by $|z| > z_0$, with $z_0 < 1$.

! **Note:** Compare this with the conclusion for the LTIC system in Section 5.6.4.

Example 6.6.2. The difference equation of the LTID system is given by:

$$y(k) + 0.2y(k-1) - 0.24y(k-2) = f(k) + f(k-1) .$$

(1) Determine the Z-transfer function and the impulse response $h(k)$.
(2) Determine whether the causal system is absolutely BIBO stable.
(3) Determine the unit step response $g(k)$.

Solution: (1) Taking the Z-transform on both sides of the difference equation yields:

$$Y(z) + 0.2z^{-1}Y(z) - 0.24z^{-2}Y(z) = F(z) + z^{-1}F(z)$$

The Z-transfer function is obtained as follows:

$$H(z) = \frac{Y(z)}{F(z)} = \frac{1 + z^{-1}}{1 + 0.2z^{-1} - 0.24z^{-2}}$$

The partial fraction expansion of the Z-transform $H(z)$ is, therefore, given by:

$$H(z) = \frac{1.4z}{z - 0.4} - \frac{0.4z}{z + 0.6}, \quad |z| > 0.6$$

Taking the inverse Z-transform, the impulse response $h(k)$ is as follows:

$$h(k) = \left[1.4(0.4)^k - 0.4(-0.6)^k\right] \varepsilon(k).$$

(2) The LITD system with Z-transfer function $H(z)$ has two poles located at $z = 0.4$, -0.6. Since all the poles lie within the unit circle, the system is absolutely BIBO stable.

(3) Taking the Z-transform of the input sequence, we obtain:

$$f(k) = \varepsilon(k) \leftrightarrow F(z) = \frac{z}{z - 1}, \quad |z| > 1$$

Note: The unit step response is the zero-state response with the step sequence as the input.

Applying the convolution property, the Z-transform of the output response is given by:

$$Y(z) = F(z)H(z) = \frac{z^2(z + 1)}{(z - 1)(z - 0.4)(z + 0.6)}$$

Taking the partial fraction expansion of $Y(z)$ yields:

$$Y(z) = \frac{2.08z}{z - 1} - \frac{0.93z}{z - 0.4} - \frac{0.15z}{z + 0.6}, \quad |z| > 1$$

Taking the inverse Z-transform, the unit step response $g(k)$ is as follows:

$$g(k) = \left[2.08 - 0.93(0.4)^k - 0.15(-0.6)^k\right] \varepsilon(k)$$

6.7 Signal flow graph and LTID system simulation

6.7.1 Block diagram representation

1. Cascaded configuration

Section 5.7.1 gave the block diagram of series or cascaded LTIC systems. In this section, the signal flow graph and system simulation is realized with a similar analysis method.

The unit impulse response is the convolution sum of that of the cascaded subsystem:

$$h(k) = h_1(k) * h_2(k) * \cdots * h_n(k)$$

The corresponding Z-transform is as follows:

$$H(z) = H_1(z) \cdot H_2(z) \cdots\cdots H_n(z). \tag{6.67}$$

The block diagram in the Z-domain is presented in Figure 6.3 (b).

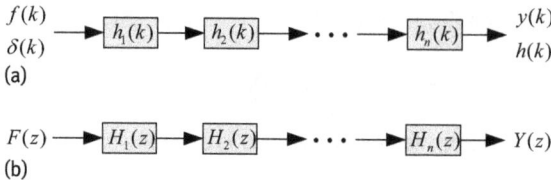

$$f(k)$$
$$\delta(k)$$ $\rightarrow$ $h_1(k)$ $\rightarrow$ $h_2(k)$ $\rightarrow$ $\cdots$ $\rightarrow$ $h_n(k)$ $\rightarrow$ $\begin{array}{l} y(k) \\ h(k) \end{array}$
(a)

$$F(z) \rightarrow H_1(z) \rightarrow H_2(z) \rightarrow \cdots \rightarrow H_n(z) \rightarrow Y(z)$$
(b)

Fig. 6.3: Block diagrams for cascaded systems, (a) Cascaded configuration for connecting LTID systems in time domain, (b) Cascaded configuration for connecting LTID systems in Z-domain

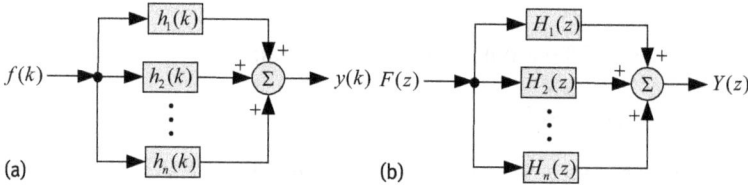

(a)   (b)

Fig. 6.4: Block diagrams for parallel systems, (a) Parallel connection in the time domain, (b) Parallel connection in the Z-domain

2. Parallel configuration

The parallel configuration between n systems is illustrated in Figure 6.4 (a). The impulse response is the sum of that of the cascaded subsystem:

$$h(k) = h_1(k) + h_2(k) + \cdots + h_n(k)$$

The corresponding Z-transform is as follows:

$$H(z) = H_1(z) + H_2(z) + \cdots + H_n(z) . \tag{6.68}$$

The block diagram in the Z-domain is presented in Figure 6.4 (b).

6.7.2 Model of basic components of LTID systems

For a linear, constant-coefficient difference equation, the model includes three basic operations: multiplication, difference and addition. These basic operations can be expressed by the ideal parts to be connected with each other, which is drawn as a block diagram. As shown in Table 6.6, the basic component units are the multiplier, delayer and adder.

! **Note:** We use the unit delay in the condition of zero state to represent the system function z^{-1}.

Tab. 6.6: Basic components of the LTID system

Components	k-domain	Z-domain
Multiplier	$f(k)\longrightarrow\boxed{a}\longrightarrow af(k)$ or $f(k)\xrightarrow{\quad a\quad} af(k)$	$F(z)\longrightarrow\boxed{a}\longrightarrow aF(z)$ or $F(z)\xrightarrow{\quad a\quad} aF(z)$
Adder	$f_1(k)\searrow^+_{\textstyle\Sigma}\nearrow f_2(k)\ _{\pm}\ \longrightarrow f_1(k)\pm f_2(k)$	$F_1(z)\searrow^+_{\textstyle\Sigma}\nearrow F_2(z)\ _{\pm}\ \longrightarrow F_1(z)\pm F_2(z)$
Delayer	$f(k)\longrightarrow\boxed{D}\longrightarrow f(k-1)$	$\begin{array}{c}\ \ \downarrow f(-1)\\[2pt]F(z)\xrightarrow{}\boxed{z^{-1}}\xrightarrow{+}\boxed{\Sigma}\ \!^+\longrightarrow z^{-1}F(z)+f(-1)\end{array}$
Delay (zero-state)	$f(k)\longrightarrow\boxed{D}\longrightarrow f(k-1)$	$F(z)\longrightarrow\boxed{z^{-1}}\longrightarrow z^{-1}F(z)$

6.7.3 Signal flow graph of LTID systems

The signal flow graph (SFG) and Mason's formula were introduced in Sections 5.7.3 and 5.7.4. The signal flow graph of an LTID system in the Z-domain is similar to that of an LTIC system, except that the integrator is replaced by a delayer. The basic concepts in the flow diagram are exactly the same, and Mason's formula is still applicable.

Example 6.7.1. The block diagram of an LTID system is given in Figure 6.5. Plot the corresponding signal flow graph.

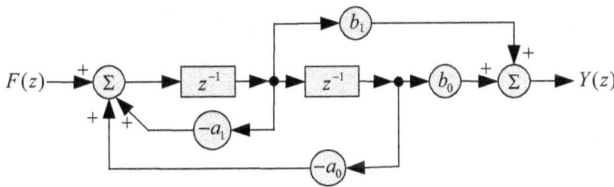

Fig. 6.5: Block diagram of the LTID system in Example 6.7.1

Solution: According to the signal flow diagram rules, the signal flow graph can be plotted directly as shown in Figure 6.6.

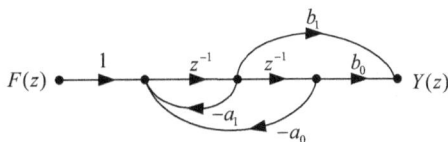

Fig. 6.6: Signal flow graph in Example 6.7.1

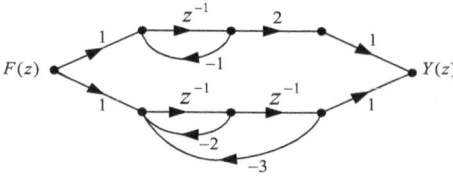

Fig. 6.7: Signal flow graph of the LTID system in Example 6.7.2

Example 6.7.2. Compute the Z-transfer function of the system in Figure 6.7 with Mason's rule.

Solution: All the loop gains are listed as follows:

$$L_1 = -z^{-1}, \quad L_2 = -2z^{-1}, \quad L_3 = -3z^{-2}$$

The determinant of the signal flow graph is calculated as follows:

$$\begin{aligned} \Delta &= 1 - (L_1 + L_2 + L_3) + (L_{12} + L_{13}) \\ &= 1 - (-z^{-1} - 2z^{-1} - 3z^{-2}) + (2z^{-2} + 3z^{-3}) \\ &= 1 + 3z^{-1} + 5z^{-2} + 3z^{-3} \end{aligned} \tag{6.69}$$

The gain of the forward path and its corresponding cofactor value are given by:

$$P_1 = 2z^{-1}, \quad \Delta_1 = 1 - (L_2 + L_3) = 1 + 2z^{-1} + 3z^{-2}$$
$$P_2 = z^{-2}, \quad \Delta_2 = 1 - L_1 = 1 + z^{-1}$$

Based on Mason's rule, the Z-transfer function is computed as follows:

$$H(z) = \frac{\sum_{i=1}^{2} P_i \Delta_i}{\Delta} = \frac{2z^{-1}(1 + 2z^{-1} + 3z^{-2}) + z^{-2}(1 + z^{-1})}{1 + 3z^{-1} + 5z^{-2} + 3z^{-3}} = \frac{2z^2 + 5z + 7}{z^3 + 3z^2 + 5z + 3} \tag{6.70}$$

6.7.4 Simulation of LTID systems

Similarly to the simulation of LTIC systems, we will discuss the realization of the LTID system. When the Z-transfer function $H(z)$ is given, the signal flow graph can also be plotted by Mason's rule to model the system. There are still three simulation methods: the direct form, the cascaded form and the parallel form.

1. Direct form structure

Example 6.7.3. The Z-transfer function is:

$$H(z) = \frac{2z + 3}{z^3 + 3z^2 + 2z + 2}.$$

Plot the signal flow graph based on Mason's rule.

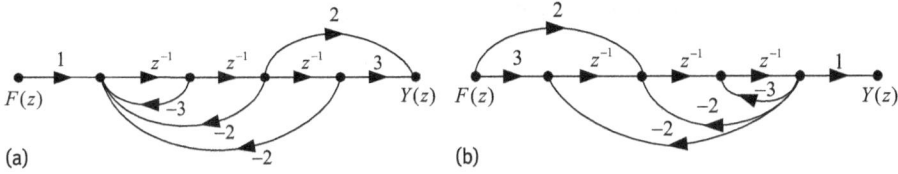

Fig. 6.8: Signal flow graph of Example 6.7.3; (a) Direct form I, (b) Direct form II

Solution: To match Mason's formula, the transfer function is reorganized as follows:

$$H(z) = \frac{2z^{-2} + 3z^{-3}}{1 - (-3z^{-1} - 2z^{-2} - 2z^{-3})} \tag{6.71}$$

By observation, the denominator of Equation (6.71) is the determinant of signal flow graph, which has three touching loops. From the numerator, we can conclude that the signal flow graph has two separated open paths.

The loop gain is given by:

$$L_1 = -3z^{-1}; \quad L_2 = -2z^{-2}; \quad L_3 = -2z^{-3}$$

The open path gain is given by:

$$P_1 = 2z^{-2}; \quad P_2 = 3z^{-3}$$

Figure 6.8 shows the signal flow diagram of two direct forms. The direct form II is the transposition of direct form I. The input and output are reversed, and all the branch directions are also reversed.

Note: For simplicity, we usually adopt the direct form I.

2. Cascaded form structure

In order to simulate a cascaded form structure, the system transfer function must be rewritten in the form of $H(z) = H_1(z) \cdot H_2(z) \cdot \ldots \cdot H_n(z)$. Each subsystem simulated in direct form is then cascaded to realize a cascaded structure. Therefore, a complicated system is divided into several first-order or second-order simple subsystems.

Example 6.7.4. The Z-transfer function is:

$$H(z) = \frac{z^2 + z}{z^2 + 5z + 6}.$$

Plot the signal flow graph in cascaded form based on Mason's rule.

Solution: The system transfer function is rewritten as:

$$H(z) = \frac{z^2 + z}{z^2 + 5z + 6} = \frac{z}{z+2} \cdot \frac{z+1}{z+3} = H_1(z) \cdot H_2(z)$$

In the above equation, $H_1(z)$ and $H_2(z)$ represent two first-order subsystems:

$$H_1(z) = \frac{z}{z+2} = \frac{1}{1 - \left(-\frac{2}{z}\right)}, \quad H_2(z) = \frac{z+1}{z+3} = \frac{1 + \frac{1}{z}}{1 - \left(-\frac{3}{z}\right)}$$

As shown in Figure 6.9, each subsystem is simulated with direct form I, and then two signal flow graphs are cascaded.

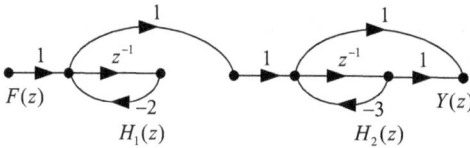

Fig. 6.9: Cascaded signal flow graph in Example 6.7.4

! **Note:** Pay attention to the simulation of $H_1(z)$.

3. Parallel form structure

In order to simulate a parallel form system, the system transfer function must be rewritten in the form of $H(z) = H_1(z) + H_2(z) + \cdots + H_n(z)$. Each subsystem simulated in direct form I is then connected in parallel to realize a parallel structure.

Example 6.7.5. The Z-transfer function is:

$$H(z) = \frac{z^2 + 4z + 4}{z^2 + 7z + 12}.$$

Plot the signal flow graph in parallel form based on Mason's rule.

Solution: The system transfer function is rewritten as:

$$H(z) = \frac{1}{z+3} + \frac{z}{z+4} = H_1(z) + H_2(z)$$

In the above equation, $H_1(z)$ and $H_2(z)$ represent two first-order subsystems:

$$H_1(z) = \frac{1}{z+3} = \frac{\frac{1}{z}}{1 - \left(-\frac{3}{z}\right)}; \quad H_2(z) = \frac{z}{z+4} = \frac{1}{1 - \left(-\frac{4}{z}\right)}$$

As shown in Figure 6.10, each subsystem is simulated with direct form I, and then two signal flow graphs are connected in parallel.

! **Note:** Pay attention to the simulation of $H_2(z)$.

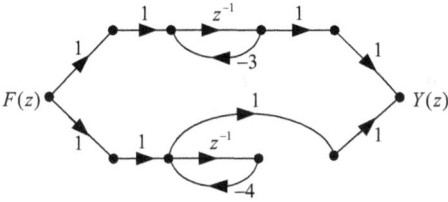

Fig. 6.10: Parallel signal flow graph in Example 6.7.5

6.8 Characteristics of frequency response

In Chapter 4, the frequency response function $H(j\omega)$ of LTIC systems was introduced. In this section, we will further discuss the frequency response function of LTID systems. The magnitude and phase spectrum can be obtained to discuss the characteristics of the system.

6.8.1 Response of LTID systems to the complex exponential sequence

In Chapter 3, we showed that the response of the LTID system is the convolution sum of the input signal and the impulse response $h(k)$. When the input signal is the complex exponential function $e^{j\Omega k}$, its response is given by:

$$y_1(k) = h(k) * e^{j\Omega k} = \sum_{m=-\infty}^{\infty} h(m)e^{j\Omega(k-m)} = e^{j\Omega k} \sum_{m=-\infty}^{\infty} h(m)(e^{j\Omega})^{-m} \tag{6.72}$$

Assuming the ROC include the unit circle, the response can be obtained from the Z-transform by substituting $z = e^{j\Omega}$, i.e.,

$$y_1(k) = e^{j\Omega k} H(z)|_{z=e^{j\Omega}} = e^{j\Omega k} H(e^{j\Omega}) \tag{6.73}$$

Similarly, the response of the input signal $e^{-j\Omega k}$ can be obtained as follows:

$$y_2(k) = h(k) * e^{-j\Omega k} = e^{-j\Omega k} H^*(z)|_{z=e^{j\Omega}} = e^{-j\Omega k} H^*(e^{j\Omega k}) \tag{6.74}$$

Note: Readers can prove it.

6.8.2 Response of LTID systems to the sinusoidal sequence

When the input signal is the sinusoidal sequence $f(k) = A \cos(\Omega k)$, its response can be computed by the linearity of the system. The input sinusoidal sequence is expressed as the sum of two complex exponential sequences:

$$f(k) = A \cos(\Omega k) = \frac{A}{2}(e^{j\Omega k} + e^{-j\Omega k}). \tag{6.75}$$

By the linear property:

$$y(k) = \frac{A}{2}[y_1(k) + y_2(k)] = \frac{A}{2}\left[e^{j\Omega k}H(e^{j\Omega}) + e^{-j\Omega k}H^*(e^{j\Omega})\right] \tag{6.76}$$

Assuming that:

$$H(e^{j\Omega T}) = |H(e^{j\Omega T})|e^{j\varphi(\Omega T)}, \quad H^*(e^{j\Omega}) = |H(e^{j\Omega})|e^{-j\varphi(\Omega)},$$

the response is given by:

$$\begin{aligned} y(k) &= \frac{A}{2}|H(e^{j\Omega})|\left[e^{j(\Omega k+\varphi(\Omega))} + e^{-j(\Omega k+\varphi(\Omega))}\right] \\ &= A|H(e^{j\Omega})|\cos(\Omega k + \varphi(\Omega)), \quad -\infty < k < \infty \end{aligned} \tag{6.77}$$

For a stable discrete-time system, the response of the sinusoidal input is still a sinusoidal sequence with the same frequency. It is referred to as sinusoidal steady-state response. The amplitude of the output is $|H(e^{j\Omega})|$ times that of the input, and the phase of the output has a phase shift $\varphi(\Omega)$.

6.8.3 Definition of frequency response of LTID systems

The condition for the existence of the frequency response function is the same as with the stability condition of the LTID system.
(i) If the ROC includes the unit circle, the frequency response function exists.
(ii) If all the poles of a causal LTID system lie within the unit circle of the Z-plane, the frequency response function exists.

The frequency response of LTID systems can be obtained from the Z-transform by substituting $z = e^{j\Omega}$:

$$H(e^{j\Omega}) = H(z)|_{z=e^{j\Omega}} = |H(e^{j\Omega})|e^{j\varphi(\Omega)}. \tag{6.78}$$

! **Note:** The frequency response function of LTIC system exists as $H(j\omega) = H(s)|_{s=j\omega}$.

The magnitude spectrum $|H(e^{j\Omega})|$ stands for the amplitude-frequency response of the system, while the phase spectrum $\varphi(\Omega)$ is referred to as the phase response of the system. The frequency response $H(e^{j\Omega})$ has the following characteristics:
(1) It is a continuous function of Ω(rad).
(2) It is a periodic function of Ω, and the period is 2π.
(3) It can be used to compute the steady-state response of sinusoidal sequences with different frequencies Ω.

Example 6.8.1. The diagram of a causal LTID system is shown in Figure 6.11. The input sequence $f(k)$ is causal, and the coefficient satisfies $0 < a < 1$. Determine the frequency response $H(e^{j\Omega})$ and draw the magnitude spectrum.

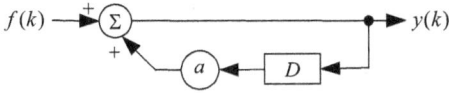

$f(k) \longrightarrow \Sigma \longrightarrow y(k)$

Fig. 6.11: Diagram of an LTID system

Solution: According to the system diagram, the difference equation is established as follows:

$$y(k) - ay(k-1) = f(k)$$

Taking the Z-transform on both sides of the equation yields:

$$Y(z) - az^{-1}Y(z) = F(z) .$$

The Z-transfer function is given by:

$$H(z) = \frac{Y(z)}{F(z)} = \frac{1}{1 - az^{-1}} = \frac{z}{z - a}, \quad |z| > a . \tag{6.79}$$

Obviously, the ROC $|z| > a$, $0 < a < 1$ contains the unit circle. Therefore, the frequency response is obtained as follows:

$$H(e^{j\Omega}) = H(z)|_{z=e^{j\Omega}} = \frac{e^{j\Omega}}{e^{j\Omega} - a} \tag{6.80}$$

The magnitude spectrum is given by:

$$\left|H(e^{j\Omega})\right| = \frac{1}{|e^{j\Omega} - a|} = \frac{1}{|\cos\Omega + j\sin\Omega - a|}$$

$$= \frac{1}{\sqrt{(\cos\Omega - a)^2 + \sin^2\Omega}} = \frac{1}{\sqrt{(1 + a^2) - 2a\cos\Omega}} \tag{6.81}$$

Assuming $a = 0.5$, we draw the magnitude spectrum in Figure 6.12:

$$\Omega = 0, \quad |H(e^{j\Omega})| = \frac{1}{1 - a} = 2$$

$$\Omega = \pi, \quad |H(e^{j\Omega})| = \frac{1}{1 + a} = \frac{2}{3}$$

$$\Omega = 2\pi, \quad |H(e^{j\Omega})| = \frac{1}{1 - a} = 2$$

Note: The magnitude spectrum is continuous and its period is 2π. ❗

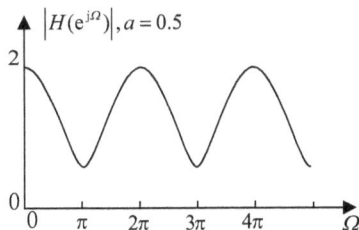

Fig. 6.12: Magnitude spectrum of the LTID system in Example 6.8.1

Example 6.8.2. Consider the system with the Z-transfer function given by:

$$H(z) = \frac{1}{2z+1}, \quad |z| > \frac{1}{2}.$$

Calculate the steady state output response of the following input sequence:

$$f(k) = 9 + 9\cos\left(\frac{\pi}{4}k\right) + 9\cos\left(\frac{\pi}{2}k + \frac{\pi}{4}\right), \quad -\infty < k < \infty.$$

Solution: For the ROC $|z| > 1/2$ including the unit circle, the frequency response is given by:

$$H(e^{j\Omega}) = \frac{1}{2e^{j\Omega} + 1} \tag{6.82}$$

(1) For the input component $f_1(k) = 9$, the frequency response is computed by substituting $\Omega = 0$ into Equation (6.82):

$$H(e^{j0}) = \frac{1}{3}$$

The response $y_1(k)$ is computed by Equation (6.77):

$$y_1(k) = 9 \times \frac{1}{3} = 3$$

(2) For the input component $f_2(k) = 9\cos((\pi/4)k)$, the response $y_2(k)$ is computed by substituting $\Omega = \pi/4$ into Equations (6.82) and (6.77):

$$H\left(e^{j\frac{\pi}{4}}\right) = \frac{1}{2e^{j\frac{\pi}{4}} + 1} = 0.36\angle -30.3°$$

$$y_2(k) = 9|H(e^{j\Omega})|\cos[\Omega k + \varphi(\Omega)] = 3.24\cos\left(\frac{\pi}{4}k - 30.3\right)$$

(3) For the input component $f_3(k) = 9\cos((\pi/2)k + \pi/4)$, the response $y_3(k)$ is computed by substituting $\Omega = \pi/2$ into Equations (6.82) and (6.77):

$$H\left(e^{j\frac{\pi}{2}}\right) = \frac{1}{2e^{j\frac{\pi}{2}} + 1} = \frac{1}{1+j2} = 0.45\angle -63.4°$$

$$y_3(k) = 9|H(e^{j\Omega})|\cos\left[\Omega k + \frac{\pi}{4} + \varphi(\Omega)\right] = 4.05\cos\left(\frac{\pi}{2}k - 18.4°\right)$$

(4) Using linearity, the steady state output response is given by:

$$y(k) = y_1(k) + y_2(k) + y_3(k)$$

$$= 3 + 3.24\cos\left(\frac{\pi}{4}k - 30.3°\right) + 4.05\cos\left(\frac{\pi}{2}k - 18.4°\right) \tag{6.83}$$

! **Note:** This is similar to CTFS analysis of

$$y(t) = \left(\frac{A_0}{2}\right)H(0) + \sum_{n=1}^{\infty} A_n|H(jn\Omega)|\cos[n\Omega t + \varphi_n + \theta(n\Omega)].$$

6.8.4 Calculation with MATLAB

MATLAB provides the function to compute and plot the zero-pole distributive diagram of the LTID system. The function is *zplane*, and its method is as follows:

```
zplane (b,a)
```

where *b* and *a* indicate the coefficient vector of Z-transfer function of the system.

MATLAB provides the function to compute the frequency response of the LTID system. The function is *freqz*, and its method is as follows:

```
freqz (num,den)
```

where *num* and *den* indicate the coefficient vector of numerator and denominator of the Z-transfer function.

Example 6.8.3. Consider the system with the Z-transfer function given by:

$$H(z) = \frac{z^{-1} + 2z^{-2} + z^{-3}}{1 - 0.5z^{-1} - 0.005z^{-2} + 0.3z^{-3}}.$$

(1) Draw the zero-pole distributive diagram of the given system.
(2) Calculate the impulse response $h(k)$ and the frequency response $H(e^{j\Omega})$.
(3) Determine the stability with MATLAB.

Solution:

```
b=[1 2 1];                 % coefficients of Numerator
a=[1 -0.5 -0.005 0.3];     % coefficients of Denominator
figure(1);zplane(b,a);     % plot the zero-pole distributive diagram
num=[0 1 2 1];
den=[1 -0.5 -0.005 0.3];
h=impz(num,den);           % impulse response
figure(2);stem(h,'.')
zlabel('k')
title('Impulse Respone')
[H,w]=freqz(num,den);      % frequency response
figure(3);plot(w/pi,abs(H))
xlabel('Fequency\omega')
title('Magnitude Response')
```

Figure 6.13 shows the results of the program. In the pole-zero plot of the system, all the poles lie within the unit circle. The system is absolutely BIBO stable. The impulse response shown in Figure 6.13 (b) is decaying and absolutely summable.

(a)

(b)

(c)

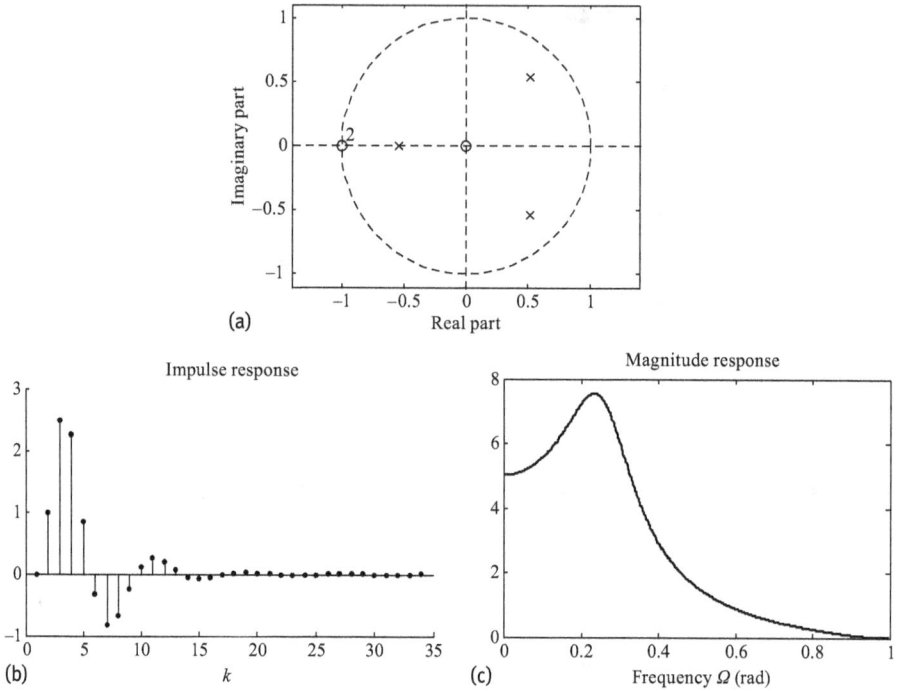

Fig. 6.13: Results of Example 6.8.3; (a) Location of poles and zeros, (b) Unit impulse response (c) Frequency response

6.9 Summary

This chapter mainly introduced Z-domain analysis methods for LTID systems. The basic signal z^k was used to obtain Z-transform and the inverse transform was defined in Section 6.3. The difference equation was solved in Section 6.5 to obtain the overall response including zero-input and zero-state responses. Section 6.6 gave the Z-transfer function $H(z) = Y(z)/F(z)$ to compute the zero-state response $y(k) = \mathcal{Z}^{-1}[F(z)H(z)]$. The frequency response $H(e^{j\Omega}) = H(z)|_{z=e^{j\Omega}}$ was analyzed in Section 6.8.

Chapter 6 problems

6.1 For the following DT sequences, calculate the bilateral Z-transforms and determine the corresponding ROC:

(1) $f(k) = \left(\dfrac{1}{2}\right)^k \varepsilon(-k-1)$

(2) $f(k) = \left(\dfrac{1}{2}\right)^{|k|}$

(3) $f(k) = \begin{cases} 2^k, & k < 0 \\ \left(\dfrac{1}{3}\right)^k, & k \geq 0 \end{cases}$

(4) $f(k) = \begin{cases} 0, & k < -4 \\ \left(\dfrac{1}{2}\right)^k, & k \geq -4 \end{cases}$

6.2 For the following DT sequences, calculate the unilateral Z-transforms and determine the corresponding ROC:

(1) $f(k) = \left(-\dfrac{1}{3}\right)^{-k} \varepsilon(k)$

(2) $f(k) = \left[\left(\dfrac{1}{2}\right)^k + \left(\dfrac{1}{3}\right)^{-k}\right] \varepsilon(k)$

(3) $f(k) = \cos\left(\dfrac{k\pi}{4}\right) \varepsilon(k)$

(4) $f(k) = \sin\left(\dfrac{k\pi}{2} + \dfrac{\pi}{4}\right) \varepsilon(k)$

6.3 Using the properties of the Z-transform, calculate the unilateral Z-transforms:

(1) $f(k) = (-1)^k k \varepsilon(k)$

(2) $f(k) = (k-1)\varepsilon(k-1)$

(3) $f(k) = k(k-1)\varepsilon(k-1)$

(4) $f(k) = (k-1)^2 \varepsilon(k-1)$

(5) $f(k) = \dfrac{a^k - b^k}{k} \varepsilon(k-1)$

(6) $f(k) = \dfrac{a^k}{k+1} \varepsilon(k)$

(7) $f(k) = \displaystyle\sum_{i=0}^{4} (-1)^i$

6.4 The Z-transform of a causal DT sequence is given below. Determine the values of $f(0), f(1), f(2)$:

$$F(z) = \dfrac{z^2}{z^2 - 3z + 2}$$

6.5 The Z-transforms of causal DT sequences are given below. Determine whether there exists $f(\infty)$ and determine the final value:

(1) $F(z) = \dfrac{z^2}{z^2 - z/6 - 1/6}$

(2) $F(z) = \dfrac{z^2}{z^2 - 5z + 6}$

6.6 The Z-transforms of DT sequences and the ROCs are given below. Calculate the inverse Z-transform in each case:

(1) $F(z) = 2z + 1 - 2z^{-2}$ $(0 < |z| < \infty)$

(2) $F(z) = \dfrac{1}{1 - az^{-1}}$ $(|z| < |a|)$

(3) $F(z) = \dfrac{z^2}{z^2 + 3z + 2}$ $(|z| > 2)$

(4) $F(z) = \dfrac{z^2}{(z - 0.5)(z - 0.25)}$ $(|z| > 0.5)$

(5) $F(z) = \dfrac{az - 1}{z - a}$ $(|z| > |a|)$

(6) $F(z) = \dfrac{z^2 + z + 1}{z^2 + z - 2}$ $(|z| > 2)$

6.7 The Z-transform of three right-sided functions is given below. Calculate the inverse Z-transform in each case:

(1) $F(z) = \dfrac{z}{z^2 - 3z + 2}$

(2) $F(z) = \dfrac{1}{(z - 0.1)(z - 0.5)(z + 0.2)}$

(3) $F(z) = \dfrac{2z(3z + 17)}{(z - 1)(z^2 - 6z + 25)}$

6.8 The Z-transform of a DT sequence is given below. Calculate the inverse Z-transform $f(k)$ in different cases of ROC:

$$F(z) = \dfrac{2z(z + 1)}{(z - 1)(z - 2)^2}$$

(1) $|z| > 2$ (2) $|z| < 1$ (3) $1 < |z| < 2$

6.9 A causal system is represented by the following difference equation:

$$y(k) - 5y(k - 1) + 6y(k - 2) = 2f(k) .$$

Calculate the zero-input response for the initial conditions $y(-1) = 1, y(-2) = -1$ and zero-state response for the input $f(k) = \varepsilon(k)$.

6.10 Consider the following difference equation of an causal system:

$$y(k) - 5y(k - 1) + 6y(k - 2) = 3f(k - 1) + 5f(k - 2) .$$

Compute the overall response for the initial conditions $y(-1) = 11/6, y(-2) = 37/36$ and the input $f(k) = \varepsilon(k)$.

6.11 The difference equation of an LTID system is given below:

$$y(k) + 4y(k - 1) + 3y(k - 2) = 4f(k) + 2f(k - 1) .$$

Determine the zero-input response and zero-state response for the input $f(k) = (-2)^k \varepsilon(k)$ and the initial conditions $y(0) = 9, y(1) = -33$.

6.12 The input–output relationship of an LTID system is given by the following difference equation:

$$y(k) - \frac{3}{4}y(k-1) + \frac{1}{8}y(k-2) = 2f(k).$$

Determine the transfer function and the impulse response of the system.

6.13 The input–output relationship of an LTID system is given by the following difference equation:

$$y(k) + \frac{1}{2}y(k-1) - \frac{1}{2}y(k-2) = f(k) - 2f(k-1).$$

(1) Determine the impulse response of the system.
(2) Calculate the zero-state response for the input

$$f(k) = \begin{cases} 2, & k = 0 \\ 1, & k = 3 \\ 0, & \text{else}. \end{cases}$$

6.14 Consider the diagram of an LTID system. Compute the zero-state response of the following inputs:

(1) $f(k) = \delta(k)$
(2) $f(k) = \varepsilon(k)$
(3) $f(k) = k\varepsilon(k)$
(4) $f(k) = \sin\left(\frac{k\pi}{3}\right)\varepsilon(k)$

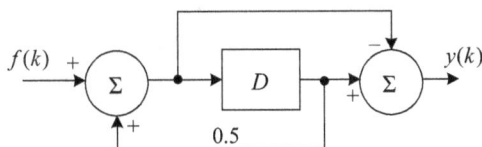

Fig. P6.1: Diagram of the LTID system in Problem 6.14

6.15 A causal LTID system with difference function is given by:

$$y(k) + y(k-1) + 0.25y(k-2) = 2f(k-1) + f(k-2)$$

(1) Calculate the transfer function $H(z)$.
(2) Compute the impulse response $h(k)$.
(3) Determine if the system is BIBO stable.

6.16 Consider the following transfer function. Find all possibilities of the ROC and determine if the system is absolutely BIBO stable in each case of ROC:

$$H_2(z) = \frac{1}{(z-0.1)(z-0.5)(z+0.2)}$$

6.17 Plot the poles and zeros of the transfer function $F(z) = -3z/(2z^2 - 5z + 2)$ and compute the inverse Z-transform in each case of ROC:

  (1) $|z| > 2$ (2) $|z| < 0.5$ (3) $0.5 < |z| < 2$

6.18 Consider the system with Z-transfer function given by:

$$H(z) = \frac{2z^2}{z^2 - \frac{3}{4}z + \frac{1}{8}}.$$

Calculate and plot the amplitude and phase spectra of the system.

6.19 The Z-transfer function is given below:

$$H(z) = \frac{2z^2 - 2z}{z^3 - 0.5z^2 + 0.25z - 0.125}.$$

Draw the signal-flow graph with direct form, cascaded form and parallel form.

6.20 The diagram of an LTID system is given below.

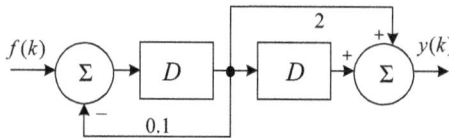

Fig. P6.2: Diagram of the LTID system in Problem 6.20

  (1) Calculate the transfer function $H(z)$.
  (2) Compute the impulse response $h(k)$.
  (3) Determine the difference equation of input-output relationship.

6.21 As is shown in Figure P6.3, a time-domain equalizer is implemented by transversal filters. When the input is $f(k) = 1/4\delta(k) + \delta(k-1) + 1/2\delta(k-2)$, the zero-state response is $y_{zs}(k)$. Given $y_{zs}(0) = 1, y_{zs}(1) = y_{zs}(3) = 0$, determine the coefficients a, b, and c.

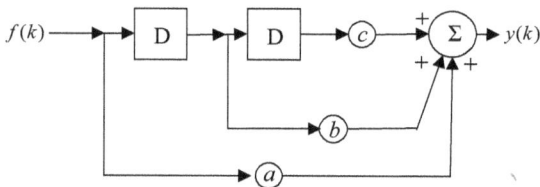

Fig. P6.3: LTID system in Problem 6.21

6.22 The signal flow graph of an LTID system is given below. Calculate the transfer function $H(z)$ using Mason's rule.

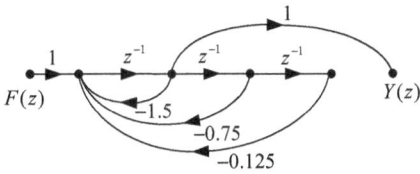

Fig. P6.4: Signal flow graph in Problem 6.22

6.23 The signal flow graph of an LTID system is given below:
   (1) Calculate the transfer function $H(z)$ using Mason's rule.
   (2) Determine the difference equation of the input–output relationship
   (3) Determine whether the system is absolutely BIBO stable.

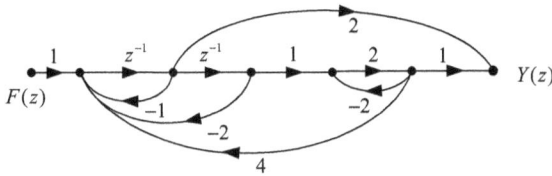

Fig. P6.5: Signal flow graph in Problem 6.23

6.24 Compute the poles and zeros and give a sketch of their locations in the complex Z-plane with MATLAB:

$$H(z) = \frac{2z(3z + 17)}{(z - 1)(z^2 - 6z + 25)}$$

6.25 Sketch the location of the zeros and poles for the given system with MATLAB:

$$H(z) = \frac{z^2 - 1}{z^2 + 0.5z + 0.5}.$$

Calculate the unit impulse response $h(k)$ and the frequency response $H(e^{j\Omega})$ with MATLAB.

7 State-space analysis of systems

Please focus on the following key questions.
1. What is the definition of the state space? What are the characteristics of state-space analysis?
2. How can we build the state equations of continuous-time and discrete-time systems?

7.0 Introduction

The methods introduced in the previous chapters deal with external descriptions for single-input single-output systems. Such methods may be inadequate in multiple-input multiple-output (MIMO) systems. It is necessary to study the internal characteristics of systems in applications of optimum or adaptive control and neural networks [26]. We need a systematic method for state-space analysis to make an internal description for systems. In state-space analysis, a set of key variables is first selected as the state variables. Then, the relationships of state variables, inputs and outputs are built as the state equation and the output equation. Lastly, the equations can be solved to obtain the system outputs.

In this chapter, we first introduce the selection of state variables in Section 7.1. The state equations of the continuous-time system and the discrete-time system are illustrated separately in Sections 7.2 and 7.3. Sections 7.4 and 7.5 give the procedure for solving the state and output equations with the Laplace transform and the Z-transform. The system transfer function and stability analysis of systems are analyzed in Section 7.6. Finally, the chapter is concluded in Section 7.7.

7.1 Basic concepts of the state space

7.1.1 State variables of systems

The state-space description of a system is introduced to realize the internal description of the system. In this approach, certain key variables are identified as the state variables of the system. These variables have the property that every possible signal in the system can be expressed as a relatively simple relationship with these state variables.

Definition. For a continuous-time system, the *state variables* $x_1(t)$, $x_2(t)$, $\ldots x_n(t)$ are the minimum number of variables to determine every possible output of the system with the given initial state and input.

https://doi.org/10.1515/9783110593907-007

For an n-th-order system, all n independent state variables are written in a column matrix form, which is called the state vector:

$$x(t) = \begin{bmatrix} x_1(t) \\ x_2(t) \\ \vdots \\ x_n(t) \end{bmatrix} \tag{7.1}$$

The state variables are the minimum number of variables of a system such that their initial values at any instant t_0 are sufficient to determine the behavior of the system for all times $t \geq t_0$ when the inputs to the system are known for $t \geq t_0$. This statement implies that an output of a system at any instant is determined completely from a knowledge of the values of the system and the input at that instant.

! **Note:** The number of variables is unique, but the state variables are not unique.

7.1.2 State equations of continuous-time and discrete-time systems

For continuous-time systems of n-th order, the state equations are n simultaneous first-order differential equations in n state variables $x_1(t)$, $x_2(t)$, ... $x_n(t)$. Given p inputs, the state function is expressed as follows:

$$\begin{cases} \dot{x}_1 = a_{11}x_1 + a_{12}x_2 + \cdots + a_{1n}x_n + b_{11}f_1 + b_{12}f_2 + \cdots + b_{1p}f_p \\ \dot{x}_2 = a_{21}x_1 + a_{22}x_2 + \cdots + a_{2n}x_n + b_{21}f_1 + b_{22}f_2 + \cdots + b_{2p}f_p \\ \vdots \\ \dot{x}_n = a_{n1}x_1 + a_{n2}x_2 + \cdots + a_{nn}x_n + b_{n1}f_1 + b_{n2}f_2 + \cdots + b_{np}f_p \end{cases} \tag{7.2}$$

These equations can be written more conveniently in matrix form:

$$\begin{bmatrix} \dot{x}_1 \\ \dot{x}_2 \\ \vdots \\ \dot{x}_n \end{bmatrix} = \begin{bmatrix} a_{11} & a_{12} & \cdots & a_{1n} \\ a_{21} & a_{22} & \cdots & a_{2n} \\ \vdots & \vdots & & \vdots \\ a_{n1} & a_{n2} & \cdots & a_{nn} \end{bmatrix} \begin{bmatrix} x_1 \\ x_2 \\ \vdots \\ x_n \end{bmatrix} + \begin{bmatrix} b_{11} & b_{12} & \cdots & b_{1p} \\ b_{21} & b_{22} & \cdots & b_{2p} \\ \vdots & \vdots & & \vdots \\ b_{n1} & b_{n2} & \cdots & b_{np} \end{bmatrix} \begin{bmatrix} f_1 \\ f_2 \\ \vdots \\ f_p \end{bmatrix} \tag{7.3}$$

or:

$$\dot{X} = AX + Bf \tag{7.4}$$

For discrete-time systems of n-th order, the state equations are n simultaneous first-order difference equations in n state variables $x_1(k)$, $x_2(k)$, ..., $x_N(k)$. Given p inputs, the state function is expressed as follows:

$$
\begin{bmatrix} x_1(k+1) \\ x_2(k+1) \\ \vdots \\ x_n(k+1) \end{bmatrix} = \begin{bmatrix} a_{11} & a_{12} & \cdots & a_{1n} \\ a_{21} & a_{22} & \cdots & a_{2n} \\ \vdots & \vdots & & \vdots \\ a_{n1} & a_{n2} & \cdots & a_{nn} \end{bmatrix} \begin{bmatrix} x_1(k) \\ x_2(k) \\ \vdots \\ x_n(k) \end{bmatrix} + \begin{bmatrix} b_{11} & b_{12} & \cdots & b_{1p} \\ b_{21} & b_{22} & \cdots & b_{2p} \\ \vdots & \vdots & & \vdots \\ b_{n1} & b_{n2} & \cdots & b_{np} \end{bmatrix} \begin{bmatrix} f_1 \\ f_2 \\ \vdots \\ f_p \end{bmatrix} \tag{7.5}
$$

These equations can be written in a convenient matrix form:

$$
\boldsymbol{X}(k+1) = \boldsymbol{A}\boldsymbol{X}(k) + \boldsymbol{B}\boldsymbol{f}(k) \tag{7.6}
$$

Note: These equations are forward difference equations. ❗

7.1.3 Output equations of continuous-time and discrete-time systems

The output equation describes the algebraic relation between the outputs, the inputs and the state variables. If there are q outputs and p inputs, the output equations for n state variables are of the form:

$$
\begin{cases} y_1 & = c_{11}x_1 + c_{12}x_2 + \cdots + c_{1n}x_n + d_{11}f_1 + d_{12}f_2 + \cdots + d_{1p}f_p \\ y_2 & = c_{21}x_1 + c_{22}x_2 + \cdots + c_{2n}x_n + d_{21}f_1 + d_{22}f_2 + \cdots + d_{2p}f_p \\ \vdots & \\ y_q & = c_{q1}x_1 + c_{q2}x_2 + \cdots + c_{qn}x_n + d_{q1}f_1 + d_{q2}f_2 + \cdots + d_{qp}f_p \end{cases} \tag{7.7}
$$

The matrix form is:

$$
\begin{bmatrix} y_1 \\ y_2 \\ \vdots \\ y_q \end{bmatrix} = \begin{bmatrix} c_{11} & c_{12} & \cdots & c_{1n} \\ c_{21} & c_{22} & \cdots & c_{2n} \\ \vdots & \vdots & & \vdots \\ c_{q1} & c_{q2} & \cdots & c_{qn} \end{bmatrix} \begin{bmatrix} x_1 \\ x_2 \\ \vdots \\ x_n \end{bmatrix} + \begin{bmatrix} d_{11} & d_{12} & \cdots & d_{1p} \\ d_{21} & d_{22} & \cdots & d_{2p} \\ \vdots & \vdots & & \vdots \\ d_{q1} & d_{q2} & \cdots & d_{qp} \end{bmatrix} \begin{bmatrix} f_1 \\ f_2 \\ \vdots \\ f_p \end{bmatrix} \tag{7.8}
$$

or:

$$
\boldsymbol{Y} = \boldsymbol{C}\boldsymbol{X} + \boldsymbol{D}\boldsymbol{f} \tag{7.9}
$$

The output equations for continuous-time and discrete-time systems are of the same form. In Equation (7.9), X, Y and f are the state vector, the output vector and the input vector, respectively.

Note: Each symbol in Equation (7.9) is a matrix. ❗

7.2 State-space description of CT systems

7.2.1 State-space description for electrical circuit systems

We now discuss a systematic procedure for determining the state-space description
of RLC networks. It is known that inductor current i_L and capacitor voltage u_C in an
RLC circuit can be used as one possible choice of state variables. Therefore, for RLC
circuits, the steps to establish state equations are summarized as follows:

(1) All the independent capacitor voltages and inductor currents are selected as state
    variables.
(2) The independent KCL and KVL equations associated with the state variables are
    listed separately.
(3) The equations are rearranged to eliminate all variables other than state variables.
    The standard form of the state equations is obtained as $\dot{X} = AX + Bf$.
(4) Observe the circuit and establish the output equations $Y = CX + Df$.

Example 7.2.1. The RLC circuit is shown in Figure 7.1, in which f_1 and f_2 are two inputs.
Taking $y_1 = u_{L2}$ and $y_2 = u_{ab}$ as outputs, find the state and output equations.

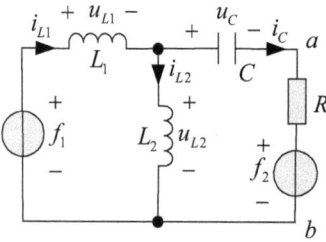

Fig. 7.1: Circuit system in Example 7.2.1

Solution: The state variables are selected as capacitor voltage and inductor currents:

$$x_1 = i_{L1}, \quad x_2 = i_{L2}, \quad x_3 = u_C$$

Then, the first-order differential equation of the first variable $x_1 = i_{L1}$ is listed with the
KVL in the outer loop formed by f_1, L_1, C, R, and f_2:

$$L_1\dot{x}_1 = u_{L1} = f_1 - f_2 - R(x_1 - x_2) - x_3$$
$$= -Rx_1 + Rx_2 - x_3 + f_1 - f_2 \tag{7.10}$$

! **Note:** To avoid introducing other variables, the capacitor current is directly expressed
as $i_C = x_1 - x_2$.

The first-order differential equation of the second variable $x_2 = i_{L2}$ is listed with the KVL in the right loop formed by L_2, C, R, and f_2:

$$L_2\dot{x}_2 = u_{L2} = x_3 + R(x_1 - x_2) + f_2$$
$$= Rx_1 - Rx_2 + x_3 + f_2 \tag{7.11}$$

The first-order differential equation of the third variable $x_3 = u_C$ is listed with the KCL:

$$C\dot{x}_3 = i_C = x_1 - x_2 \tag{7.12}$$

Therefore, the state equations can be obtained as follows:

$$\begin{cases} \dot{x}_1 = -\frac{R}{L_1}x_1 + \frac{R}{L_1}x_2 - \frac{1}{L_1}x_3 + \frac{1}{L_1}f_1 - \frac{1}{L_1}f_2 \\ \dot{x}_2 = \frac{R}{L_2}x_1 - \frac{R}{L_2}x_2 + \frac{1}{L_2}x_3 + \frac{1}{L_2}f_2 \\ \dot{x}_3 = \frac{1}{C}x_1 - \frac{1}{C}x_2 \end{cases} \tag{7.13}$$

Note: Pay attention to the standard arrangement order of state variables and inputs. ❗

This can be written in the following matrix form:

$$\begin{bmatrix} \dot{x}_1 \\ \dot{x}_2 \\ \dot{x}_3 \end{bmatrix} = \begin{bmatrix} -\frac{R}{L_1} & \frac{R}{L_1} & -\frac{1}{L_1} \\ \frac{R}{L_2} & -\frac{R}{L_2} & \frac{1}{L_2} \\ \frac{1}{C} & -\frac{1}{C} & 0 \end{bmatrix} \begin{bmatrix} x_1 \\ x_2 \\ x_3 \end{bmatrix} + \begin{bmatrix} \frac{1}{L_1} & -\frac{1}{L_1} \\ 0 & \frac{1}{L_2} \\ 0 & 0 \end{bmatrix} \begin{bmatrix} f_1 \\ f_2 \end{bmatrix} \tag{7.14}$$

The output equations can now be expressed as a linear combination of state variables and inputs. From the circuit, we have:

(**Note:** All variables other than state variables should be eliminated.) ❗

$$\begin{cases} y_1 = u_{L2} = L_2\dot{x}_2 \\ \quad = Rx_1 - Rx_2 + x_3 + f_2 \\ y_2 = u_{ab} = Ri_C + f_2 = RC\dot{x}_3 + f_2 \\ \quad = Rx_1 - Rx_2 + f_2 \end{cases} \tag{7.15}$$

or:

$$\begin{bmatrix} y_1 \\ y_2 \end{bmatrix} = \begin{bmatrix} R & -R & 1 \\ R & -R & 0 \end{bmatrix} \begin{bmatrix} x_1 \\ x_2 \\ x_3 \end{bmatrix} + \begin{bmatrix} 0 & 1 \\ 0 & 1 \end{bmatrix} \begin{bmatrix} f_1 \\ f_2 \end{bmatrix} \tag{7.16}$$

7.2.2 State-space description from differential equations

An LTIC system described by a linear constant-coefficient differential equation can be changed directly to obtain state and output equations. The following examples are given to illustrate the detailed process of establishing the state-space description from differential equations in two cases.

Example 7.2.2. The differential equation of an LTIC system is given by:

$$y'''(t) + a_2 y''(t) + a_1 y'(t) + a_0 y(t) = b_0 f(t).$$

Find the state and output equations.

Solution:

(1) Select the state variables: $x_1 = y$, $x_2 = y'$, $x_3 = y''$.

(2) It is easy to obtain the state and output equations as follows:

$$\begin{cases} \dot{x}_1 = x_2 \\ \dot{x}_2 = x_3 \\ \dot{x}_3 = -a_0 x_1 - a_1 x_2 - a_2 x_3 + b_0 f \end{cases} \tag{7.17}$$

Note: The third equation is obtained by rearranging the given differential equation:

$$y = x_1 \tag{7.18}$$

(3) The equations are written in matrix form:

$$\begin{bmatrix} \dot{x}_1 \\ \dot{x}_2 \\ \dot{x}_3 \end{bmatrix} = \begin{bmatrix} 0 & 1 & 0 \\ 0 & 0 & 1 \\ -a_0 & -a_1 & -a_2 \end{bmatrix} \begin{bmatrix} x_1 \\ x_2 \\ x_3 \end{bmatrix} + \begin{bmatrix} 0 \\ 0 \\ b_0 \end{bmatrix} f \tag{7.19}$$

$$y = \begin{bmatrix} 1 & 0 & 0 \end{bmatrix} \begin{bmatrix} x_1 \\ x_2 \\ x_3 \end{bmatrix} \tag{7.20}$$

Example 7.2.3. The differential equation of an LTIC system is given by:

$$y'''(t) + a_2 y''(t) + a_1 y'(t) + a_0 y(t) = b_1 f'(t) + b_0 f(t). \tag{7.21}$$

Find the state and output equations.

Solution: The differential equation in Equation (7.21) is decomposed into two equations by assuming an auxiliary variable $q(t)$:

$$q'''(t) + a_2 q''(t) + a_1 q'(t) + a_0 q(t) = f(t) \tag{7.22}$$

$$y(t) = b_1 q'(t) + b_0 q(t) \tag{7.23}$$

Note: A similar method of combining two equations was introduced in Section 2.1.2.

For the differential equation in Equation (7.22), the establishment process of the state equations is the same in Example 7.2.2.

The state variables are selected as:

$$x_1 = q, \quad x_2 = q', \quad x_3 = q''. \tag{7.24}$$

By observing Equation (7.22), the state equations are listed as:

$$\begin{cases} \dot{x}_1 = x_2 \\ \dot{x}_2 = x_3 \\ \dot{x}_3 = -a_0 x_1 - a_1 x_2 - a_2 x_3 + f. \end{cases} \tag{7.25}$$

By observing Equations (7.23) and (7.24), the output equations are listed as:

$$y = b_1 x_2 + b_0 x_1 \tag{7.26}$$

The state and output equations are expressed in a matrix form as follows:

$$\begin{bmatrix} \dot{x}_1 \\ \dot{x}_2 \\ \dot{x}_3 \end{bmatrix} = \begin{bmatrix} 0 & 1 & 0 \\ 0 & 0 & 1 \\ -a_0 & -a_1 & -a_2 \end{bmatrix} \begin{bmatrix} x_1 \\ x_2 \\ x_3 \end{bmatrix} + \begin{bmatrix} 0 \\ 0 \\ 1 \end{bmatrix} f \tag{7.27}$$

$$y = \begin{bmatrix} b_0 & b_1 & 0 \end{bmatrix} \begin{bmatrix} x_1 \\ x_2 \\ x_3 \end{bmatrix} \tag{7.28}$$

7.2.3 State-space description from the system diagram and the flow graph

Essentially, the system block diagram and the signal flow graph are consistent. When given either of them, the integrator output can be selected directly as the state variable to establish state equations.

Example 7.2.4. The system block diagram and flow graph of an LTIC system are given in Figure 7.2. Find the state and output equations.

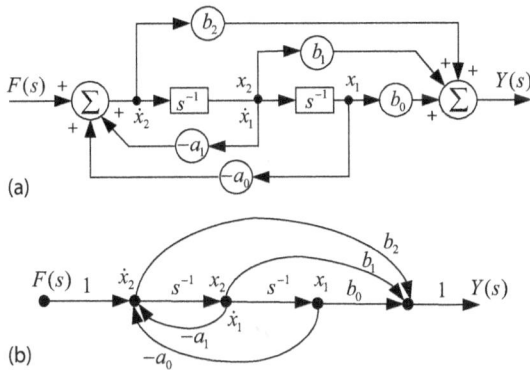

Fig. 7.2: System diagram and flow graph in Example 7.2.4; (a) System block diagram, (b) Signal flow graph

Solution: The state variables x_1, x_2 are selected as outputs of two integrators. By observation, the state equations are given by:

$$\begin{cases} \dot{x}_1 = x_2 \\ \dot{x}_2 = -a_0 x_1 - a_1 x_2 + f \end{cases} \tag{7.29}$$

! **Note:** How can we obtain the differential equation from the diagram and graph in the S-domain?

Also, the output y is given by:

$$\begin{aligned} y &= b_0 x_1 + b_1 x_2 + b_2(-a_0 x_1 - a_1 x_2 + f) \\ &= (b_0 - a_0 b_2)x_1 + (b_1 - a_1 b_2)x_2 + f \end{aligned} \tag{7.30}$$

The state equations and the output equation are written in matrix form:

$$\begin{bmatrix} \dot{x}_1 \\ \dot{x}_2 \end{bmatrix} = \begin{bmatrix} 0 & 1 \\ -a_0 & -a_1 \end{bmatrix} \begin{bmatrix} x_1 \\ x_2 \end{bmatrix} + \begin{bmatrix} 0 \\ 1 \end{bmatrix} f \tag{7.31}$$

$$y = \begin{bmatrix} b_0 - a_0 b_2 & b_1 - a_1 b_2 \end{bmatrix} \begin{bmatrix} x_1 \\ x_2 \end{bmatrix} + f \tag{7.32}$$

7.2.4 State-space description with MATLAB

In MATLAB, the tool function *tf2ss* can be used to obtain the state-space equation from the system equation. The function format is as follows:

```
[A, B, C, D]=tf2ss(num, den)
```

The input variables *num* and *den* denote the coefficients of numerator and denominator polynomial of the transfer function, respectively. The outputs A, B, C, D are the matrices of the state-space equation.

Example 7.2.5. The system transfer function is given below. Find the state and output equations:

$$H(s) = \frac{1}{s^2 + 5s + 10}$$

Solution:

```
num = [0 0 1];            % coefficients of numerator
den = [1 5 10];           % coefficients of denominator
[A,B,C,D]=tf2ss(num, den)
```

The operation result is:

$$A = \begin{bmatrix} -5 & -10 \\ 1 & 0 \end{bmatrix}, \quad B = \begin{bmatrix} 1 \\ 0 \end{bmatrix}, \quad C = \begin{bmatrix} 0 & 1 \end{bmatrix}, \quad D = 0$$

Therefore, the state-space equations of the system are given by:

$$\begin{bmatrix} \dot{x}_1 \\ \dot{x}_2 \end{bmatrix} = \begin{bmatrix} -5 & -10 \\ 1 & 0 \end{bmatrix} \begin{bmatrix} x_1 \\ x_2 \end{bmatrix} + \begin{bmatrix} 1 \\ 0 \end{bmatrix} f \tag{7.33}$$

$$y = \begin{bmatrix} 0 & 1 \end{bmatrix} \begin{bmatrix} x_1 \\ x_2 \end{bmatrix} \tag{7.34}$$

7.3 State-space description of DT systems

7.3.1 State-space description from difference equations

The state equation of LTID systems is in the form of $X(k + 1) = AX(k) + Bf(k)$. The output equation is in the form of $Y(k) = CX(k) + Df(k)$. The following examples are given to illustrate the detailed process of establishing the state-space description from difference equations.

Example 7.3.1. The difference equation of an LTID system is given by:

$$y(k) + a_1 y(k - 1) + a_0 y(k - 2) = bf(k) . \tag{7.35}$$

Find the state and output equations.

Solution: Select the state variables: $x_1(k) = y(k - 2)$, $x_2(k) = y(k - 1)$
It is easy to obtain state equations and output equations as follows:

$$\begin{cases} x_1(k + 1) = x_2(k) \\ x_2(k + 1) = -a_0 x_1(k) - a_1 x_2(k) + bf(k) \end{cases} \tag{7.36}$$

Note: The second equation is obtained by rearranging the given difference equation:

$$y(k) = -a_0 x_1(k) - a_1 x_2(k) + bf(k) \tag{7.37}$$

The equations are written in matrix form as:

$$\begin{bmatrix} x_1(k + 1) \\ x_2(k + 1) \end{bmatrix} = \begin{bmatrix} 0 & 1 \\ -a_0 & -a_1 \end{bmatrix} \begin{bmatrix} x_1(k) \\ x_2(k) \end{bmatrix} + \begin{bmatrix} 0 \\ b \end{bmatrix} f(k) \tag{7.38}$$

$$y(k) = \begin{bmatrix} -a_0 & -a_1 \end{bmatrix} \begin{bmatrix} x_1(k) \\ x_2(k) \end{bmatrix} + bf(k) \tag{7.39}$$

Example 7.3.2. The forward-difference equation of an LTID system is given by:

$$y(k + 2) + a_1 y(k + 1) + a_0 y(k) = bf(k) . \tag{7.40}$$

Find the state and output equations.

Solution: Select the state variables: $x_1(k) = y(k)$, $x_2(k) = y(k + 1)$

! **Note:** Find the rule of choosing the state variables from the above examples.

The state equations and output equations are given by:

$$\begin{cases} x_1(k + 1) = x_2(k) \\ x_2(k + 1) = -a_0 x_1(k) - a_1 x_2(k) + bf(k) \end{cases} \tag{7.41}$$

$$y(k) = x_1(k) \tag{7.42}$$

or:

$$\begin{bmatrix} x_1(k + 1) \\ x_2(k + 1) \end{bmatrix} = \begin{bmatrix} 0 & 1 \\ -a_0 & -a_1 \end{bmatrix} \begin{bmatrix} x_1(k) \\ x_2(k) \end{bmatrix} + \begin{bmatrix} 0 \\ b \end{bmatrix} f(k) \tag{7.43}$$

$$y(k) = \begin{bmatrix} 1 & 0 \end{bmatrix} \begin{bmatrix} x_1(k) \\ x_2(k) \end{bmatrix} \tag{7.44}$$

Example 7.3.3. The forward-difference equation of an LTID system is given by:

$$y(k + 2) + a_1 y(k + 1) + a_0 y(k) = b_2 f(k + 2) + b_1 f(k + 1) + b_0 f(k) . \tag{7.45}$$

Find the state and output equations.

Solution: The difference equation in Equation (7.45) is decomposed into two equations by assuming an auxiliary variable $q(k)$:

$$q(k + 2) + a_1 q(k + 1) + a_0 q(k) = f(k) \tag{7.46}$$

$$y(k) = b_2 q(k + 2) + b_1 q(k + 1) + b_0 q(k) \tag{7.47}$$

! **Note:** A similar method of combining two equations was introduced in Section 3.1.2.

For the difference equation in Equation (7.46), the establishment process of the state equations is the same as in Example 7.3.2.

The state variables are selected as:

$$x_1(k) = q(k) \quad , \quad x_2(k) = q(k + 1) \tag{7.48}$$

By observing Equation (7.46), the state equations are listed as:

$$\begin{cases} x_1(k+1) = x_2(k) \\ x_2(k+1) = -a_0 x_1(k) - a_1 x_2(k) + f(k) \end{cases} \tag{7.49}$$

By observing Equations (7.47) and (7.48), the output equations are listed as:

$$\begin{aligned} y(k) &= b_2 q(k+2) + b_1 q(k+1) + b_0 q(k) \\ &= b_2[-a_0 x_1(k) - a_1 x_2(k) + f(k)] + b_1 x_2(k) + b_0 x_1(k) \\ &= (b_0 - a_0 b_2)x_1(k) + (b_1 - a_1 b_2)x_2(k) + b_2 f(k) \end{aligned} \tag{7.50}$$

Note: All variables other than state variables should be eliminated. !

The state and output equations are expressed in a matrix form as follows:

$$\begin{bmatrix} x_1(k+1) \\ x_2(k+1) \end{bmatrix} = \begin{bmatrix} 0 & 1 \\ -a_0 & -a_1 \end{bmatrix} \begin{bmatrix} x_1(k) \\ x_2(k) \end{bmatrix} + \begin{bmatrix} 0 \\ 1 \end{bmatrix} f(k) \tag{7.51}$$

$$y(k) = [b_0 - a_0 b_2 \quad b_1 - a_1 b_2] \begin{bmatrix} x_1(k) \\ x_2(k) \end{bmatrix} + b_2 f(k) \tag{7.52}$$

7.3.2 State-space description from system diagrams and flow graphs

Section 7.2.3 shows the procedure of establishing state equations for continuous-time systems. For discrete-time systems, the stable variables are directly selected at the de-layer output to establish the state-space equations.

Example 7.3.4. The system diagram and flow graph of an LTID system are given in Figure 7.3. Find the state and output equations.

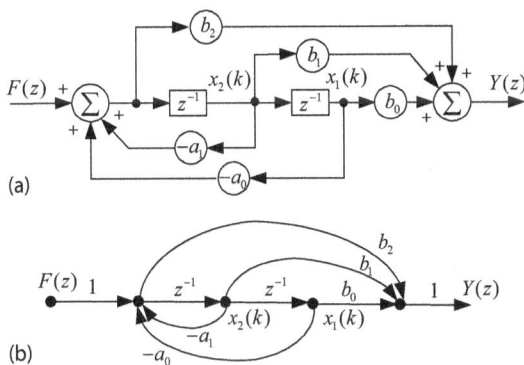

Fig. 7.3: System diagram and flow graph in Example 7.3.4;
(a) System block diagram, (b) Signal flow graph

Solution: The state variables $x_1(k)$, $x_2(k)$ are selected as two delayer outputs. By observation, the state equations are given by:

$$\begin{cases} x_1(k+1) = x_2(k) \\ x_2(k+1) = -a_0 x_1(k) - a_1 x_2(k) + f(k) \end{cases} \tag{7.53}$$

! **Note:** How can we obtain the difference equation from the diagram and graph in the Z-domain?

The output $y(k)$ is given by:

$$\begin{aligned} y(k) &= b_0 x_1(k) + b_1 x_2(k) + b_2[-a_0 x_1(k) - a_1 x_2(k) + f(k)] \\ &= (b_0 - a_0 b_2)x_1(k) + (b_1 - a_1 b_2)x_2(k) + f(k) \end{aligned} \tag{7.54}$$

The state equations and the output equation are written in matrix forms:

$$\begin{bmatrix} x_1(k+1) \\ x_2(k+1) \end{bmatrix} = \begin{bmatrix} 0 & 1 \\ -a_0 & -a_1 \end{bmatrix} \begin{bmatrix} x_1(k) \\ x_2(k) \end{bmatrix} + \begin{bmatrix} 0 \\ 1 \end{bmatrix} f(k) \tag{7.55}$$

$$y(k) = \begin{bmatrix} b_0 - a_0 b_2 & b_1 - a_1 b_2 \end{bmatrix} \begin{bmatrix} x_1(k) \\ x_2(k) \end{bmatrix} + f(k) \tag{7.56}$$

In conclusion, the methods of establishing the state equations from system diagrams or flow graphs are applicable for the system transfer function. For LTIC and LTID systems, we can draw the system flow graph in direct form I using Mason's formula. Then, the state variables are selected as the outputs of each integrator s^{-1} or delayer z^{-1}.

! **Note:** How can we determine the state equations of the given transfer function $H(s) = 1/(s+a)$?

7.4 Solution of state-space equations of LTIC systems

7.4.1 Laplace transform solution of state equations

The state equations of an LTIC system are N simultaneous linear differential equations of the first order. These equations can be solved in both the time and frequency domains (Laplace transform). The former is complex and is not discussed here. The latter is relatively easier to deal with than the time-domain solution.

The state equation is of the matrix form:

$$\dot{X}(t) = AX(t) + Bf(t) \tag{7.57}$$

Taking the Laplace transform on both sides yields:

$$sX(s) - X(0_-) = AX(s) + BF(s) \tag{7.58}$$

Rearranging and collecting the terms corresponding to $X(s)$ on the left-hand side of the equation results in the following:

$$(sI - A)X(s) = X(0_-) + BF(s) \tag{7.59}$$

where I is the $N \times N$ identity matrix. From Equation (7.59), we have:

$$X(s) = (sI - A)^{-1}X(0_-) + (sI - A)^{-1}BF(s) \tag{7.60}$$

where:

$$\Phi(s) = (sI - A)^{-1} \tag{7.61}$$

Thus, Equation (7.60) reduces to:

$$X(s) = \Phi(s)X(0_-) + \Phi(s)BF(s) \tag{7.62}$$

Finally, the time-domain expression can be obtained by the inverse Laplace transform:

$$x(t) = \underbrace{\mathcal{L}^{-1}[\Phi(s)X(0_-)]}_{\text{zero-input component}} + \underbrace{\mathcal{L}^{-1}[\Phi(s)BF(s)]}_{\text{zero-state component}} \tag{7.63}$$

The above formula gives the desired solution. Observe the two components of the solution. The first component is the zero-input component when the input $f(t) = 0$. In a similar manner, the second component is the zero-state component.

7.4.2 Laplace transform solution of output equations

The output equation of an LTIC system is given by:

$$Y(t) = CX(t) + Df(t) \tag{7.64}$$

Taking the Laplace transform on both sides yields:

$$Y(s) = CX(s) + DF(s) \tag{7.65}$$

Substituting Equation (7.62) into this equation, we have:

$$Y(s) = C\Phi(s)X(0_-) + [C\Phi(s)B + D]F(s) \tag{7.66}$$

Similarly, the first component is the zero-input response, and the second component is the zero-state response in the S-domain:

$$Y_{zi}(s) = C\Phi(s)X(0_-)$$
$$Y_{zs}(s) = [C\Phi(s)B + D]F(s) \tag{7.67}$$

The time-domain solution can be obtained by using the inverse Laplace transform on the two above equations:

$$y_{zi}(t) = \mathcal{L}^{-1}[C\Phi(s)X(0_-)]$$
$$y_{zs}(t) = \mathcal{L}^{-1}\{[C\Phi(s)B + D]F(s)\}$$
(7.68)

From Equation (7.67), the zero-state output in the S-domain is obtained as follows:

$$Y_{zs}(s) = [C\Phi(s)B + D]F(s) = H(s) \cdot F(s)$$
(7.69)

Note that the transfer function of a system is defined as $H(s) = Y_{zs}(s)/F(s)$. The transfer function matrix of the system is given by:

$$H(s) = C\Phi(s)B + D = C(sI - A)^{-1}B + D$$
(7.70)

❗ **Note:** The ij-th element $H_{ij}(s)$ of the $H(s)$ matrix is the transfer function that relates the output $y_i(t)$ to the input $x_j(t)$.

Example 7.4.1. The state equation and output equation of an LTIC system are as follows:

$$\begin{cases} x_1'(t) = x_1(t) + f(t) \\ x_2'(t) = x_1(t) - 3x_2(t) \end{cases} \quad , \quad y(t) = -\frac{1}{4}x_1(t) + x_2(t)$$

The initial conditions are $x_1(0_-) = 1$, $x_2(0_-) = 1$. Determine the zero-input response and zero-state response of the output.

Solution: Let us first find $\Phi(s) = (sI - A)^{-1}$. We have:

$$[sI - A] = s\begin{bmatrix} 1 & 0 \\ 0 & 1 \end{bmatrix} - \begin{bmatrix} 1 & 0 \\ 1 & -3 \end{bmatrix} = \begin{bmatrix} s-1 & 0 \\ -1 & s+3 \end{bmatrix}$$

and:

$$(sI - A)^{-1} = \frac{1}{(s-1)(s+3)}\begin{bmatrix} s+3 & 0 \\ 1 & s-1 \end{bmatrix} = \begin{bmatrix} \frac{1}{s-1} & 0 \\ \frac{1}{(s-1)(s+3)} & \frac{1}{s+3} \end{bmatrix}$$

The initial condition is given as:

$$X(0_-) = \begin{bmatrix} 1 \\ 2 \end{bmatrix}$$

Therefore, the zero-input and zero-state component in the S-domain can be obtained by Equation (7.67):

$$Y_{zi}(s) = C(sI - A)^{-1}X(0_-)$$

$$= \begin{bmatrix} -\frac{1}{4} & 1 \end{bmatrix}\begin{bmatrix} \frac{1}{s-1} & 0 \\ \frac{1}{(s-1)(s+3)} & \frac{1}{s+3} \end{bmatrix}\begin{bmatrix} 1 \\ 2 \end{bmatrix} = \frac{7}{4} \cdot \frac{1}{(s+3)}$$

$$Y_{zs}(s) = [C(sI - A)^{-1}B + D]F(s)$$

$$= \begin{bmatrix} -\frac{1}{4} & 1 \end{bmatrix}\begin{bmatrix} \frac{1}{s-1} & 0 \\ \frac{1}{(s-1)(s+3)} & \frac{1}{s+3} \end{bmatrix}\begin{bmatrix} 1 \\ 0 \end{bmatrix} \cdot \frac{1}{s} = \frac{1}{12} \cdot \left(\frac{1}{(s+3)} - \frac{1}{s}\right)$$

The responses in the time domain are given by the inverse Laplace transform:

$$y_{zi}(t) = \frac{7}{4}e^{-3t}\varepsilon(t), \quad y_{zs}(t) = \frac{1}{12}(e^{-3t} - 1)\varepsilon(t)$$

Note: We should obtain the matrices A, B, C, D correctly and grasp the basic matrix ❗
operations, including inversion and multiplication.

7.4.3 Calculation with MATLAB

In MATLAB, the tool function *lsim* can be used to compute zero-input and zero-state
responses from the state-space matrices. The function format is as follows:

```
[y,v]=lsim(A,B,C,D,X,t,v0)
```

The inputs A, B, C, D are coefficient matrices; X is the input, t is the time interval, and
$v0$ is the initial condition. The output y is the response.

Example 7.4.2. Find the zero-input and zero-state response for the system. Its state
equation and output equation are as follows:

$$\begin{bmatrix} \dot{x}_1(t) \\ \dot{x}_2(t) \end{bmatrix} = \begin{bmatrix} -2 & -2 \\ 1 & 0 \end{bmatrix} \begin{bmatrix} x_1(t) \\ x_2(t) \end{bmatrix} + \begin{bmatrix} 10 \\ 0 \end{bmatrix} f(t), \quad y(t) = \begin{bmatrix} 1 & 0 \end{bmatrix} \begin{bmatrix} x_1(t) \\ x_2(t) \end{bmatrix}$$

where the input is $f(t) = t\varepsilon(t)$. The initial conditions are given by:

$$\begin{bmatrix} x_1(0) \\ x_2(0) \end{bmatrix} = \begin{bmatrix} 5 \\ 0 \end{bmatrix}.$$

Solution:

```
% Compute zero-input response
A=[-2 -2;1 0]; B=[10;0]; C=[1 0]; D=[0];
v0=[5;0]                           % initial condition
t=0:.01:5;
X=[0*ones(size(t))]';              % zero-input
[y,v]=lsim(A,B,C,D,X,t,v0);        % calculate zero-input response
plot(t,y);grid;xlabel('t');ylabel('y');title('zero-input response');
% Compute zero-state response
v0=[0;0];                          % zero state
X=[1*t]';                          % input signal t
[y,v]=lsim(A,B,C,D,X,t,v0);        % calculate zero-state response
plot(t,y);grid;xlabel('t');ylabel('y');title('zero-state');
```

The operation result is shown in Figure 7.4.

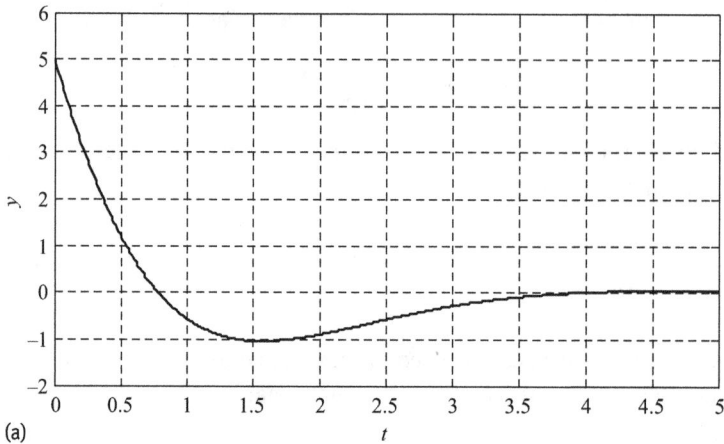

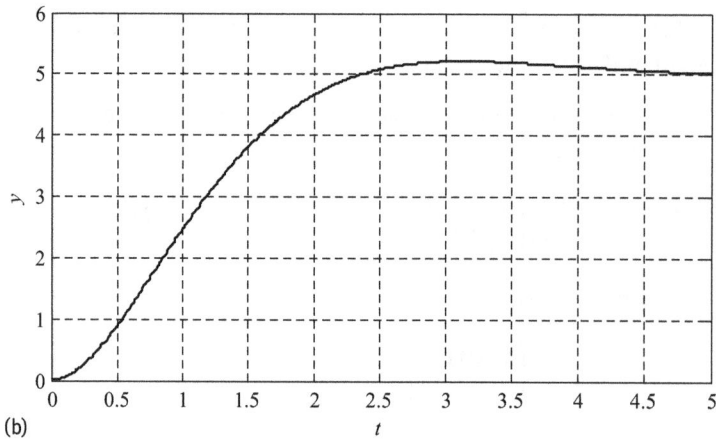

Fig. 7.4: Zero-input response and zero-state response; (a) Zero-input response, (b) Zero-state response

7.5 Solution of state-space equations of LTID systems

7.5.1 Z-transform solution of state equations

The solution of the state equations of the LTID system is similar to that of those of the LTIC system. The state equations of an LTID system are N simultaneous linear difference equations of the first order. These equations can be solved in both the time and Z-domains. We only illustrate the Z-domain solution.

The state equation is of the matrix form:

$$X(k + 1) = AX(k) + Bf(k) \tag{7.71}$$

Taking the Z-transform on both sides yields:

$$zX(z) - zX(0) = AX(z) + BF(z) \tag{7.72}$$

Rearranging and collecting the terms corresponding to $X(z)$ on the left-hand side of the equation results in the following:

$$(zI - A)X(z) = zX(0) + BF(z) \tag{7.73}$$

where I is the $N \times N$ identity matrix. From Equation (7.73), we have:

$$X(z) = (zI - A)^{-1}zX(0) + (zI - A)^{-1}BF(z) \tag{7.74}$$

where:

$$\Phi(z) = (zI - A)^{-1} \tag{7.75}$$

Thus, Equation (7.74) reduces to:

$$X(z) = \Phi(z)zX(0_-) + \Phi(z)BF(z) \tag{7.76}$$

Finally, the time-domain expression can be obtained by the inverse Z-transform:

$$x(k) = \underbrace{Z^{-1}[\Phi(z)zX(0_-)]}_{\text{zero-input component}} + \underbrace{Z^{-1}[\Phi(z)BF(z)]}_{\text{zero-state component}} \tag{7.77}$$

Note: Pay attention to the difference between Equations (7.77) and (7.63). ❗

The above formula gives the desired solution. Observe the two components of the solution. The first component is the zero-input component when the input $f(k) = 0$. In a similar manner, the second component is the zero-state component.

7.5.2 Z-transform solution of output equations

The output equation of an LTID system is given by:

$$Y(k) = CX(k) + Df(k) \tag{7.78}$$

Taking the Z-transform on both sides yields:

$$Y(z) = CX(z) + DF(z) \tag{7.79}$$

Substituting Equation (7.76) into this equation, we have:

$$Y(z) = C\Phi(z)zX(0) + [C\Phi(z)B + D]F(z) \tag{7.80}$$

Similarly, the first component is the zero-input component, and the second component is the zero-state component in the S-domain:

$$\begin{aligned} Y_{zi}(z) &= C\Phi(z)zX(0_-) \\ Y_{zs}(z) &= [C\Phi(z)B + D]F(z) \end{aligned} \tag{7.81}$$

The time-domain solution can be obtained by using the inverse Laplace transform on the two above equations:

$$y_{zi}(k) = Z^{-1}[C\Phi(z)zX(0_-)]$$
$$y_{zs}(k) = Z^{-1}\{[C\Phi(z)B + D]F(z)\}$$

(7.82)

From Equation (7.81), the zero-state response in the Z-domain is obtained as follows:

$$Y_{zs}(z) = [C\Phi(z)B + D]F(z) = H(z) \cdot F(z)$$

(7.83)

Note that the transfer function of a system is defined as $H(z) = Y_{zs}(z)/F(z)$. The transfer function matrix of the LTID system is given by:

$$H(z) = C(zI - A)^{-1}B + D$$

(7.84)

! **Note:** The ij-th element $H_{ij}(z)$ of the $H(z)$ matrix is the transfer function that relates the output $y_i(k)$ to the input $x_j(k)$.

Example 7.5.1. Consider the system block diagram in Figure 7.5. Find the zero-state output of the given LTID system with two inputs $f_1(k) = \delta(k)$ and $f_2(k) = \varepsilon(k)$.

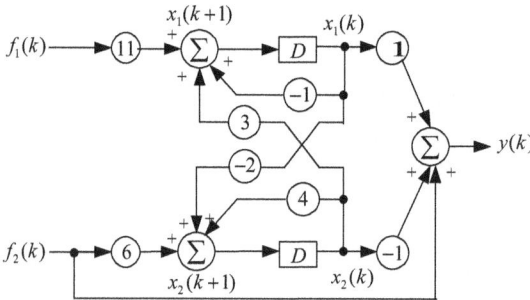

Fig. 7.5: System diagram in Example 7.5.1

Solution:
(1) The state variables are selected as the outputs of two delayers. The state-space equations are established as follows:

$$\begin{cases} x_1(k + 1) = -x_1(k) + 3x_2(k) + 11f_1(k) \\ x_2(k + 1) = -2x_1(k) + 4x_2(k) + 6f_2(k) \\ \quad\quad y(k) = x_1(k) - x_2(k) + f_2(k) \end{cases}$$

The four matrices are:

$$A = \begin{bmatrix} -1 & 3 \\ -2 & 4 \end{bmatrix}, \quad B = \begin{bmatrix} 11 & 0 \\ 0 & 6 \end{bmatrix}, \quad C = \begin{bmatrix} 1 & -1 \end{bmatrix}, \quad D = \begin{bmatrix} 0 & 1 \end{bmatrix}.$$

(2) Under the zero-state condition, taking the Z-transform on state equation $X(k+1) = AX(k) + Bf(k)$ yields:

$$zX(z) = AX(z) + BF(z) .$$

Therefore, we have:

$$X(z) = (zI - A)^{-1} BF(z)$$

$$= \frac{1}{(z+1)(z-4)+6} \begin{bmatrix} z-4 & 3 \\ -2 & z+1 \end{bmatrix} \begin{bmatrix} 11 & 0 \\ 0 & 6 \end{bmatrix} \begin{bmatrix} 1 \\ \frac{z}{z-1} \end{bmatrix}$$

$$= \frac{1}{(z-1)(z-2)} \begin{bmatrix} 11(z-4) + \frac{18z}{z-1} \\ -22 + \frac{6z(z+1)}{z-1} \end{bmatrix}$$

$$= \begin{bmatrix} \frac{-3}{z-1} + \frac{14}{z-2} - \frac{18}{(z-1)^2} \\ \frac{-8}{z-1} + \frac{14}{z-2} - \frac{12}{(z-1)^2} \end{bmatrix}$$

(3) From the output equation $Y(z) = CX(z) + DF(z)$, we can obtain:

$$Y(z) = \begin{bmatrix} 1 & -1 \end{bmatrix} \begin{bmatrix} \frac{-3}{z-1} + \frac{14}{z-2} - \frac{18}{(z-1)^2} \\ \frac{-8}{z-1} + \frac{14}{z-2} - \frac{12}{(z-1)^2} \end{bmatrix} + \begin{bmatrix} 0 & 1 \end{bmatrix} \begin{bmatrix} 1 \\ \frac{z}{z-1} \end{bmatrix} = 1 + \frac{6}{z-1} + \frac{-6}{(z-1)^2}$$

(4) Taking the inverse Z-transform, the zero-state response is given by:

$$y(k) = \delta(k) + (12 - 6k)\varepsilon(k-1)$$

Note: Try to compute the transfer function $H(z)$. !

7.5.3 Calculation with MATLAB

In MATLAB, the tool function *dlsim* can be used to compute zero-input and zero-state responses from the state-space matrices. The function format is as follows:

```
[y,v]=dlsim(A,B,C,D,X,v0)
```

The inputs A, B, C, D are coefficient matrices; X is the input and $v0$ is the initial condition. The output y is the response.

Example 7.5.2. The state equation and the output equation of an LTID system are as follows:

$$\begin{bmatrix} x_1(k+1) \\ x_2(k+1) \end{bmatrix} = \begin{bmatrix} 0 & 1 \\ -1 & 1.9021 \end{bmatrix} \begin{bmatrix} x_1(k) \\ x_2(k) \end{bmatrix} + \begin{bmatrix} 1 \\ 0 \end{bmatrix} [f(k)] , \quad y(k) = \begin{bmatrix} -1 & 1 \end{bmatrix} \begin{bmatrix} x_1(k) \\ x_2(k) \end{bmatrix}$$

(1) The initial state is:

$$\begin{bmatrix} x_1(0) \\ x_2(0) \end{bmatrix} = \begin{bmatrix} -11.7557 \\ -6.1803 \end{bmatrix},$$

compute the zero-input response.

(2) The initial state is

$$\begin{bmatrix} x_1(0) \\ x_2(0) \end{bmatrix} = \begin{bmatrix} -10 \\ -4 \end{bmatrix}$$

and the input is $f(k) = \varepsilon(k)$. Compute the entire response.

Solution:

```
% Compute zero-input response
A=[0 1;-1 1.9021];B=[1;0];C=[-1 1];D=[0];
v0=[-11.7557;-6.1803];              % initial condition
k=0:1:40;
X=[0*k]';                           % zero-input
[y,v]=dlsim(A,B,C,D,X,v0);          % zero-input response
stem(k,y);xlabel('k');ylabel('y');
% Compute the whole response
v0=[-10;-4];                        % initial condition
X=[1*ones(size(k))]';               % input sequence
[y,v]=dlsim(A,B,C,D,X,v0);          % entire response
stem(k,y);xlabel('k');ylabel('y');
```

The operation result is shown in Figure 7.6.

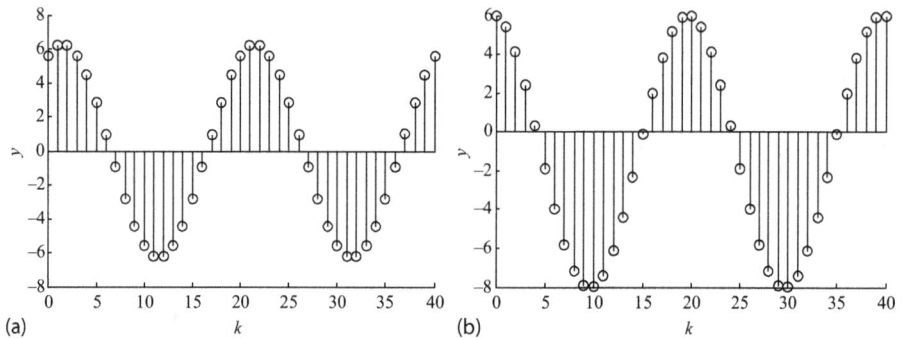

(a)

(b)

Fig. 7.6: Zero-input response and entire response; (a) Zero-input response (b) Entire response

7.6 Stability analysis from the transfer function matrix

7.6.1 Stability condition

In Sections 5.6 and 6.6, we stated that the stability of the system can be determined by the poles' position of the system transfer function. A causal LTIC system is BIBO stable if all the poles lie in the left half of the S-plane. A causal LTID system is BIBO stable if all the poles lie within the unit circle of the Z-plane. In this section, the stability of the system can be determined directly from the coefficient matrix of the state-space equations.

In Section 7.4.2, the Laplace transfer function matrix is given by:

$$H(s) = C\Phi(s)B + D = C(sI - A)^{-1}B + D \qquad (7.85)$$

To find the poles of $H(s)$, we have to compute the denominator of $(sI - A)^{-1}$, i.e., $\det(sI - A)$. The poles are the eigenvalues (characteristic roots) of matrix A. If all the eigenvalues lie in the left half of the S-plane, a causal LTIC system is stable.

In Section 7.5.2, the Z-transfer function matrix is given by:

$$H(z) = C(zI - A)^{-1}B + D \qquad (7.86)$$

Similarly, the stability of a causal LTID system can be determined by judging whether all the eigenvalues of matrix A lie within the unit circle of the Z-plane.

Note: The stability of a causal system only depends on matrix A. ❗

Example 7.6.1. Find the state equations of the given LTID system in Figure 7.7 with $a = -1$, $-1 < b < 1$ and determine the stability.

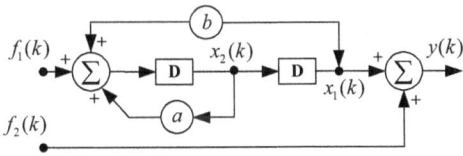

Fig. 7.7: System diagram in Example 7.6.1

Solution: The state variables $x_1(k)$, $x_2(k)$ are selected as two delayer outputs. By observation, the state equations are given by:

$$\begin{cases} x_1(k+1) = x_2(k) \\ x_2(k+1) = bx_1(k) + ax_2(k) + f_1(k) \end{cases} \qquad (7.87)$$

It is written in matrix form:

$$\begin{bmatrix} x_1(k+1) \\ x_2(k+1) \end{bmatrix} = \begin{bmatrix} 0 & 1 \\ b & a \end{bmatrix} \begin{bmatrix} x_1(k) \\ x_2(k) \end{bmatrix} + \begin{bmatrix} 0 & 0 \\ 1 & 0 \end{bmatrix} \begin{bmatrix} f_1(k) \\ f_2(k) \end{bmatrix}$$

Where

$$A = \begin{bmatrix} 0 & 1 \\ b & a \end{bmatrix}$$

is the coefficient matrix.

The characteristic equation of A is computed to find the characteristic roots:

$$\det(zI - A) = \begin{vmatrix} z & -1 \\ -b & z+1 \end{vmatrix} = z^2 + z - b = 0$$

$$z_{1,2} = -\frac{1}{2} \pm \frac{1}{2}\sqrt{1+4b}$$

According to the range of $-1 < b < 1$, the complex and real roots are discussed separately.

(1) $1 + 4b < 0$ or $-1 < b < -1/4$:

The roots must satisfy the following condition:

$$\sqrt{\left(\frac{1}{2}\right)^2 + \left(\frac{1}{2}\sqrt{1+4b}\right)^2} < 1$$

We obtain $b < 1/2$, which lies in range of $-1 < b < -1/4$.

(2) $1 + 4b \geq 0$ or $-1/4 \leq b < 1$:

The roots must satisfy the following condition:

$$\left| -\frac{1}{2} \pm \frac{1}{2}\sqrt{1+4b} \right| < 1 \rightarrow b < 0$$

Combining $b < 0$ with $-1/4 \leq b < 1$ yields $-1/4 \leq b < 0$.

In conclusion, the system is stable when all the eigenvalues lie within the unit circle of the Z-plane if $-1 < b < 0$.

7.6.2 Calculation with MATLAB

In MATLAB, the tool function *ss2tf* can be used to compute the transfer function matrix from the state-space equations. The function format is as follows:

```
[num, den]=ss2tf(A, B, C, D, k)
```

The inputs A, B, C, D are coefficient matrices, and k is the column number of $H(s)$; *num* is the numerator polynomial of column k in $H(s)$ and *den* is the denominator polynomial of $H(s)$.

Example 7.6.2. The state-space description of an LTIC system is given as follows:

$$\begin{bmatrix} \dot{x}_1(t) \\ \dot{x}_2(t) \end{bmatrix} = \begin{bmatrix} 2 & 3 \\ 0 & -1 \end{bmatrix} \begin{bmatrix} x_1(t) \\ x_2(t) \end{bmatrix} + \begin{bmatrix} 0 & 1 \\ 1 & 0 \end{bmatrix} \begin{bmatrix} f_1(t) \\ f_2(t) \end{bmatrix}$$

$$\begin{bmatrix} y_1(t) \\ y_2(t) \end{bmatrix} = \begin{bmatrix} 1 & 1 \\ 0 & -1 \end{bmatrix} \begin{bmatrix} x_1(t) \\ x_2(t) \end{bmatrix} + \begin{bmatrix} 1 & 0 \\ 1 & 0 \end{bmatrix} \begin{bmatrix} f_1(t) \\ f_2(t) \end{bmatrix}$$

Determine the system transfer function matrix $H(s)$ with MATLAB.

Solution:

```
A=[2 3;0 -1];B=[0 1;1 0];C=[1 1;0 -1];D=[1 0;1 0];
[num1,den1]=ss2tf(A,B,C,D,1)
[num2,den2]=ss2tf(A,B,C,D,2)
```

The operation result is as follows:

```
num1 = 1   0 -1      den1 = 1 -1 -2
       1  -2  0
num2 = 0   1  1      den2 = 1 -1 -2
       0   0  0
```

Note: The symbols *num1* and *num2* are the coefficients of columns 1 and 2, respectively. !

Therefore, the system transfer function matrix is given by:

$$H(s) = \frac{1}{s^2 - s - 2} \begin{bmatrix} s^2 - 1 & s + 1 \\ s^2 - 2s & 0 \end{bmatrix} = \begin{bmatrix} \frac{s+1}{s-2} & \frac{1}{s-2} \\ \frac{s}{s+1} & 0 \end{bmatrix}$$

7.7 Summary

This chapter mainly introduced the state-space description methods. In Section 7.2 and 7.3, we illustrated the state-space equations to establish internal description of CT and DT systems. Their corresponding solutions $H(s) = C(sI - A)^{-1}B + D$ and $H(z) = C(zI - A)^{-1}B + D$ were detailed in Section 7.4 and 7.5, respectively. Section 7.6 presented the stability condition by judging the eigenvalues of matrix A.

Chapter 7 problems

7.1 An RLC circuit is shown in Figure P7.1. Taking $y_1 = i_c$ and $y_2 = u$ as outputs, find the state and output equations.

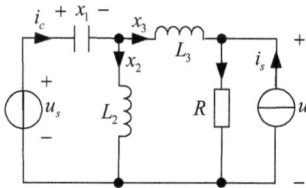

Fig. P7.1: RLC circuit in Problem 7.1

7.2 Consider the circuit in Figure P7.2. Taking $u_c(t)$ and $i_L(t)$ as the state variables, determine the state and output equations.

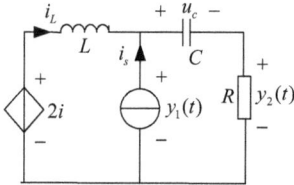

Fig. P7.2: RLC circuit in Problem 7.2

7.3 Consider the differential equation of an LTIC system; determine the state and output equations:

$$y'''(t) + 5y''(t) + 2y'(t) + y(t) = f(t)$$

7.4 Consider the differential equation of an LTIC system; determine the state and output equations:

$$2y'''(t) + 5y''(t) + 3y'(t) + 2y(t) = 2f'''(t) + 7f''(t) + 2f'(t)$$

7.5 Consider the difference equation of an LTID system:

$$y(k) + 4y(k-1) + 3y(k-2) = f(k-1) + 2f(k-2)$$

(1) Determine the Z-transfer function $H(z)$.
(2) Draw the system flow graph.
(3) Determine the state and output equations.

7.6 The system flow graph is shown in Figure P7.3. Write the state equation and the output equation.

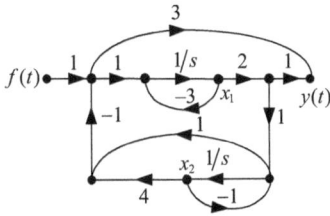

Fig. P7.3: System flow graph in Problem 7.6

7.7 The system diagram is shown in Figure P7.4. Write the state equation and the output equation.

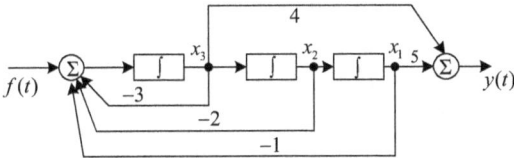

Fig. P7.4: The system diagram in Problem 7.7

7.8 Consider the Laplace transfer function. Determine the state equation and the output equation.

$$H(s) = \frac{2s^2 + 9s}{s^2 + 4s + 12}$$

7.9 Consider the state and the output equations. Determine the system flow graph and the differential equations:

$$\begin{bmatrix} \dot{x}_1 \\ \dot{x}_2 \end{bmatrix} = \begin{bmatrix} -4 & 1 \\ -3 & 0 \end{bmatrix} \begin{bmatrix} x_1 \\ x_2 \end{bmatrix} + \begin{bmatrix} 1 \\ 1 \end{bmatrix} f$$

$$y = \begin{bmatrix} 1 & 0 \end{bmatrix} \begin{bmatrix} x_1 \\ x_2 \end{bmatrix}$$

7.10 The CT system flow graph is shown in Figure P7.5. Write the state equation and the output equation.

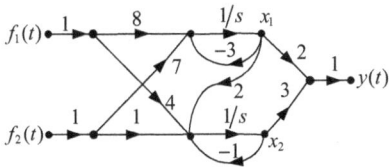

Fig. P7.5: System diagram in Problem 7.10

7.11 The DT system flow graph is shown in Figure P7.6. Write the state equation and the output equation.

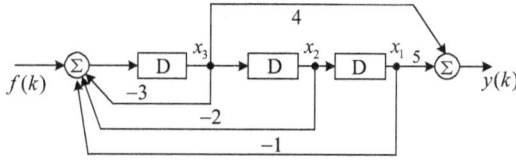

Fig. P7.6: The system diagram in Problem 7.11

7.12 The state-space equations and the initial conditions are given below. Determine the output with the input $f(t) = e^{2t}\varepsilon(t)$ using the Laplace transform:

$$\begin{bmatrix} \dot{x}_1 \\ \dot{x}_2 \end{bmatrix} = \begin{bmatrix} -1 & 0 \\ 1 & 0 \end{bmatrix} \begin{bmatrix} x_1 \\ x_2 \end{bmatrix} + \begin{bmatrix} 1 \\ 0 \end{bmatrix} f$$

$$\begin{bmatrix} y_1 \\ y_2 \end{bmatrix} = \begin{bmatrix} 1 & 0 \\ 0 & 1 \end{bmatrix} \begin{bmatrix} x_1 \\ x_2 \end{bmatrix} + \begin{bmatrix} 6 \\ 0 \end{bmatrix} f \quad ; \quad \begin{bmatrix} x_1(0) \\ x_2(0) \end{bmatrix} = \begin{bmatrix} 6 \\ 6 \end{bmatrix}$$

7.13 The state-space equations of a DT system are given below. Determine the unit impulse response $h(k)$ using the Z-transform:

$$\begin{bmatrix} x_1(k+1) \\ x_2(k+1) \end{bmatrix} = \begin{bmatrix} 0 & 1 \\ -6 & 5 \end{bmatrix} \begin{bmatrix} x_1(k) \\ x_2(k) \end{bmatrix} + \begin{bmatrix} 0 \\ 1 \end{bmatrix} f$$

$$\begin{bmatrix} y_1(k) \\ y_2(k) \end{bmatrix} = \begin{bmatrix} 1 & 1 \\ 2 & -1 \end{bmatrix} \begin{bmatrix} x_1(k) \\ x_2(k) \end{bmatrix}$$

7.14 Consider the state and output equations of an LTID system:

$$\begin{bmatrix} x_1(k+1) \\ x_2(k+1) \end{bmatrix} = \begin{bmatrix} -2 & -3 \\ 2 & 1 \end{bmatrix} \begin{bmatrix} x_1(k) \\ x_2(k) \end{bmatrix} + \begin{bmatrix} 1 \\ 0 \end{bmatrix} f(k)$$

$$y(k) = \begin{bmatrix} 3 & 2 \end{bmatrix} \begin{bmatrix} x_1(k) \\ x_2(k) \end{bmatrix}$$

(1) Draw the system flow graph and mark the state variables on it.
(2) Determine the Z-transfer function $H(z)$.

7.15 The state-space equations of an LTIC system are given below:

$$\begin{bmatrix} \dot{x}_1 \\ \dot{x}_2 \end{bmatrix} = \begin{bmatrix} -4 & 1 \\ -3 & 0 \end{bmatrix} \begin{bmatrix} x_1 \\ x_2 \end{bmatrix} + \begin{bmatrix} 1 \\ 1 \end{bmatrix} f$$

$$y(t) = \begin{bmatrix} 1 & 0 \end{bmatrix} \begin{bmatrix} x_1 \\ x_2 \end{bmatrix}$$

(1) Determine the transfer function $H(s)$ using the Laplace transform.
(2) Determine the differential equation.

7.16 The state-space equations of an LTIC system are given below:

$$\begin{bmatrix} \dot{x}_1 \\ \dot{x}_2 \end{bmatrix} = \begin{bmatrix} -4 & 1 \\ -3 & 0 \end{bmatrix} \begin{bmatrix} x_1 \\ x_2 \end{bmatrix} + \begin{bmatrix} 1 \\ 1 \end{bmatrix} f$$

$$y(t) = \begin{bmatrix} 1 & 0 \end{bmatrix} \begin{bmatrix} x_1 \\ x_2 \end{bmatrix}$$

(1) Determine the transfer function $H(s)$ using the Laplace transform.
(2) Determine the differential equation.

7.17 The state equation of an LTIC system is given below. Determine whether the system is BIBO stable:

$$\begin{bmatrix} \dot{x}_1 \\ \dot{x}_2 \end{bmatrix} = \begin{bmatrix} 4 & 3 \\ -3 & 4 \end{bmatrix} \begin{bmatrix} x_1 \\ x_2 \end{bmatrix} + \begin{bmatrix} 1 \\ 1 \end{bmatrix} f$$

7.18 The coefficient matrix A in the state equations of an LTID system is given below:

$$A = \begin{bmatrix} 1 & b \\ 2 & 0.5 \end{bmatrix}$$

Determine the requirements of parameter b if the system is BIBO stable.

7.19 The state-space equations and the initial conditions are given below. Determine the zero-state response with the input $f(t) = 2\varepsilon(t)$ and the zero-input response with MATLAB:

$$\dot{x}(t) = \begin{bmatrix} -2 & 1 \\ 0 & -1 \end{bmatrix} x(t) + \begin{bmatrix} 1 \\ 0 \end{bmatrix} f(t) \quad y(t) = \begin{bmatrix} 1 & 0 \end{bmatrix} x(t) ; \quad x(0) = \begin{bmatrix} 2 \\ 2 \end{bmatrix}$$

7.20 The state-space description of an LTIC system is given as follows:

$$\dot{x}(t) = \begin{bmatrix} -4 & 5 \\ 0 & 1 \end{bmatrix} x(t) + \begin{bmatrix} 0 \\ 1 \end{bmatrix} f(t)$$

$$y(t) = \begin{bmatrix} 1 & 0 \end{bmatrix} x(t)$$

Determine the system transfer function matrix $H(s)$ with MATLAB.

8 Applications of system analysis

Please focus on the following key questions.
1. How is the signal modulated in communication systems?
2. What is the basic theory of controlling a linear system? Can any arbitrary system be observed or controlled?
3. How can a digital filter be designed?
4. What is the Kalman filter and how can it be used to make predictions?

8.0 Introduction

This chapter introduces some applications of signal processing. Our aim is to motivate readers' interest to explore the applications. Section 8.1 introduces signal modulation in communication systems. Section 8.2 briefly introduces the discrete-time Fourier transform for digital signal processing. Control system analysis is considered in Section 8.3. For digital signal filtering, classic FIR and IIR filter design methods are introduced in Section 8.4. Sections 8.5 and 8.6 give the applications of the state-space analysis. Digital image processing by convolution is illustrated in Section 8.7. Finally, Section 8.8 concludes the chapter with a summary of important concepts.

8.1 Application of the Fourier transform in communication systems

Modulation and demodulation are important techniques in communication systems. In some cases, the signal is modulated to a high frequency, which is suitable for transmission in channels. Modulation will result in spectrum shifting of the signal [9]. Generally, the signal of the low-frequency voice or video signal in communication systems is called the baseband signal. The baseband signal is multiplied by a high-frequency carrier to obtain the modulated signal. Modulation can be achieved by changing the high-frequency component of the carrier wave along with the signal amplitude, so as to change the amplitude, phase, or frequency of the carrier wave. Demodulation is to recover the original signal from the modulated one at the receiver, that is, to extract the baseband signal from the carrier wave for the receiver to manipulate and understand it.

https://doi.org/10.1515/9783110593907-008

8.1.1 Double-sideband suppressed-carrier amplitude modulation (DSB-SC-AM)

The diagram of the double-sideband suppressed-carrier amplitude modulation (DSB-SC-AM) is shown in Figure 8.1. The modulated signal $x(t) = m(t)c(t) = m(t)\cos(\omega_c t)$ is achieved by multiplying the carrier wave $c(t)$ and the modulation signal $m(t)$.

! **Note:** Here, we suppose $\omega_c \gg \omega_m$.

The spectrum of the carrier wave $c(t) = \cos(\omega_c t)$ is:

$$C(j\omega) = \pi\left[\delta(\omega - \omega_c) + \delta(\omega + \omega_c)\right] . \tag{8.1}$$

Supposing the spectrum of the modulation signal $m(t)$ is $M(j\omega)$, the spectrum of the modulated signal $x(t) = m(t)c(t)$ is obtained by:

$$X(j\omega) = \frac{1}{2\pi}M(j\omega) * C(j\omega) = \frac{1}{2}\{M\left[j(\omega - \omega_c)\right] + M\left[j(\omega + \omega_c)\right]\} \tag{8.2}$$

! **Note:** Refer to the frequency-convolution property of CTFT.

The amplitude frequency characteristics of the modulation signal and the modulated signal are presented in Figure 8.2. The spectrum $X(j\omega)$ of the modulated signal is achieved by splitting $M(j\omega)$ into two parts and shifting by ω_c units to the left and right, respectively. The parts exceeding ω_c are called the upper sideband (USB), and the parts below ω_c are called the lower sideband (LSB).

Figure 8.3 shows the process of demodulation. The modulated signal $x(t)$ is multiplied with the sine signal $\cos(\omega_c t)$ generated by the local oscillator. The sinusoidal signal has the same frequency as the carrier. The spectrum $X(j\omega)$ of the modulated

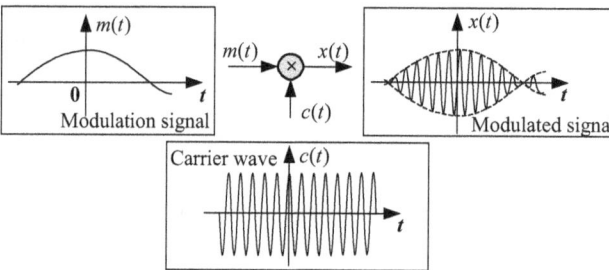

Fig. 8.1: Diagram of the DSB-SC-AM system

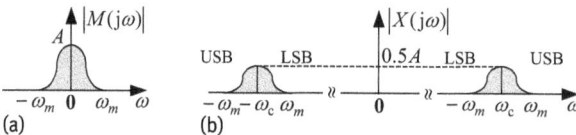

Fig. 8.2: Spectrum of modulation and modulated signals

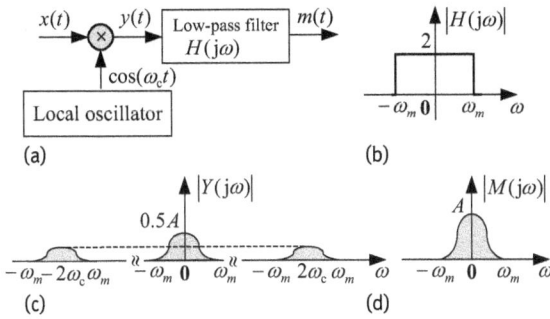

Fig. 8.3: Demodulation of the DSB-SC-AM system

signal will be shifted by ω_c units along the ω axis to the left and right sides (with half amplitude), as is shown in Figure 8.3 (c). The low-pass filter in Figure 8.3 (b) is implemented on the spectrum $Y(j\omega)$ to obtain the spectrum of the modulation signal $M(j\omega)$.

8.1.2 Amplitude modulation (AM)

Amplitue modulation (AM) is used in commercial AM broadcasting, especially medium-wave broadcasting. It can be achieved by adding a carrier wave $A_c \cos(\omega_c t)$ of certain intensity to the transmitted signal, as is shown in Figure 8.4. It can be seen that the envelope of the modulated signal $s(t)$ is the modulation signal $m(t)$. The coefficient k_a of amplitude modulation satisfies:

$$1 + k_a m(t) > 0 \tag{8.3}$$

The modulated signal can be denoted as follows:

$$s(t) = A_c[1 + k_a m(t)] \cos(\omega_c t) \tag{8.4}$$

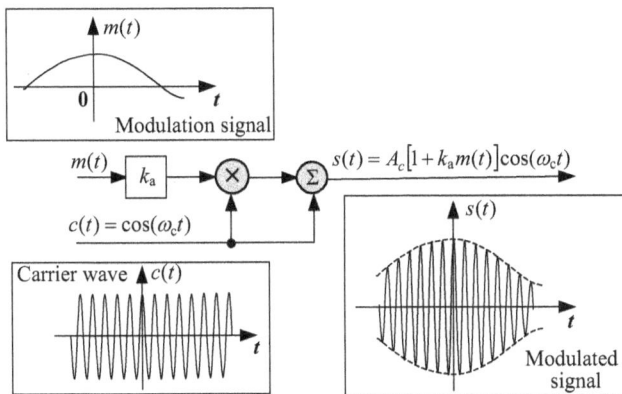

Fig. 8.4: Illustration of amplitude modulation

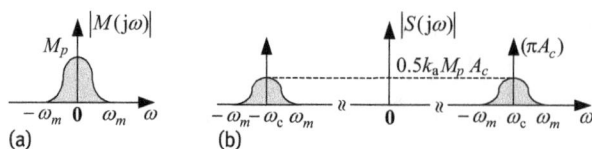

Fig. 8.5: Spectrum of the modulation and modulated signal; (a) Spectrum of the modulation signal, (b) Spectrum of the modulated signal

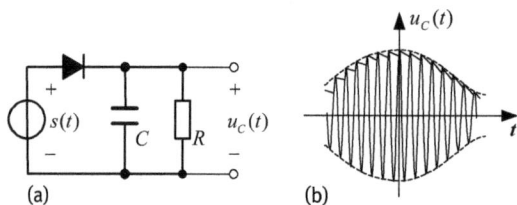

Fig. 8.6: Envelope detection; (a) Envelope detector, (b) The detected envelope

Take the CTFT on both sides of the equation to obtain its spectrum as follows:

$$s(j\omega) = A_c \pi \left[\delta(\omega - \omega_c) + \delta(\omega + \omega_c) \right]$$
$$+ \frac{k_a A_c}{2} \{ M \left[j(\omega - \omega_c) \right] + M \left[j(\omega + \omega_c) \right] \} \tag{8.5}$$

The amplitude spectra of the modulation signal $m(t)$ and the modulated signal $s(t)$ are shown in Figure 8.5. The spectrum $|S(j\omega)|$ indicates both the components of the modulation signal and frequency-shifting of the carrier wave.

The demodulation of the AM signal is realized by envelope detection. The envelope detector in Figure 8.6 (a) is composed of a diode, resistor and capacitor. The envelope extracted from the modulated signal $s(t)$ is shown in Figure 8.6 (b). The modulation signal $m(t)$ can be recovered by DC-isolation on $u_C(t)$ using high-pass filter and low-pass smoothing.

8.1.3 Pulse-amplitude modulation (PAM)

The pulse-amplitude modulation (PAM) is realized by using the periodic rectangular pulse carrier wave. As is shown in Figure 8.7, the sine carrier wave $c(t)$ is replaced by $p(t)$.

In Chapter 4, the exponential CTFS expansion of the periodic rectangular pulse carrier $p(t)$ with amplitude 1, pulse width τ and period T_c is:

$$p(t) = \sum_{n=-\infty}^{\infty} F_n e^{jn\omega_c t} = \frac{\tau}{T_c} \sum_{n=-\infty}^{\infty} \text{Sa}\left(\frac{n\omega_c \tau}{2} \right) e^{jn\omega_c t} \tag{8.6}$$

(Note: Refer to Section 4.2.2.)

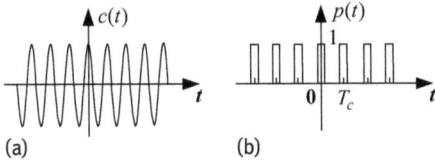

Fig. 8.7: The carrier waves, (a) Sine carrier wave, (b) Rectangular pulse carrier wave

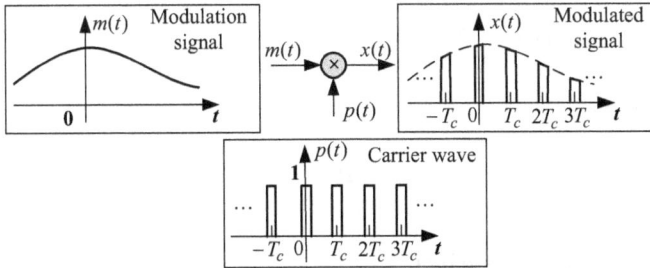

Fig. 8.8: IPulse-amplitude modulation

Where $\omega_c = 2\pi/T_c$ is the fundamental angular frequency of the carrier wave. Taking CTFT on both sides yields:

$$P(j\omega) = 2\pi \sum_{n=-\infty}^{\infty} \frac{\tau}{T_c} \mathrm{Sa}\left(\frac{n\omega_c\tau}{2}\right) \delta(\omega - n\omega_c) \tag{8.7}$$

It can be seen from Figure 8.8 that the output of the PAM modulator is $x(t) = m(t)p(t)$. The spectra of the modulation signal and the modulated signal are shown in Figure 8.9. The spectrum of the modulation signal is $m(t) \leftrightarrow M(j\omega)$. Combining $P(j\omega)$ in Equation (8.7) with $M(j\omega)$, the spectrum of the modulated signal $x(t) = m(t)p(t)$ is given by:

$$X(j\omega) = \sum_{n=-\infty}^{\infty} \frac{\tau}{T_c} \mathrm{Sa}\left(\frac{n\omega_c\tau}{2}\right) M\left[j(\omega - n\omega_c)\right] . \tag{8.8}$$

Note: Readers can prove it using the frequency-convoluiton property of CTFT. !

The spectrum $X(j\omega)$ is the convolution of the spectrum $M(j\omega)$ of the modulation signal and the spectrum $P(j\omega)$ of the periodic rectangular pulse carrier. It can be seen from Figure 8.9 that $X(j\omega)$ can be considered as the periodic extension of $M(j\omega)$. As long as the carrier fundamental frequency satisfies $\omega_c \geq 2\omega_m$, the modulation signal $M(j\omega)$ can be recovered from the modulated signal $X(j\omega)$ by a low-pass filter.

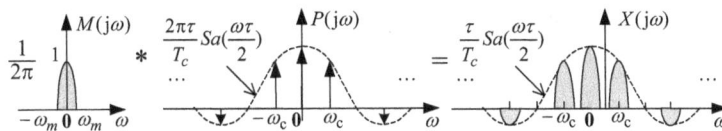

Fig. 8.9: Spectra of pulse-amplitude modulation

8.2 Fast Fourier transform

8.2.1 Discrete-time Fourier series (DTFS)

In Chapter 4, the exponential CTFS representation of $f_T(t)$ is defined as follows:

$$f_T(t) = \sum_{n=-\infty}^{\infty} F_n e^{jn\Omega t}, \tag{8.9}$$

where the exponential CTFS coefficients F_n are calculated as:

$$F_n = \frac{1}{T} \int_{(T)} f_T(t) e^{-jn\Omega t} \, dt, \tag{8.10}$$

Ω being the fundamental frequency given by $\Omega = 2\pi/T$.

For the discrete-time sequence $f_N(k)$, the corresponding basis functions are $\{e^{jn\Omega k}, n \in Z\}$, where Ω is the fundamental frequency given by $\Omega = 2\pi/N$. An important difference between DT and CT complex exponential functions lies in the frequency-periodicity property of DT exponential sequences.

Since:

$$e^{jn\frac{2\pi}{N}k} = e^{j(n+N)\frac{2\pi}{N}k} = e^{jn\frac{2\pi}{N}k} \cdot e^{j2\pi k},$$

the period of the exponential sequence is N. The Fourier series coefficients are periodic with a period of N.

! **Note:** The CT exponentials do not have this periodicity property.

The DT exponential expansion can be expressed as:

$$f_N(k) = \sum_{n=0}^{N-1} C_n e^{jn\Omega k} = \sum_{n=0}^{N-1} C_n e^{jn\frac{2\pi}{N}k} \tag{8.11}$$

Where

$$C_n = \frac{1}{N} \sum_{k=0}^{N-1} f_N(k) e^{-jn\Omega k} = \frac{1}{N} F_N(n).$$

The *DTFS coefficients* $F_N(n)$ are defined as:

$$F_N(n) = \sum_{k=0}^{N-1} f_N(k) e^{-jn\Omega k}. \tag{8.12}$$

The *DTFS expansion* of DT periodic sequence $f_N(k)$ is defined as:

$$f_N(k) = \frac{1}{N} \sum_{n=0}^{N-1} F_N(n) e^{jn\Omega k}. \tag{8.13}$$

For simplicity, by substituting $W = e^{-j\Omega} = e^{-j(2\pi/N)}$ into the above equations, the DTFS pair for a periodic sequence $f_N(k)$ can be expressed as follows:

$$\text{DTFS}[f_N(k)] = F_N(n) = \sum_{k=0}^{N-1} f_N(k) W^{nk}$$

$$\text{IDTFS}[F_N(n)] = f_N(k) = \frac{1}{N} \sum_{n=0}^{N-1} F_N(n) W^{-nk}$$

(8.14)

Note: The period of $f_N(k)$ and $F_N(n)$ is N. !

8.2.2 Discrete-time Fourier transform (DTFT)

In this section, the frequency representation for an aperiodic sequence is considered. When the period of $f_N(k)$ is $N \to \infty$, the angular frequency $\Omega = 2\pi/N$ takes a very small value, say $d\theta$. Substituting $n\Omega = \theta$ and applying the limit $N \to \infty$ into Equation (8.12), the DT Fourier transform (DTFT) can be obtained as follows:

$$F(e^{j\theta}) = \lim_{N \to \infty} \sum_{k=\langle N \rangle} f_N(k) e^{-jn\frac{2\pi}{N}k} = \sum_{k=-\infty}^{\infty} f(k) e^{-jk\theta}$$

(8.15)

The DTFT of the aperiodic sequence $f(k)$ is a continuous function of θ and its period is 2π.

Substituting $n\Omega = \theta$ and applying the limit $N = 2\pi/\Omega \to \infty$, Equation (8.13) reduces to the following integral:

$$f(k) = \frac{1}{2\pi} \int_{-\pi}^{\pi} F(e^{j\theta}) e^{jk\theta} \, d\theta$$

(8.16)

Note: Readers can prove it with $1/N = \Omega/2\pi \to (d\theta)/(2\pi)$. !

The *DTFT pair* for an aperiodic sequence $f(k)$ is given by:

$$\text{DTFT}[f(k)] = F(e^{j\theta}) = \sum_{k=-\infty}^{\infty} f(k) e^{-jk\theta}$$

$$\text{IDTFT}[F(e^{j\theta})] = f(k) = \frac{1}{2\pi} \int_{-\pi}^{\pi} F(e^{j\theta}) e^{jk\theta} \, d\theta$$

(8.17)

Note: The sufficient condition for the existence of DTFT is the absolutely summable condition as: $\sum_{k=-\infty}^{\infty} |f(k)| < \infty$. !

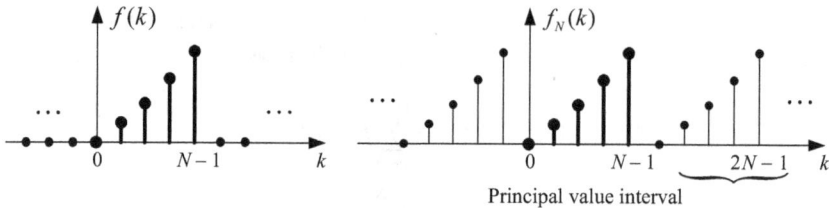

Fig. 8.10: Relationship of aperiodic sequence $f(k)$ with periodic sequence $f_N(k)$

8.2.3 Discrete Fourier transform (DFT)

With the increased use of digital computers and specialized hardware in digital signal processing (DSP), interest has focused around transforms that are suitable for digital computations. Because of the continuous nature of θ, direct implementation of the DTFT is not suitable on such digital devices. The discrete Fourier transform (DFT) is analyzed to be computed efficiently on digital computers and other DSP boards.

As Figure 8.10 shows, a time-limited sequence $f(k)$, which is non-zero within the limits $0 \le k \le N - 1$, can be considered as a cycle of the periodic sequence $f_N(k)$:

$$f(k) = \begin{cases} f_N(k), & (0 \le k \le N - 1) \\ 0, & \text{else} \end{cases}$$

The forward DFT and the inverse DFT of $f(k)$ are defined as follows:

$$F(n) = \text{DFT}[f(k)] = \sum_{k=0}^{N-1} f(k)e^{-jn\frac{2\pi}{N}k} = \sum_{k=0}^{N-1} f(k)W^{nk} \quad (0 \le n \le N - 1)$$

$$\text{(8.18)}$$

$$f(k) = \text{IDFT}[F(n)] = \frac{1}{N}\sum_{n=0}^{N-1} F(n)e^{jn\frac{2\pi}{N}k} = \frac{1}{N}\sum_{n=0}^{N-1} F(n)W^{-nk} \quad (0 \le k \le N - 1)$$

! **Note:** Pay attention to the difference with DTFS in Equation (8.14).

8.2.4 Relationship between Fourier transforms

Figure 8.11 gives the characteristics of five forms of Fourier transform pairs in time and frequency domains. If $f(k)$ and $F(n)$ are considered as the principal value interval of $f_N(k)$ and $F_N(n)$, respectively, the DFT transform pair is exactly the same as the DTFS transform pair. Comparing the spectrum of DTFT and DFT, the relationship can be obtained as follows:

$$F(n) = F(e^{j\theta})\big|_{\theta=\frac{2\pi}{N}n}$$

The DFT $F(n)$ is the sampled value of DTFT $F(e^{j\theta})$ at discrete frequencies $\theta = 2\pi n/N$ for $0 \le n < N$.

! **Note:** Find out all the relationships in Figure 8.11.

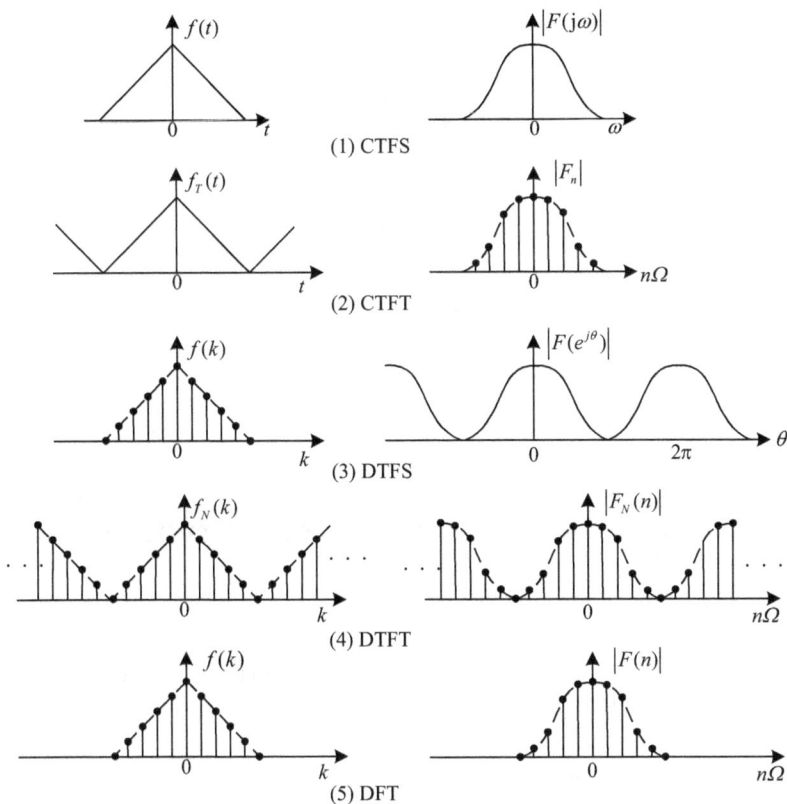

Fig. 8.11: Characteristics of Fourier transforms in time and frequency domains

8.2.5 Fast Fourier transform (FFT)

There are several well-known techniques, including the radix-2, radix-4, split radix, Winograd and prime factor algorithms that are used for computing the DFT. These algorithms are referred to as fast Fourier transform (FFT) algorithms. In this section, we illustrate the radix-2 decimation-in-time FFT algorithm [23].

Figure 8.12 shows a *butterfly* flow graph for the 8-point DFT. For the sequence with length $N = 2^M$, the computational complexity of DFT is $O(N^2)$. The FFT needs M times butterfly operations. Each butterfly operation includes one complex multiplication and two complex additions. The total number of complex multiplications is $0.5N \log_2 N$, and the total number of complex additions is $N \log_2 N$. Therefore, the complexity of FFT can be expressed as $O(N \log_2 N)$. The difference in computational complexity is obvious when the length of the sequence is big [27].

Note: More details can be seen in Reference [21].

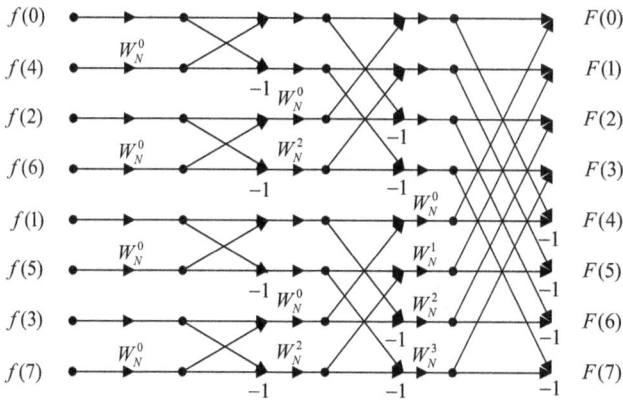

Fig. 8.12: Decimation-in-time implementation of an 8-point DFT

8.3 Application of the Laplace transform in control systems

8.3.1 Diagram of the closed-loop feedback system

A feedback controlling system is designed to produce the desired output $y(t)$ for a given input $f(t)$. The closed-loop feedback system can detect the difference between the actual output and the expected output to adjust the inputs for correction [28]. It can solve many problems caused by disturbance, such as random noise in electronic systems, wild wind affecting the precision of antennas and the movement of aircraft affecting observation systems. Figure 8.13 (a) is the block diagram of an open-loop system, and Figure 8.13 (b) is the block diagram of a closed-loop system.

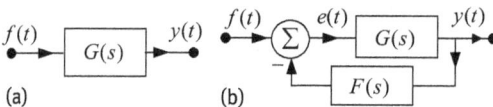

Fig. 8.13: (a) Open-loop system and (b) closed-loop (feedback) system

Note: Feedback systems have advantages of obtaining a more stable output.

8.3.2 Analysis of an automatic position control system

Figure 8.14 gives an illustration of controlling the angular position of an electrome-chanical system, such as a tracking antenna, a telescope platform, or an antiaircraft weapon. The expected angular position is θ_i, and the actual angular position mea-

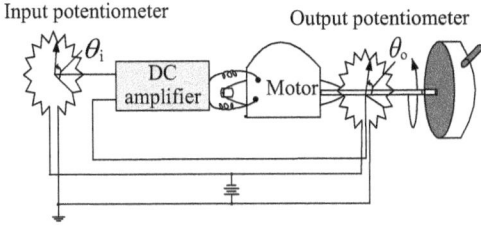

Fig. 8.14: Automatic position control system

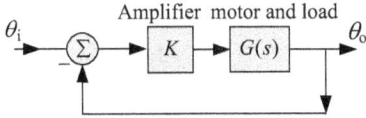

Fig. 8.15: Block diagram of system

sured by a potentiometer is θ_o. The sliding arm of the potentiometer is fixed on the output axis. The difference of the expected input θ_i and the actual measured output θ_o is amplified and applied on the input terminal of a motor. If $\theta_i - \theta_o = 0$, the actual output equals the expected value. No input will be applied on the motor. If $\theta_i - \theta_o \neq 0$, a nonzero input will be applied on the electric motor to rotate the axis until $\theta_i - \theta_o = 0$.

The block diagram of the system is shown in Figure 8.15. The amplitude gain K is adjustable. Assuming $G(s)$ is the transfer function between the output angle θ_o and the input voltage of motor, the transfer function $H(s)$ relating the output $y(t) = \theta_o$ and input $f(t) = \theta_i$ to the closed-loop system is:

$$H(s) = \frac{Y(s)}{F(s)} = K\frac{G(s)}{1 + KG(s)} \tag{8.19}$$

Note: The transfer function can be obtained by Mason's rule.

Based on the system transfer function, the unit step response is analyzed. The unit step function means an instantaneous or abrupt change. It is very difficult to track a step input. Therefore, the unit step response is frequently applied to evaluate the control ability of the adaptive system.

For a unit step input $\theta_i(t) = \varepsilon(t)$, the Laplace transform is $F(s) = 1/s$. According to Equation (8.19), the Laplace transform of the output is as follows:

$$Y(s) = H(s) \cdot F(s) = \frac{KG(s)}{1 + KG(s)} \cdot \frac{1}{s} \tag{8.20}$$

Assuming $G(s) = 1/(s(s + 8))$, we can obtain:

$$Y(s) = H(s) \cdot F(s) = \frac{1}{s} \cdot \frac{\frac{K}{s(s+8)}}{1 + \frac{K}{s(s+8)}} = \frac{K}{s(s^2 + 8s + K)} \tag{8.21}$$

Now the features of system are analyzed based on three different values of the gain K.

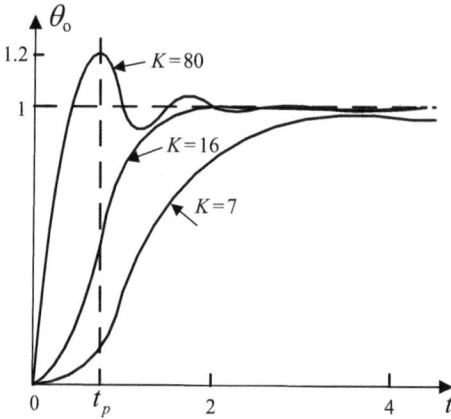

Fig. 8.16: Waves of the output with different gains

(1) $K = 7$:

$$Y(s) = \frac{7}{s(s^2 + 8s + 7)} = \frac{7}{s(s + 1)(s + 7)} = \frac{1}{s} - \frac{7/6}{s + 1} + \frac{1/6}{s + 7} \tag{8.22}$$

The inverse Laplace transform is computed to obtain the output:

$$\theta_o(t) = \left(1 - \frac{7}{6}e^{-t} + \frac{1}{6}e^{-2t}\right)\varepsilon(t) \tag{8.23}$$

The corresponding wave is plotted in Figure 8.16, which indicates that the response reaches the expected value in a very slow trend. In order to accelerate this procedure, the gain is enhanced to $K = 80$.

(2) $K = 80$:

$$Y(s) = \frac{80}{s(s^2 + 8s + 80)} = \frac{80}{s(s + 4 - j8)(s + 4 + j8)} = \frac{1}{s} + \frac{\sqrt{5}/4 \cdot e^{j153°}}{s + 4 - j8} + \frac{\sqrt{5}/4 \cdot e^{-j153°}}{s + 4 + j8} \tag{8.24}$$

Calculating the inverse Laplace transform, we obtain the output as follows:

$$\theta_o(t) = \left[1 + \frac{\sqrt{5}}{2}e^{-4t}\cos(8t + 153°)\right]\varepsilon(t) \tag{8.25}$$

The output is also plotted in Figure 8.16. The approaching speed is increased at the cost of vibration. The stable value of the response is 1, and the steady-state error is $e_r = 0$. The time for arrival of the peak value is $t_p = 0.393$ s. The rise time $t_r = 0.175$ s is defined as the period from 10% to 90% of the steady-state value, which represents the response speed. A well-designed system should have small amplitude vibration, small rise time t_r, and small steady-state error e_r.

! **Note:** How can we calculate the peak time $t_p = 0.393$ s?

(3) $K = 16$:

In order to avoid the vibration with large amplitude, the system should be designed with real roots. In this example, the roots are complex eigenvalues when $K > 16$. So, we choose $K = 16$ and compute the transfer function:

$$Y(s) = \frac{16}{s(s^2 + 8s + 16)} = \frac{16}{s(s + 4)^2} = \frac{1}{s} - \frac{1}{s + 4} - \frac{4}{(s + 4)^2} \tag{8.26}$$

Calculating the inverse Laplace transform yields:

$$\theta_o(t) = [1 - (4t + 1)e^{-4t}]\varepsilon(t) \tag{8.27}$$

The wave of response is also plotted in Figure 8.16. Hence, selecting $K = 16$ can lead to the fastest response and no vibration. In conclusion, the state of this system with $K > 16$ is called under damping, the state with $K < 16$ is over damping, and the state with $K = 16$ is critical damping.

8.4 Digital filters

8.4.1 Filter classification

In this section, we will give a brief introduction to digital filters. A digital filter is designed to transform an input sequence by changing its frequency characteristics in a predefined manner. Digital filters can be classified into four important categories: low-pass, high-pass, bandpass and bandstop, based on the magnitude response in the frequency domain. In the case of the ideal filter, the shape of the magnitude spectrum is rectangular with a sharp transition between the range of frequencies passed and the range of frequencies blocked by the filter. The range of frequencies passed by the filter is referred to as the passband of the filter, while the range of blocked frequencies is referred to as the stopband.

Figure 8.17 gives the magnitude responses of ideal filters. The low-pass filter removes the higher frequencies in the range of $\Omega_C \le |\Omega| \le \pi$. It is observed that the low-pass filter has a unity gain in the passband and zero gain in the stopband. The high-pass filter has a passband of $\Omega_C \le |\Omega| \le \pi$ and a stopband of $|\Omega| < \Omega_C$. It blocks the lower frequencies $|\Omega| < \Omega_C$, while the higher frequencies $\Omega_C \le |\Omega| \le \pi$ are passed with a unity gain. The ideal bandpass filter has a passband of $\Omega_{C1} \le |\Omega| \le \Omega_{C2}$ and a stopband of $|\Omega| \le \Omega_{C1}$ and $\Omega_{C2} \le |\Omega| \le \pi$. The ideal bandstop filter has a passband of $|\Omega| \le \Omega_{C1}$ and $\Omega_{C2} \le |\Omega| \le \pi$ and a stopband of $\Omega_{C1} \le |\Omega| \le \Omega_{C2}$.

Note: The magnitude spectrum is an even function.

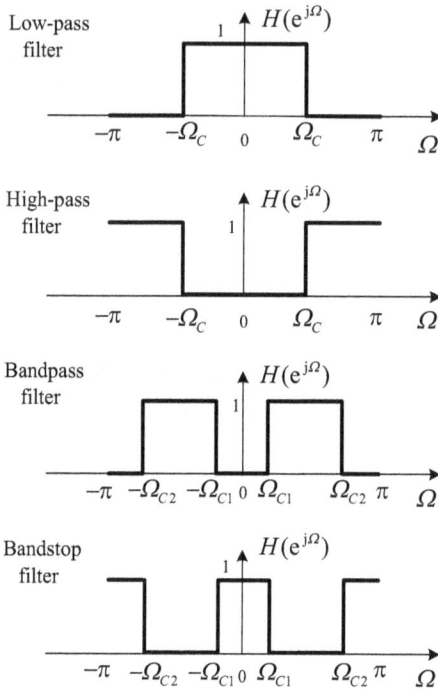

Low-pass filter

High-pass filter

Bandpass filter

Bandstop filter

Fig. 8.17: Magnitude responses of ideal filters

8.4.2 FIR and IIR filters

A second classification of digital filters is made on the length of their impulse response $h(k)$. The length (or width) of a digital filter is the number N of samples k beyond which the impulse response $h(k)$ is zero in both directions along the k-axis. A finite impulse response (FIR) filter is defined as a filter whose length N is finite. On the other hand, if the length N of the filter is infinite, the filter is called an infinite impulse response (IIR) filter.

Since the output response is obtained by the convolution of the impulse response and the input sequence, the output of an FIR filter is finite in length if the input sequence itself is finite in length. On the other hand, an IIR filter produces an output response that is always infinite in length.

The stability characteristics of FIR and IIR filters are now studied. Recall that an LTID system with impulse response function $h(k)$ is BIBO stable if:

$$\sum_{k=-\infty}^{\infty} |h(k)| < \infty .$$ (8.28)

! **Note:** The time-domain stability condition is that the impulse response is absolutely summable.

Since the FIR filter is nonzero for only a limited number of samples k, it always satisfies the stability criterion. For IIR filters with infinite length, the summation $\sum h(k)$ may not be finite, even if the amplitudes of the impulse functions are finite. In other words, it is not guaranteed that an IIR filter will always be stable.

In designing digital filters, we should measure the implementation cost. The number of delay elements used is an important criterion. IIR filters are implemented using a feedback loop, in which the number of delay elements is determined by the order of the IIR filter. The number of delay elements used in FIR filters depends on their length, and so the implementation cost of such filters increases with the number of filter taps. A FIR filter with a large number of taps may, therefore, be computationally infeasible.

8.4.3 IIR filter design using the impulse-invariance method

In this section, we briefly discuss the impulse-invariance transformation to design an IIR filter. It provides a linear transformation between the DT and CT frequency domains. To derive the impulse invariance transformation, the impulse response $h_a(t)$ of a CT filter is sampled to obtain the DT impulse response $h(k)$:

$$h(k) = h_a(t)|_{t=kT} = h_a(kT) \tag{8.29}$$

Suppose the Laplace transform $H_a(s)$ of $h_a(t)$ is a rational proper fraction with N first-order poles; the partial fraction expansion is as follows:

$$H_a(s) = \sum_{i=1}^{N} \frac{A_i}{s - s_i} \tag{8.30}$$

Calculating the inverse Laplace transformation yields:

$$h_a(t) = \sum_{i=1}^{N} A_i e^{s_i t} \varepsilon(t) \tag{8.31}$$

It is evenly sampled, and its Z-transformation is given by Equation (8.33):

$$h(k) = h_a(kT) = \sum_{i=1}^{N} A_i e^{s_i kT} \varepsilon(k) \tag{8.32}$$

$$H(z) = \sum_{k=0}^{\infty} \left(\sum_{i=1}^{N} A_i e^{s_i kT} \right) z^{-k} = \sum_{i=1}^{N} A_i \sum_{k=0}^{\infty} (e^{s_i T} z^{-1})^k = \sum_{i=1}^{N} \frac{A_i}{1 - e^{s_i T} z^{-1}} \tag{8.33}$$

It can be seen by comparing Equation (8.33) with Equation (8.30) that the partial fraction $1/(s - s_i)$ of $H_a(s)$ is replaced by $1/(1 - e^{s_i T} z^{-1})$ to obtain the transfer function $H(z)$ to design the IIR filter.

Note: we only need to compute the poles of $H_a(s)$ to obtain $H(z)$.

Example 8.4.1. Consider the following filter:

$$H_a(s) = \frac{1}{(s+1)(s+2)}$$

Use the impulse invariance transformation to derive the transfer function of the equivalent digital filter.

Solution: Express the transfer function of the CT filter as follows:

$$H_a(s) = \frac{1}{(s+1)(s+2)} = \frac{1}{s+1} - \frac{1}{s+2} \tag{8.34}$$

Using Equation (8.33), the transfer function of the DT filter is given by:

$$H(z) = \frac{1}{1-e^{-T}z^{-1}} - \frac{1}{1-e^{-2T}z^{-1}} = \frac{z}{z-e^{-T}} - \frac{z}{z-e^{-2T}} = \frac{(e^{-T}-e^{-2T})z}{z^2-(e^{-T}+e^{-2T})z+e^{-3T}} \tag{8.35}$$

Substituting $s = j\omega$ and $z = e^{j\Omega}$ into Equations (8.34) and (8.35) yields:

$$H_a(j\omega) = \frac{1}{(j\omega+1)(j\omega+2)}$$

$$H(e^{j\Omega}) = \frac{(e^{-T}-e^{-2T})e^{j\Omega}}{e^{j2\Omega}-(e^{-T}+e^{-2T})e^{j\Omega}+e^{-3T}}$$

Figure 8.18 shows the magnitude spectra with different sampling periods of $T = 1$ s, 0.2 s, and 0.1 s. Because the DT impulse response $h(k)$ is the sampled version of $h_a(t)$, the magnitude spectrum of the digital filter is the periodic extension of that of the analog filter. The period of $|H(e^{j\Omega})|$ is 2π. From Figure 8.18 (b), it can be seen that the magnitude spectrum of the DT filter with $T = 0.1$ is very close to that of the original CT filter.

! **Note:** Review the periodicity in Section 8.2.2.

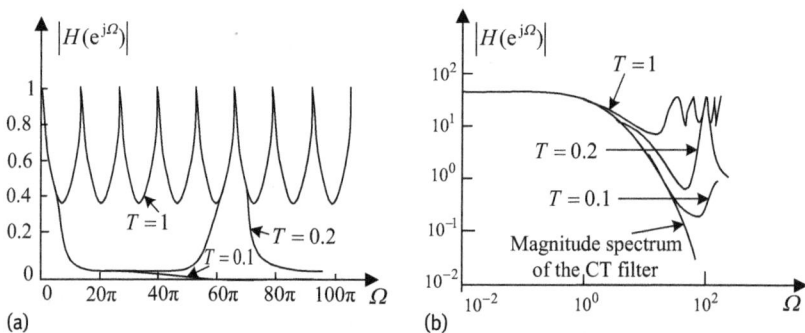

(a)

(b)

Fig. 8.18: Magnitude spectra of Example 8.4.1; (a) Magnitude spectra, (b) Magnitude spectra in logarithm coordinate

8.4.4 IIR filter design using bilinear transformation

At times, the impulse invariance transformation suffers from aliasing, which may lead to deviations from the original DT specifications. An alternative to the impulse invariance transformation is the bilinear transformation, which is a nonlinear mapping between the CT and DT frequency domains. It provides a one-to-one mapping from the S-plane to the Z-plane. The mapping equation is given by:

$$s = \frac{2}{T}\left(\frac{z-1}{z+1}\right), \quad z = \frac{1+sT/2}{1-sT/2} \tag{8.36}$$

where $2/T$ is the normalization constant and T is the sampling interval. To derive the frequency characteristics of the bilinear transformation, we substitute $s = j\omega$ and $z = e^{j\Omega}$ in Equation (8.36):

$$j\omega = \frac{2}{T}\frac{e^{j\Omega}-1}{e^{j\Omega}+1} = \frac{2}{T}\frac{e^{j\Omega/2}-e^{-j\Omega/2}}{e^{j\Omega/2}+e^{-j\Omega/2}} = \frac{2}{T}\frac{j\sin\left(\frac{\Omega}{2}\right)}{\cos\left(\frac{\Omega}{2}\right)} = j\frac{2}{T}\tan\left(\frac{\Omega}{2}\right) \tag{8.37}$$

Note: Readers can prove it. ❗

The resulting expression is given by:

$$\omega = \frac{2}{T}\tan\left(\frac{\Omega}{2}\right), \quad \Omega = 2\arctan\left(\frac{\omega T}{2}\right) \tag{8.38}$$

The CT frequencies $\omega = [0, \infty]$ are mapped to the DT frequencies $\Omega = [0, \pi]$, and the negative CT frequencies $\omega = [-\infty, 0]$ are mapped to the DT frequencies $\Omega = [-\pi, 0]$. Since the CT frequency range $[-\infty, \infty]$ is mapped on to the DT frequency range $[-\pi, \pi]$, there is no overlap between adjacent replicas constituting the magnitude response of the digital filter. Frequency warping, therefore, eliminates the undesirable effects of aliasing from the transformed digital filter.

Example 8.4.2. Bilinear transformation is used to design a low-pass Butterworth filter with the following specifications:

$$\text{passband } (0 \le |\Omega| \le 60\pi\,\text{rad}) \qquad 0.9 \le |H(e^{j\Omega})| \le 1\,;$$
$$\text{stopband } (|\Omega| \ge 320\pi\,\text{rad}) \qquad |H(e^{j\Omega})| \le 0.1\,.$$

The sampling interval is $T = 0.002$ s.

Note: More details can be found in Reference [6]. ❗

Solution: Use Equation (8.38) to calculate the CT frequency domain:

$$\omega_1 = \frac{2}{0.002} \tan\left(\frac{60\pi \times 0.002}{2}\right) = 190.76 \, \text{rad/s}$$

$$\omega_2 = \frac{2}{0.002} \tan\left(\frac{320\pi \times 0.002}{2}\right) = 1575.75 \, \text{rad/s}$$

The gain terms G_p and G_s are given by:

$$G_p = \frac{1}{(1-0.1)^2} - 1 = 0.2346, \quad G_s = \frac{1}{(0.1)^2} - 1 = 99$$

The order of the Butterworth filter is given by:

$$N = \frac{1}{2} \times \frac{\ln(0.2346/99)}{\ln(190.76/1575.75)} = 1.43,$$

which is rounded up to $N = 2$. According to the specifications, the frequency characteristics $|H(j\omega)| = 1/(\sqrt{1+(\omega/\omega_c)^{2N}})$ of the Butterworth filter must satisfy the following equations [29]:

$$
\begin{cases}
|H(j\omega_1)| = \dfrac{1}{\sqrt{1+\left(\frac{190.76}{\omega_c}\right)^{2N}}} \geq 0.9 \\[4mm]
|H(j\omega_2)| = \dfrac{1}{\sqrt{1+\left(\frac{1575.75}{\omega_c}\right)^{2N}}} \leq 0.1
\end{cases}
$$

The cut-off frequency ω_c of the Butterworth filter is obtained as:

$$274 \leq \omega_c \leq 499.5 \tag{8.39}$$

The cut-off frequency ω_c is selected as $\omega_c = 100\pi$. The Laplace transfer function can be obtained as follows:

$$H_a(s) = \frac{10,000\pi^2}{\left(s - 100\pi e^{j\frac{3}{4}\pi}\right)\left(s - 100\pi e^{-j\frac{3}{4}\pi}\right)} = \frac{10,000\pi^2}{s^2 - 200\pi \cos\left(\frac{3}{4}\pi\right)s + 10,000\pi^2}$$

$$= \frac{98,696}{s^2 + 444.28s + 98,696}$$

$$\tag{8.40}$$

Substituting $s = 2/T((z-1)/(z+1)) = 1000((z-1)/(z+1))$ into Equation (8.40) yields:

$$H(z) = \frac{0.064(z^2 + 2z + 1)}{z^2 - 1.168z + 0.424} \tag{8.41}$$

Figure 8.19 shows the magnitude spectra of the original CT filter and the DT filter designed by bilinear transformation. We can observe that the magnitude spectrum of the DT filter satisfies the specified band requirements, and it is better than the original CT filter.

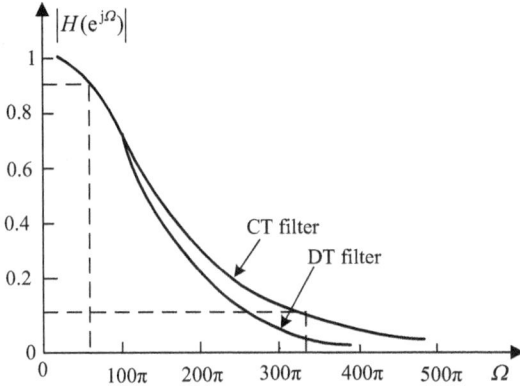

Fig. 8.19: Magnitude spectra of Example 8.4.2

8.4.5 FIR filter design using the windowing method

A finite impulse response (FIR) filter is defined as a filter whose length N is finite. A causal FIR filter is represented by the following transfer function:

$$H(z) = \sum_{k=0}^{N-1} h_N(k)z^{-k} = h(0) + h(1)z^{-1} + h(2)z^{-2} + \cdots + h(N-1)z^{-(N-1)} \qquad (8.42)$$

The design of FIR filters can be realized by the windowing approach. We first illustrate the windowing principle by a simple average filter. A simple average filter is described by its impulse response $h_N(k)$ and difference equation:

$$h_N(k) = \begin{cases} 1/N, & 0 \le k \le N-1 \\ 0, & k < 0, \ k \ge N \end{cases} \qquad (8.43)$$

$$y(k) = \frac{1}{N} \sum_{i=0}^{N-1} f(k-i) \qquad (8.44)$$

Its frequency response is given by:

$$H_N(e^{j\Omega}) = \frac{1}{N} \frac{\sin(N\Omega/2)}{\sin(\Omega/2)} e^{-j\frac{N\Omega}{2}}. \qquad (8.45)$$

Figure 8.20 shows the impulse response $h_N(k)$ and the logarithmic amplitude spectrum for a simple 33-point average filter. Because the cut-off frequency only depends on the window length N, the steep transition characteristics and the passband width cannot be taken into account simultaneously.

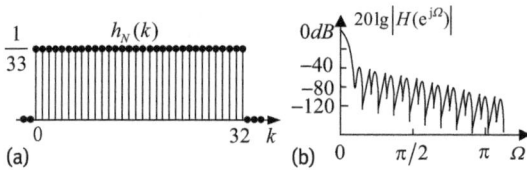

(a)

(b)

Fig. 8.20: Impulse response and amplitude spectrum of a simple average filter; (a) Impulse response, (b) Logarithmic amplitude spectrum

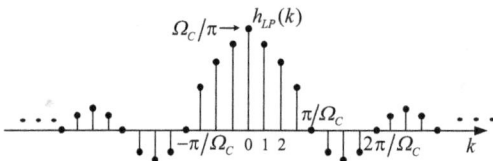

Fig. 8.21: Impulse response of an ideal low-pass filter

The drawback of this average filter is that all coefficients are $1/N$. If different weights are assigned in Equation (8.43), the filtering characteristics of weighted low-pass filter will be much better. To realize the approximation of the ideal low-pass FIR filter, the impulse response $h_{LP}(k)$ is designed as follows:

$$h_{LP}(k) = \frac{\Omega_c}{\pi} \text{Sa}(\Omega_c k) = \frac{\sin(\Omega_c k)}{\pi k} \tag{8.46}$$

Figure 8.21 shows the impulse response $h_{LP}(k)$ with $\Omega_c = \pi/4$. If it is truncated symmetrically and shifted $(N-1)/2$ to the right, the impulse response of the FIR low-pass filter $h_N(k)$ can be obtained as follows:

(Note: We suppose that the length N is odd.)

$$h_N[k] = h_{LP}[k - (N-1)/2] \cdot w_{rect}[k] = \frac{\Omega_c \sin\{\Omega_c[k - (k-1)/2]\}}{\pi[k - (k-1)/2]} \cdot w_{rect}(k) \tag{8.47}$$

Where Ω_c is the cut-off frequency and $w_{rect}[k] = \varepsilon[k] - \varepsilon[k - N]$ is the rectangular window. Figure 8.22 (a) shows the impulse response $h_N(k)$. The magnitude spectrum

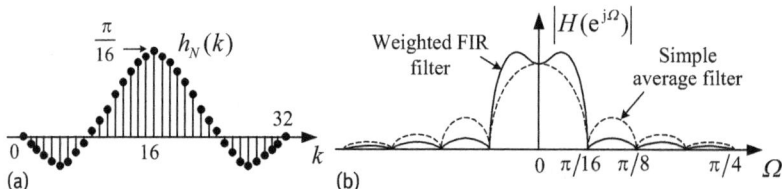

(a)

(b)

Fig. 8.22: Impulse response and amplitude spectrum of a FIR filter; (a) Impulse response, (b) Magnitude spectrum

is shown in Figure 8.22 (b), where we observe that the designed FIR low-pass filter is better than the simple average filter.

Note: The center lobe is the main lobe, and the lobes its side are the side lobes. !

There are a number of alternatives to the rectangular window. Commonly used windows are the Bartlett (triangular) window, the Hamming window, the Hanning window and the Blackman window. If we choose a smoother window, the Gibbs phenomenon is greatly improved.

The Bartlett (triangular) window

$$w_{\text{bart}}(k) = \begin{cases} \frac{2k}{N-1}, & 0 \le k \le (N-1)/2 \\ 2 - \frac{2k}{N-1}, & (N-1)/2 < k \le N-1 \\ 0, & \text{otherwise}. \end{cases}$$

The Hamming window

$$w_{\text{hamm}}(k) = \begin{cases} 0.54 - 0.46 \cos\left(\frac{2\pi k}{N-1}\right), & 0 \le k \le N-1 \\ 0, & \text{otherwise}. \end{cases}$$

The Hanning window

$$w_{\text{hann}}(k) = \begin{cases} 0.5 - 0.5 \cos\left(\frac{2\pi k}{N-1}\right), & 0 \le k \le N-1 \\ 0, & \text{otherwise}. \end{cases}$$

The Blackman window

$$w_{\text{blac}}(k) = \begin{cases} 0.42 - 0.5 \cos\left(\frac{2\pi k}{N-1}\right) + 0.08 \cos\left(\frac{4\pi k}{N-1}\right), & 0 \le k \le N-1 \\ 0, & \text{otherwise}. \end{cases}$$

By choosing different windows, we prefer to design a FIR filter by minimizing the transition band and reducing the strength of the ripples.

8.5 Controllability and observability of linear systems

In modern control theory, the controllability and observability of the system can be analyzed by the state equations. In the process of launching satellites, the states of the satellite are adjusted according to the initial state and excitation inputs. This requires that the system be completely controllable and, therefore, the satellite can go smoothly into orbit. Otherwise, the ground control center cannot control the satellite system into its designated working state. In addition, the states of position, speed and acceleration need to be observed. In real applications, the position in space can be observed, but the velocity and acceleration cannot be directly observed. The observability of the system is to study whether the internal states of the system can be calculated by observing the outputs of the system.

8.5.1 Controllability of linear systems

The concept of controllability refers to the ability of a controller to arbitrarily alter the functionality of the system. If the system can be controlled from an initial state $x(0) = x_0$ to zero state $x(T) = 0$ by applying an input signal $f(t)$ after a period of time, the state x_0 is called a controllable state. Complete state controllability (all states are controllable) describes the ability of an external input to change the internal state of a system from any initial state to any other final state in a finite time interval.

! **Note:** More details can be seen in Reference [11].

The precise concept of controllability is defined as follows: A state x_0 is controllable at time t_0 if for some finite time t_1 there exists an input $f(t)$ that transfers the state $x(t)$ from x_0 to the origin at time t_1. A system is called completely controllable at time t_0 if every state x_0 in the state space is controllable. If only a part of the states is controllable, the system is partially controllable.

The state-space equations of continuous time and discrete time are given as follows, respectively:

$$\dot{X} = AX + Bf$$

$$X(k + 1) = AX(k) + Bf(k)$$

For LTI systems, a system is controllable if and only if its controllability matrix has a full row rank. The *controllability matrix* M_C of a system is defined as follows:

$$M_C = [B, AB, A^2B, \dots, A^{n-1}B] \tag{8.48}$$

! **Note:** We should know how to compute the rank of a matrix.

Example 8.5.1. Given the state equations of an LTIC system, determine the controllability:

$$\begin{bmatrix} \dot{x}_1(t) \\ \dot{x}_2(t) \end{bmatrix} = \begin{bmatrix} -2 & 1 \\ 1 & -2 \end{bmatrix} \begin{bmatrix} x_1(t) \\ x_2(t) \end{bmatrix} + \begin{bmatrix} 1 \\ 1 \end{bmatrix} f(t)$$

Solution: From the state equations, the matrices A and B are obtained as follows:

$$A = \begin{bmatrix} -2 & 1 \\ 1 & -2 \end{bmatrix}, \quad B = \begin{bmatrix} 1 \\ 1 \end{bmatrix}$$

We obtain

$$AB = \begin{bmatrix} -1 \\ -1 \end{bmatrix}.$$

The controllability matrix is given by:

$$M_C = [B, AB] = \begin{bmatrix} 1 & -1 \\ 1 & -1 \end{bmatrix}$$

Obviously, the corresponding determinant of the matrix Mc is equal to zero. The rank is 1, and the matrix is not full rank. Therefore, the corresponding system is not controllable.

8.5.2 Observability of linear systems

The observability describes whether the internal state variables of the system can be externally measured. The state variables of a system might not be measurable for any of the following reasons:
1. The location of the particular state variable might not be physically accessible (a capacitor or a spring, for instance).
2. There are no appropriate instruments to measure the state variable.
3. The state variable is a derived "dummy" variable that has no physical meaning.

If state variables cannot be directly observed, for any of the reasons above, the values of the internal state variables must be calculated or estimated by using the input/output relation of the system and the output history from the starting time. In other words, we have to find out what the inside of the system (the internal system states) is like, by only observing the outside performance of the system (input and output). The precise definition of observability is given as follows.

A system with an initial state $x(t_0)$ is observable if, and only if, the value of the initial state can be determined from the system output $y(t)$ that has been observed through the time interval $t_0 < t < t_1$. A system is said to be completely observable if all the possible initial states of the system can be observed. If only some of the initial states can be determined, then the system is not completely observable.

The observability of the system is dependent only on the system states and the system output, so we can simplify the state equations to remove the input terms:

$$\dot{X} = AX$$

$$Y = CX$$

Therefore, the observability of the system is dependent only on the coefficient matrices A and C. The observability matrix M_O is defined as the following to precisely determine whether a system is observable:

$$M_O = [C, CA, CA^2, \ldots, CA^{n-1}]^T \tag{8.49}$$

The system is observable if, and only if, the Q matrix is a full rank matrix.

Note: For discrete-time systems, the controllability and observability criteria are the same.

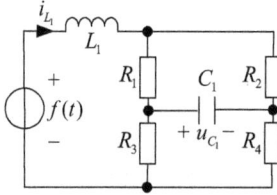

Fig. 8.23: Circuit system of Example 8.5.2

Example 8.5.2. Given the bridge circuit shown in Figure 8.23, the input is $f(t)$ and the output is voltage $u_{C_1}(t)$ of the capacitor. Determine the controllability and observability of the system.

Solution: The current $i_{L_1}(t)$ of the inductor and the voltage $u_{C_1}(t)$ of the capacitor are selected as state variables; the state-space equations are obtained as follows:

$$\begin{cases} \frac{d}{dt} i_{L_1}(t) = -\frac{1}{L_1}\left(\frac{R_1 R_2}{R_1+R_2} + \frac{R_3 R_4}{R_3+R_4}\right) i_{L_1}(t) + \frac{1}{L_1}\left(\frac{R_1}{R_1+R_2} - \frac{R_3}{R_3+R_4}\right) u_{C_1}(t) + \frac{f(t)}{L_1} \\ \frac{d}{dt} u_{C_1}(t) = -\frac{1}{C_1}\left(\frac{R_1}{R_1+R_2} - \frac{R_3}{R_3+R_4}\right) i_{L_1}(t) - \frac{1}{C_1}\left(\frac{1}{R_1+R_2} + \frac{1}{R_3+R_4}\right) u_{C_1}(t) \end{cases} \tag{8.50}$$

$$y(t) = u_{C_1}(t) \tag{8.51}$$

! **Note:** Readers can prove it.

The corresponding coefficient matrixes can be obtained:

$$A = \begin{bmatrix} -\frac{1}{L_1}\left(\frac{R_1 R_2}{R_1+R_2} + \frac{R_3 R_4}{R_3+R_4}\right) & \frac{1}{L_1}\left(\frac{R_1}{R_1+R_2} - \frac{R_3}{R_3+R_4}\right) \\ -\frac{1}{C_1}\left(\frac{R_1}{R_1+R_2} - \frac{R_3}{R_3+R_4}\right) & -\frac{1}{C_1}\left(\frac{1}{R_1+R_2} + \frac{1}{R_3+R_4}\right) \end{bmatrix}$$

$$B = \begin{bmatrix} \frac{1}{L_1} \\ 0 \end{bmatrix}, \quad C = \begin{bmatrix} 0 \\ 1 \end{bmatrix}, \quad D = 0$$

So, the controllability matrix and the observability matrix are given by:

$$M_C = [B \quad AB] = \begin{bmatrix} \frac{1}{L_1} & -\frac{1}{L_1^2}\left(\frac{R_1 R_2}{R_1+R_2} + \frac{R_3 R_4}{R_3+R_4}\right) \\ 0 & -\frac{1}{L_1 C_1}\left(\frac{R_1}{R_1+R_2} - \frac{R_3}{R_3+R_4}\right) \end{bmatrix} \tag{8.52}$$

$$M_O = \begin{bmatrix} C \\ CA \end{bmatrix} = \begin{bmatrix} 0 & 1 \\ -\frac{1}{C_1}\left(\frac{R_1}{R_1+R_2} - \frac{R_3}{R_3+R_4}\right) & -\frac{1}{C_1}\left(\frac{1}{R_1+R_2} + \frac{1}{R_3+R_4}\right) \end{bmatrix} \tag{8.53}$$

It can be seen from the matrix expression that both the necessary and sufficient conditions of full rank are:

$$\frac{R_1}{R_1 + R_2} - \frac{R_3}{R_3 + R_4} \neq 0 \tag{8.54}$$

or:

$$R_1 R_4 \neq R_2 R_3 \tag{8.55}$$

This is the condition of the bridge imbalance. Therefore, when the bridge is unbalanced in condition of $R_1 R_4 = R_2 R_3$, the system is controllable and observable.

8.5.3 Calculation with MATLAB

In MATLAB, the tool functions *ctrb* and *obsv* can be used to compute the controllability matrix and the observability matrix.

Example 8.5.3. The state-space equations of a second-order LTIC system are given by:

$$\begin{bmatrix} \dot{x}_1(t) \\ \dot{x}_2(t) \end{bmatrix} = \begin{bmatrix} -3 & 1 \\ 1 & -3 \end{bmatrix} \begin{bmatrix} x_1(t) \\ x_2(t) \end{bmatrix} + \begin{bmatrix} 1 & 1 \\ 1 & 1 \end{bmatrix} \begin{bmatrix} f_1(t) \\ f_2(t) \end{bmatrix}$$

$$\begin{bmatrix} y_1(t) \\ y_2(t) \end{bmatrix} = \begin{bmatrix} 1 & 1 \\ 1 & -1 \end{bmatrix} \begin{bmatrix} x_1(t) \\ x_2(t) \end{bmatrix}.$$

Determine the controllability and observability of the system with MATLAB.

Solution:

```
a=[-3,1;1,-3];
b=[1,1;1,1];
c=[1,1;1,-1];
d=[0];
cam=ctrb(a,b);          % compute the controllability matrix
rcam=rank(cam)          % compute the rank
oam=obsv(a,c);          % compute the observability matrix
roam=rank(oam)          % compute the rank
```

The operation result is *rcam =1* and *roam =2*. Therefore, this system is an uncontrollable but observable system.

Note: Repeat Example 8.5.1 using MATLAB.   !

8.6 Applications of the Kalman filter

8.6.1 Basic principles of the Kalman filter

The Kalman filter was proposed by Rudolf Emil Kalman, a Hungarian mathematician. The Kalman filter is an optimal recursive data processing algorithm [30]. In this section, we will briefly describe the Kalman filtering process.

A discrete-time control process is described by a linear difference equation as follows:

$$S(k) = \Phi \cdot S(k-1) + w(k) \tag{8.56}$$

$$Z(k) = H \cdot S(k) + v(k) \tag{8.57}$$

where $S(k)$ is the system state at time k, $Z(k)$ is the measured value at time k, Φ is the state transition matrix, and H is the observation matrix. The process noise vector $w(k)$

and the observation noise vector $v(k)$ are uncorrelated normal white noise sequences with zero mean. Let Q and R be the covariance matrices of dynamic noise $w(k)$ and $v(k)$, respectively.

According to the system model, the prediction is made based on the previous state $S(k-1|k-1)$ to obtain $S(k|k-1)$:

$$S(k|k-1) = \Phi \cdot S(k-1|k-1) \tag{8.58}$$

In order to update system state, the covariance matrix P of $S(k|k-1)$ is predicted as follows:

$$P(k|k-1) = \Phi \cdot P(k-1|k-1)\Phi^{T} + Q \tag{8.59}$$

Where $P(k|k-1)$ is the covariance matrix of $S(k|k-1)$, $P(k-1|k-1)$ is the covariance matrix of $S(k-1|k-1)$, Φ^{T} is the transpose matrix of Φ, and Q is the covariance matrix of system process noise.

The optimal estimate $S(k|k)$ is obtained by the prediction result $S(k|k-1)$ of the state and the measured value $Z(k)$:

$$S(k|k) = \Phi \cdot S(k|k-1) + K_g(k) \cdot (Z(k) - H \cdot \Phi \cdot S(k|k-1)) \tag{8.60}$$

where K_g is the Kalman gain:

$$K_g(k) = P(k|k-1) \cdot H^{T} \cdot (H \cdot P(k|k-1) \cdot H^{T} + R)^{-1} \tag{8.61}$$

In order to continue the recursive operation of the Kalman filter, the covariance matrix of $S(k|k)$ is updated as follows:

$$P(k|k) = (I - K_g(k) \cdot H) \cdot P(k|k-1) \tag{8.62}$$

Where I is a unit matrix.

8.6.2 Temperature prediction simulation with MATLAB

Supposing that the temperature of a room is constant, we have $\Phi = 1$. Equations (8.58) and (8.59) can be simplified as follows:

$$S(k|k-1) = S(k-1|k-1)$$
$$P(k|k-1) = P(k-1|k-1) + Q$$

Since the measured temperature value is one-dimensional data, the observation matrix is $H = 1$. Equations (8.60)–(8.62) can be simplified as follows:

$$S(k|k) = S(k|k-1) + K_g(k) \cdot (Z(k) - S(k|k-1))$$
$$K_g(k) = P(k|k-1) \cdot (P(k|k-1) + R)^{-1}$$
$$P(k|k) = (1 - K_g(k)) \cdot P(k|k-1)$$

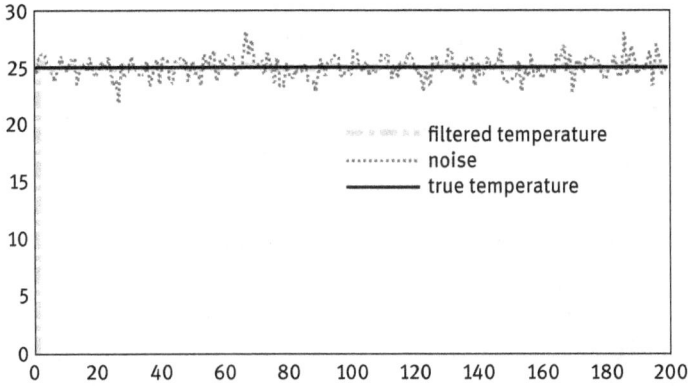

Fig. 8.24: Results of Kalman filtering

The real temperature of the room is assumed to be 25 °C. A number of 200 measurements are simulated. The average of these measurements is 25 °C. Gaussian white
noise with a standard deviation of several degrees is added. The simulation results
are shown in Figure 8.24. The optimization result of the Kalman filter is close to the
room's true temperature of 25 °C. The MATLAB simulation is detailed as follows:

```
N=200;                              % simulating 200 measurements
w=0.1*randn(1,N);                   % process noise
x(1)=25;                            % true temperature
V=randn(1,N);                       % observation noise
q1=std(V);                          % standard deviation of V
Rvv=q1.^2;                          % covariance of measurement process
q2=std(x);                          % standard deviation of x
Rxx=q2.^2;
q3=std(w);                          % standard deviation of w
Qww=q3.^2;
Y=x+V;                              % signals with noise
p(1)=1;                             % initial covariance
Bs(1)=0;                            % optimal estimated initial value
for t=2:N;                          % start filtering
x(t)=x(t-1);
p1(t)=p(t-1)+Qww;                   % covariance estimation
Kg(t)=p1(t)/(p1(t)+Rvv);           % find the Kalman gain
Bs(t)=x(t)+Kg(t)*(Y(t)-x(t-1));    % the optimal value at time t
p(t)=p1(t)-Kg(t)*p1(t);            % find the deviation of the optimal state
end
```

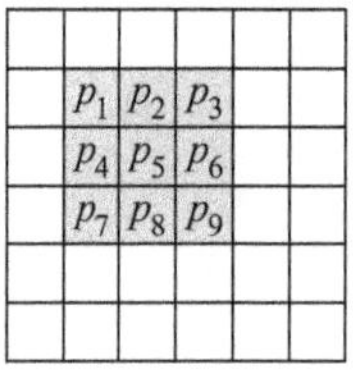

Fig. 8.25: Illustration of an image

8.7 Applications of convolution in image processing

In image processing, the 2D convolution operation is used to realize image filtering. The 2D convolution kernel is a small matrix, which is also referred to as a mask. Image filtering is accomplished by convolving a kernel and an image. Figure 8.25 is an image piece and Equation (8.63) is an illustration of convolution kernel.

Note: The size of the kernel can be designed as 3×3, 5×5, or another size according to different requirements.

$$G = \begin{bmatrix} g_1 & g_2 & g_3 \\ g_4 & g_5 & g_6 \\ g_7 & g_8 & g_9 \end{bmatrix} \tag{8.63}$$

Convolution is the process of flipping both the rows and columns of the kernel and then multiplying locally similar entries and summing [31]. The central element p_5' of the resulting image would be a weighted combination of all the entries of the image matrix:

$$p_5' = \sum_{i=1}^{9} g_i p_i = g_1 p_1 + g_2 p_2 + \cdots + g_9 p_9 \tag{8.64}$$

The values of each pixel of the output image are obtained by summing the multiplication of each kernel value and the corresponding input image pixel value.

Image filtering allows us to apply various effects on images, such as blurring, sharpening, embossing and edge detection. Three kernels A1, A2 and A3 are given for realizing mean filtering and edge detecting:

$$A1 = \frac{1}{25} \begin{bmatrix} 1 & 1 & 1 & 1 & 1 \\ 1 & 1 & 1 & 1 & 1 \\ 1 & 1 & 1 & 1 & 1 \\ 1 & 1 & 1 & 1 & 1 \\ 1 & 1 & 1 & 1 & 1 \end{bmatrix}, \quad A2 = \frac{1}{100} \begin{bmatrix} 1 & 1 & 1 & \cdots & 1 \\ 1 & 1 & 1 & \cdots & 1 \\ \vdots & \vdots & \vdots & & \vdots \\ 1 & 1 & 1 & \cdots & 1 \\ 1 & 1 & 1 & \cdots & 1 \end{bmatrix},$$

$$A3 = \begin{bmatrix} 0 & -1 & 0 \\ -1 & 4 & -1 \\ 0 & -1 & 0 \end{bmatrix}$$

Note: The kernel can be considered as a 2D impulse response.

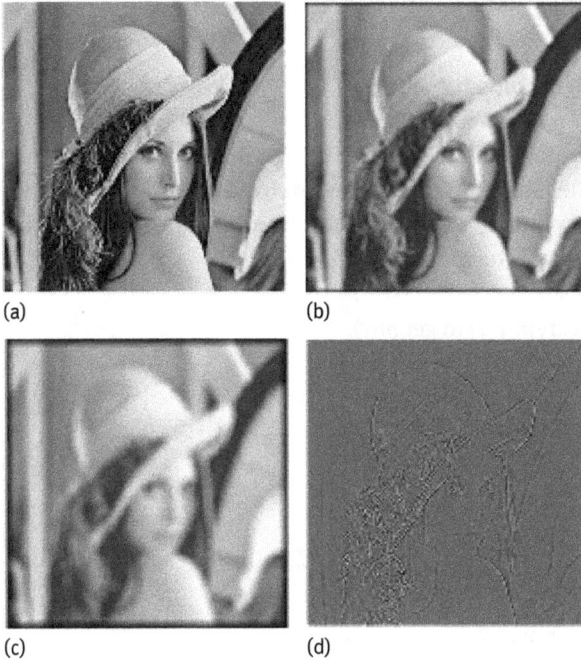

Fig. 8.26: Results of image filtering; (a) Original image, (b) Result of A1 kernel convolution, (c) Result of A2 kernel convolution, (d) Result of A3 kernel convolution

Figure 8.26 gives the original image and the resulting images by kernel convolution. The first two kernels are used to remove noise from an image by computing the average of the current pixel and its neighbors. From Figures 8.26 (b) and (c), we can observe that larger size of the kernel will have a better denoising effect, but the resulting image is blurred with a lower quality. Figure 8.26 (d) gives the results of edge detecting. The kernel A3 is used to detect edges in the horizontal and vertical directions. The resulting image is a new image, in which the edges are enhanced to make it look sharper.

8.8 Summary

This chapter mainly presented applications of system analysis. The signal was modulated by multiplying the carrier wave in Section 8.1. Section 8.2 detailed that the DFT was the sampled value of DTFT at discrete frequencies. The IIR and FIR digital filters were introduced in Section 8.4. In Section 8.5, the controllability and observability was judged if the matrix M_C or M_O had a full rank. The Kalman fiter was used in data prediction and 2D convolution was applied in image processing.

⚡ Chapter 8 problems

8.1 Consider the baseband signal $m(t) = \cos 1000t$.
(1) Draw the spectrum of $m(t) = \cos 1000t$.
(2) Draw the spectrum of the DSB-SC modulated signal $m(t)\cos 10{,}000t$ and identify the upper sideband (USB) and lower sideband (LSB).

8.2 The system for signal encryption (scrambling) is given in Figure P8.1. The input $m(t)$ is encrypted to obtain the output signal $y(t)$.
(1) Draw the spectrum of $y(t)$.
(2) Give the method to decrypt $y(t)$ to get $m(t)$.

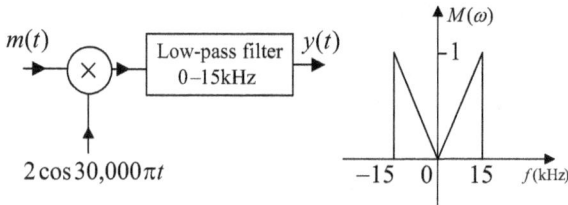

Fig. P8.1: System diagram in Problem 8.2

8.3 For the following DT sequences, calculate the DTFTs:

(1) $f(k) = \delta(k)$
(2) $f(k) = \varepsilon(k)$
(3) $f(k) = \mathrm{sgn}(k)$

8.4 Design a low-pass IIR filter with the following specifications using impulsive invariance transformation:

(1) passband $(0 \le |\Omega| \le 0.25\pi \text{ rad/s})$ $0.8 \le |H(e^{j\Omega})| \le 1$
(2) stopband $(0.75\pi \le |\Omega| \le \pi \text{ rad/s})$ $|H(e^{j\Omega})| \le 0.2$

8.5 Repeat Problem 8.4 using bilinear transformation.

8.6 The sampling frequency is $4\,\text{kHz}$, and the window length is $N = 10$. Design a low-pass FIR filter using windowing to approximate the ideal filter. The transfer function of the ideal low-pass filter is given by:

$$H(e^{j\Omega}) = \begin{cases} e^{-j2\Omega}, & |\Omega| < 2000\pi \\ 0, & \text{else} \end{cases}$$

8.7 Repeat Problem 8.6 using MATLAB.

8.8 Given the parameter matrices of the state equations:

$$A = \begin{bmatrix} 1 & 0 & 0 & 0 \\ 0 & 2 & 0 & 0 \\ -6 & -2 & 3 & 0 \\ -3 & -2 & 0 & 4 \end{bmatrix}, \quad B = \begin{bmatrix} 1 \\ 0 \\ 3 \\ 2 \end{bmatrix}, \quad C = \begin{bmatrix} -4 & -3 & 1 & 1 \end{bmatrix}$$

(1) determine the controllability and the observability;

(2) compute the Laplace transfer function $H(s)$.

8.9 Given the parameter matrixes of the state equations, analyze the controllability and observability of the system:

$$A = \begin{bmatrix} 0 & 1 & 0 \\ 0 & 0 & 1 \\ -6 & -11 & -6 \end{bmatrix}, \quad B = \begin{bmatrix} 0 \\ 0 \\ 1 \end{bmatrix}, \quad C = [4 \quad 5 \quad 1], \quad D = 0$$

8.10 High-pass filtering is used to detect the edges or suppress low-frequency noise in an image. At times, high-pass filters are also used to sharpen the edges of an image.

Given the filter with the impulse response as follows, illustrate the application of extracting the edges of an image using MATLAB:

$$h(m, n) = \frac{1}{9} \begin{bmatrix} -1 & -1 & -1 \\ -1 & 8 & -1 \\ -1 & -1 & -1 \end{bmatrix}$$

References

[1]   Alan V. Oppenheim, Alan S. Willsky and S. Hamid Nawab. Signals and systems [M]. Second
      Edition. Beijing: Publishing House of Electronics Industry, 2015.
[2]   Guo Baolong, Yan Yunyi and Zhu Juanjuan. Signals and systems engineering [M]. Beijing:
      Higher Education Press, 2014 (only available in Chinese).
[3]   Zheng Junli, Yang Weili and Ying Qihang. Signal and system [M]. Beijing: Higher Education
      Press, 1981 (only available in Chinese).
[4]   Alan V. Oppenheim. Signals and systems. Translated by Liu Shutang [M]. Xi'an: Xi'an Jiao Tong
      University press, 1985.
[5]   Guan Zhizhong and Xia Gongke. Signal and linear system. [M]. Third Edition. Beijing: Higher
      Education Press, 1992 (only available in Chinese).
[6]   Mrinal Mandal and Amir Asif. Continuous and discrete time signals and systems [M]. Beijing:
      Posts and Telecom Press, 2010.
[7]   Simon Haykin and Barry Van Veen. Signals and systems [M]. Second Edition. New York, Chich-
      ester, Weinheim, Brisbane, Singapore, Toronto: John Wiley & Sons. Inc., 2003.
[8]   Chen Shengtan, Guo Baolong and Li Xuewu. Signals and systems [M].Third Edition. Xi'an: Xid-
      ian University Press, 2008 (only available in Chinese).
[9]   Paolo Prandoni and Martin Vetterli. Signal processing for communications [M]. Lausanne: EPFL
      Press, 2008.
[10] Guo Lei and Guo Baolong. The visual system with distributed reasoning theory [M]. Xi'an: Xid-
      ian University Press, 1995.
[11] Bhagawandas Pannalal Lathi. Linear systems and signals. Translated by Liu Shutang [M].
      Xi'an: Xi'an Jiao Tong University press, 2006.
[12] Edward A. Lee and Pravin Varaiya. Structure and interpretation of signals and systems [M].
      Second Edition. Reading: Addison Wesley, 2011.
[13] Zhang Yongrui. The short version of the signal and system [M]. Xi'an: Xidian University Press,
      2014.
[14] Wu Dazheng, Yang Linyao, Zhang Yongrui and Guo Baolong. Signal and linear system analysis
      [M]. Beijing: Higher Education Press, 2005 (only available in Chinese).
[15] Srdjan Stanković, Irena Orović and Ervin Sejdić. Multimedia signals and systems: basic and ad-
      vanced algorithms for signal processing [M]. Cham, Heidelberg, New York, Dordrecht, London:
      Springer International Publishing, 2016.
[16] Simon Haykin and Barry Van Veen. Signals and systems [M]. Second Edition. Beijing: Publish-
      ing House of Electronics Industry, 2012.
[17] Chen Houjin, Hu Jian and Xue Jian. Signal and system [M]. Second Edition. Beijing: Higher
      Education Press, 2015 (only available in Chinese).
[18] Ruthber Rodriguez. Audio signals processing with digital filters implementation using Mydsp
      [J]. Journal of Engineering & Applied Sciences, 2017, 12(1).
[19] Simon Haykin. Adaptive filter theory. Translated by Zheng Baoyu [M]. Fourth Edition. Beijing:
      Publishing House of Electronics Industry, 2003.
[20] Simon Haykin. Adaptive filter theory [M]. Fifth Edition. Beijing: Publishing House of Electronics
      Industry, 2017.
[21] John G. Proakis and Dimitris K. Manolakis. Digital signal processing: principles, algorithms and
      applications [M]. London: Pearson Press, 2013.
[22] Alan V. Oppenheim, Ronald W. Schafer and John R. Buck. Discrete-time signal processing [M].
      Beijing: Tsinghua University Press, 2003.

https://doi.org/10.1515/9783110593907-009

[23] Harish Parthasarathy. Textbook of signals and systems [M]. London: I K International Publishing House Pvt. Ltd, 2008.

[24] Hwei P. Hsu. Schaum's outline of signals and systems [M]. Third Edition. New York, Chicago, San Francisco, Athens, London, Madrid, Mexico City, Milan, New Delhi, Singapore, Sydney, Toronto: McGraw-Hill Education Press, 2013.

[25] Luis Chaparro. Signals and systems using MATLAB [M]. Oxford: Academic Press, 2010.

[26] William A. Gardner. Introduction to random processes. With applications to signals and systems [M]. Second Edition. New York, Chicago, San Francisco, Athens, London, Madrid, Mexico City, Milan, New Delhi, Singapore, Sydney, Toronto: McGraw-Hill Education Press, 1986.

[27] Srdjan Stanković, Irena Orović and Ljubiša Stanković. Polynomial fourier domain as a domain of signal sparsity [J]. Signal Processing, 2017, 130:243–253.

[28] Marco F. Duarte and Yonina C. Eldar. Structured compressed sensing: from theory to applications [J].IEEE Transactions on Signal Processing, 2011, 59(9):4053–4085.

[29] Ronald Bracewell. The Fourier transform and its applications [M]. Third Edition. New York, Chicago, San Francisco, Athens, London, Madrid, Mexico City, Milan, New Delhi, Singapore, Sydney, Toronto: McGraw-Hill Science/Engineering/Math Press, 1999.

[30] Jian Pan, Xinhua Yang, Huafeng Cai and Bingyian Mu. Image noise smoothing using a modified Kalman filter [J]. Neurocomputing, 2016, 173(P3):1625–1629.

[31] Deyun Wei, Qiwen Ran and Yuanmin Li. A convolution and correlation theorem for the linear canonical transform and its application [J]. Circuits Systems & Signal Processing, 2012, 31(1):301–312.

Index

https://doi.org/10.1515/9783110593907-010